Maintenance Fundamentals for Wind Technicians

Wayne Kilcollins

Associate of
**Northern Maine Community College,
Presque Isle, Maine**

DELMAR
CENGAGE Learning·

Australia • Brazil • Canada • Mexico • Singapore • United Kingdom • United States

Maintenance Fundamentals for Wind Technicians, 1E
Wayne Kilcollins

Vice President, Careers & Computing: Dave Garza

Director of Learning Solutions: Sandy Clark

Associate Acquisitions Editor: Nicole Sgueglia

Executive Editor: Dave Boelio

Managing Editor: Larry Main

Senior Product Manager: Sharon Chambliss

Editorial Assistant: Courtney Troeger

Vice President, Marketing: Jennifer Baker

Marketing Director: Deborah Yarnell

Associate Marketing Manager: Jillian Borden

Production Director: Wendy Troeger

Production Manager: Mark Bernard

Senior Content Project Manager: Cheri Plasse

Art Director: GEX, Inc.

Media Editor: Debbie Bordeaux

For product information and technology assistance, contact us at
Cengage Learning Customer & Sales Support, 1-800-354-9706
For permission to use material from this text or product,
submit all requests online at **www.cengage.com/permissions.**
Further permissions questions can be e-mailed to
permissionrequest@cengage.com

Library of Congress Control Number: 2011944382

ISBN-13: 978-1-111-30774-5

ISBN-10: 1-111-30774-1

Delmar
5 Maxwell Drive
Clifton Park, NY 12065-2919
USA

Cengage Learning is a leading provider of customized learning solutions with office locations around the globe, including Singapore, the United Kingdom, Australia, Mexico, Brazil, and Japan. Locate your local office at **www.cengage.com/global**

To learn more about Delmar, visit **www.cengage.com/delmar**

Purchase any of our products at your local college store or at our preferred online store **www.cengagebrain.com**

Notice to the Reader

Printed at CLDPC, USA, 09-21

Dedication

This book is dedicated to my wonderful wife, Terry, and my daughter, Kristen, for their patience and support throughout this project.

Table of Contents

Preface

Maintenance Fundamentals for Wind Technicians presents information related to the maintenance and operation of wind-turbine systems. Wind-turbine systems represent a major investment for project developers and community groups in providing an electrical power resource to utility customers or to supplement local municipal needs. An organization's commitment to a major capital investment in equipment also means a related investment in personnel and resources to continue the safe and efficient operation of its equipment. Modern wind turbines bring together electrical, electronic, mechanical, and fluid power systems to produce a self-contained power production plant.

Unlike traditional large electrical power plants, wind-turbine generators are smaller plants that may be located in remote locations where wind resources are available. This type of setting requires technicians to be proficient over a wide range of disciplines. Wind-energy technicians must be as comfortable working with electrical systems as they are with mechanical and fluid power systems. Developing dedicated teams for providing only electrical or mechanical support would not be cost-effective for these organizations. Developing a workforce that is proficient working across these disciplines has been the goal of wind-turbine manufacturers and wind-project developers over the past decade. To this end, community colleges and career-training organizations have developed programs to provide students with an understanding of these disciplines along with an educational bridge between these disciplines for existing technicians. This textbook is intended as a resource for colleges and certification programs developing our wind-energy workforce. *Maintenance Fundamentals for Wind Technicians* brings together technical concepts and safety information necessary for technicians to understand and work effectively with wind-energy systems.

TEXTBOOK LAYOUT

This text book has been developed with several sections that provide the reader with concepts that build toward an overall understanding of the maintenance requirements for wind-turbine systems. The introduction develops a background for the reader to become familiar with the current state of the wind industry along with training and career opportunities that may be of interest for further research. Section I develops the reader's understanding of safety requirements for working at heights, in remote areas, offshore, and with mechanical and electrical equipment. Section II provides the reader with information on wind-turbine subsystems, test equipment, and testing techniques, as well as discussions on maintenance activities that ensure the optimum performance of the overall plant. This section also includes technical discussions and calculations that are used to further develop the reader's understanding of wind-turbine subsystems.

Section III presents each of the wind-turbine areas that a technician would be involved in during maintenance or service work. These chapters provide the reader with discussions of typical work activities encountered along with the safety information necessary to complete assigned tasks. Discussions in this section also focus on technical information and respective calculations necessary for the reader to further understand the subsystems. Section IV, the final section of the text, discusses methods technicians and managers may use to understand the mountain of data gathered by typical wind-farm management systems. Tracking and analyzing these data enables the development of maintenance strategies and a means to determine the effectiveness of those strategies.

GOAL

The goal of *Maintenance Fundamentals for Wind Technicians* is to provide valuable reference materials and technical information for wind-energy students and technicians alike and to help them understand the activities they may be required to perform in the field. Wind-energy technicians need to be comfortable using an electrical test meter as well as a grease gun and working at ground level or on the rotor assembly 80 meters above the ground. Each activity has its own challenges and rewards. Preparation before beginning any activity is the key to success.

REVIEW QUESTIONS

Each chapter includes a set of review questions that highlight topics from the text. Review questions are a good means for readers to determine their understanding of the material and what concepts may need further study. This is also a valuable tool for the instructor to gauge student knowledge gained from the reading exercise.

SUPPLEMENTAL MATERIALS

A carefully prepared Instructor Resources CD is available to accompany the text. The Components of the CD include:

- An Instructor's Guide which includes review questions from each chapter along with solutions and suggestions for supplemental classroom discussions

- A PowerPoint Presentation is available to aid in the presentation of material and provide discussion points for subtopics within each chapter. Presentations also include images from the text to assist with discussion points for key concepts.

- A Test Bank in ExamView format has been developed for each chapter. This will help instructors spend less time preparing exams and offer more contact time with students to improve the learning process.

- An Image Libary that contains the images found in the text.

Acknowledgments

Developing a textbook such as *Maintenance Fundamentals for Wind Technicians* requires research into as many wind-turbine systems as possible to give the reader a flavor for the maintenance activities necessary with each system. Gathering information for this project has been an interesting challenge for wind-farm operators and manufacturer representatives who have assisted with the research process. My sincerest thanks to each person who has been willing to provide images and technical advice to complete this project.

I offer special thanks to the following people:

Chad Allen, Cianbro Corporation, Wind Energy Services

Neil Browne, Northeastern Junior College, Sterling, Colorado

Bruce Chapman, First Wind LLC

Donnin Custer, Western Iowa Tech Community College, Sioux City, Iowa

Terrence (Mike) Daigle, Jr., Vestas—American Wind Technology, Incorporated

Craig Evert, Iowa Lakes Community College, Estherville, Iowa

Bruce Graham, Cloud County Community College, Junction City, Kansas

Petra Hemming, Maschinenfabrik Wagner GmbH & Co. KG (PLARAD)

George Lister, Texas State Technical College—West Texas, Abilene, Texas

Phil Parks, University of Maine at Presque Isle

Dave St Peter, University of Maine at Presque Isle

Caroline Zimmermann, REpower Systems SE

Mary Barton Akeley Smith

Victoria Blackwood, Maine Maritime Academy, Castine, Maine

Finally, I am indebted to the class of 2011 in Wind-Power Technology at Northern Maine Community College.

Wayne H. Kilcollins
March 2012

Introduction

KEY TERMS

American Wind Energy Association (AWEA)
availability
blade
Canadian Wind Energy Association (CanWEA)
control system
department of environmental protection (DEP)
direct drive
doubly fed asynchronous induction generator
down tower assembly (DTA)
drive system
drive train
Environmental Protection Agency (EPA)
Federal Aviation Administration (FAA)
foundation
gearbox
generator
horizontal axis wind turbine (HAWT)
hub
inverter
land use and regulation commission (LURC)
lattice towers

main shaft
nacelle
National Occupational Competency Testing Institute (NOCTI)
Organization of the Petroleum Exporting Countries (OPEC)
pad-mount transformer
permanent magnet generator
pitch system
production tax credit (PTC)
rectifier
return on investment (ROI)
rotor assembly
synchronous induction generator
transformer
tube towers
utility grid
vertical axis wind turbine (VAWT)
wild alternating current (wild AC)
wind-resource studies

OBJECTIVES

After reading this chapter and completing the review questions, you should be able to:

- Explain trends in the wind-energy industry.
- Identify some wind-turbine manufacturers.
- Describe some of the differences among utility-scale wind turbines.
- Identify the components of a wind-turbine system.
- Describe some of the steps in developing a wind farm.
- Describe the role of a control center used by a wind-turbine manufacturer.

- Identify the skills necessary for a wind-energy technician.
- Describe some of the wind-technician specialty tracks.
- Describe some of the career paths available to wind-energy technicians.
- Describe the role of a wind-turbine preventative-maintenance program.

INTRODUCTION

Wind-power and other alternative-energy developments have taken hold in the United States and many other countries over the last two decades. Wind-generated electricity dates back to the 1860s when Charles Brush developed a

12-kilowatt (kW), direct current (DC) dynamo turbine to supply power for his domestic lighting. Not until the 1930s and 1940s did wind-generated electricity become popular for rural applications. Wind generators were available with voltages ranging from 6 to 110 volts that could be used to charge batteries for lighting or to operate home appliances such as refrigerators. Some radio manufacturers such as Zenith even gave away wind-generator systems to promote the sale of their products. Wind-generated electricity is not a new concept, and many of the technologies have been around for years, but innovations in recent years have made the systems less expensive to manufacture due to economies of scale. It is less expensive per wind-turbine unit to produce when the manufacturers are set up to mass produce the systems. Henry Ford promoted the concept of economy of scale with his automobile production lines in the early 1900s. It is easy to see what this philosophy did for the automobile industry. Manufacturing advancements have also led to the larger, more efficient systems available today. Government subsidies have been another motivating factor in the growing wind-energy industry. Subsidies in alternative-energy projects over the past decade have enabled wind-energy developers to leverage their investment and ensure a profit. **Figure 1-1** shows the correlation between government incentives and wind-energy development stability. Companies would not be able to make large investments and stay in business without a profit margin. One U.S. incentive for wind energy is the **production tax credit (PTC)** paid per kilowatt-hour generated for the first 10 years of service. The current PTC was increased from 1.9 cents to 2.1 cents per kilowatt-hour and extended through 2012 as part of the American Recovery and Reinvestment Act of 2009. Several other federal and state programs are available, including tax incentives for homeowners to install residential alternative-energy sources such as wind, solar thermal, and photovoltaic systems. Utility-scale electrical generation projects that use natural gas, coal, and oil along with nuclear energy have enjoyed government subsidies for many years. Many of the incentive programs and technology innovations have been motivated by instability in foreign oil supplies. One leading factor in the instability was brought about by the 1973 oil crisis.

The 1970s Oil Crises

In 1973, Arab members of the **Organization of the Petroleum Exporting Countries** (OPEC) proclaimed an oil embargo in response to American military supplies being provided to Israel during the 1973 Arab–Israeli War. In the developed world, the embargo created wide-scale fuel shortages and inflated energy costs for several years. During this period, OPEC members realized that they could increase their profits through a collective effort to control member countries' oil production. This instability showed most oil-importing nations of the world that a better way was needed to ensure their national energy security. European countries such as Denmark, Germany, and Spain promoted community- and commercial-scale wind-energy projects to promote homegrown energy resources and reduce their reliance on foreign petroleum. The United States promoted alternative-energy projects and energy-conservation programs during the 1970s but drifted away from these technologies when foreign petroleum resources and imports seemed to stabilize in the 1980s. The turn of the 21st century, however, brought more instability and rising energy costs to the petroleum market.

Renewed Interest in Wind Energy

Energy instability and possible climate change issues brought on a renewed desire to establish homegrown alternative-energy resources in the United States. Wind-energy technology had matured to the point that large-scale installations were becoming competitive with other energy sources. Wind-energy production costs decreased significantly from the 1980s to the early 1990s: the estimated cost of wind-energy production in 1993 was 7.5 cents per kilowatt-hour. Current estimates have wind-energy production costs around 3.5 cents per kilowatt-hour. This decrease in production cost has established the viability of utility-scale wind power.

Installed wind capacity in the United States has increased dramatically since the pioneering days of the 1980s. California was the first state to develop large wind farms; by the mid-1990s, these farms produced roughly one third of the global wind-generated electricity. Since the 1990s, wind farms have been developed in several locations throughout the United States and around the world. Several European countries have had community and utility-scale projects in place since the late 1970s and early 1980s. Germany led the way until 2008 with the most installed wind power of any nation until the United States surpassed it with 25,369 megawatts (MW) to Germany's 23,902 MW. China closed in on the United States in 2009 and surpassed it by the end of the year. **Figure 1-2** shows installed wind power of several countries at the end of 2009. The top five countries with the

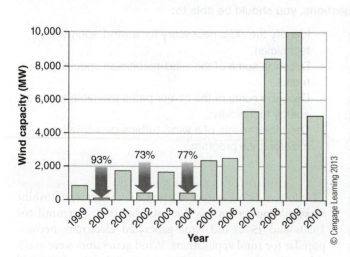

FIGURE 1-1 *Correlation between government incentives and wind-energy development stability*

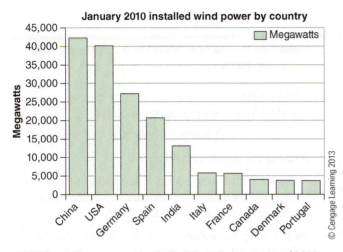

January 2010 installed wind power by country

FIGURE 1-2 *Top 10 countries for installed wind power as of 2010*

most installed wind power by the end of 2009 were China, the United States, Germany, Spain, and India. As of the end of 2009, there were 188.2 gigawatts (GW) of global installed wind power compared to 2008 with 122.2 GW.

Wind power has made great strides over the past several years, and this enthusiasm has brought several equipment manufacturers to the U.S. market. Wind-turbine manufacturers with a presence in the United States include General Electric (GE), Vestas, Clipper Wind Power, Siemens, and Gamesa. Foreign manufacturers locating in the United States have been able to take advantage of available material, technical, and labor resources. This is a familiar trend for other import markets in the United States such as automobiles. Reductions in transportation and elimination of import tariffs make foreign manufacturers more competitive with domestic manufacturers in the bidding process for U.S. wind-farm projects. The presence of these manufacturers in the United States has enabled the boom in wind-farm development. Many of the wind-farm projects are so large that one single supplier cannot provide enough turbines to complete the projects on schedule. An example would be the Horse Hollow Wind Energy Center project in Texas. This project has 291 GE 1.5-MW wind turbines and 130 Siemens 2.3-MW wind turbines to put out 735.5 MW at full production. This is one of many wind projects in the United States today.

Wind-Farm Projects

Utility-scale wind projects have been producing power in several countries in Europe such as Denmark, Germany, and Spain since the late 1970s. California was the first U.S. state to develop utility-scale projects in the mid-1980s. California has strong summer winds that blow through mountain passes and create ideal areas in which to experiment with wind-energy technologies. Areas such as Tehachapi, Altamont Pass, and San Gorgonio Pass were home to the first U.S. wind-farm

projects. Since the late 1980s, Texas has developed the most installed wind power in the United States with projects such as Big Spring, Roscoe, Delaware Mountain, Sweetwater, and Horse Hollow. Estimated total installed nameplate capacity for Texas projects exceeded 10,135 MW by August 2011 (AWEA). The wind-energy boom did not stop in California and Texas. Since the turn of the 21st century, states such as Iowa, Minnesota, and Washington, among others, have developed utility-scale wind projects. As the trend to develop more wind programs continues, projects are underway to install wind turbines off the coast of Massachusetts, Maine, and New Jersey. Wind turbines have increased in size and efficiency since the first utility-scale units were installed in California. Turbine size has grown with technology advances in materials and blade development.

WIND-ENERGY DEVELOPMENT IN THE PAST DECADE

Over the past decade, wind-turbine systems have changed in size and complexity to meet the ever-increasing demand for renewable or so-called green energy. In the late 1990s and early 2000s, wind turbines typically ran around 250 kW. These turbines were large in comparison to the 25-kW units that were being installed in California and parts of Europe during the 1980s. Increasing wind-farm output then required constructing more and more turbines over limited geographical areas. **Figure 1-3** shows a ridge in Tehachapi, California, crowded with hundreds of wind turbines mounted on **lattice towers**. The effect of this crowded farm is not visually appealing. Reducing the number of turbines was inevitable to improve the use of limited land

FIGURE 1-3 *Hundreds of wind turbines on lattice towers crowd ridge near Tehachapi, California*

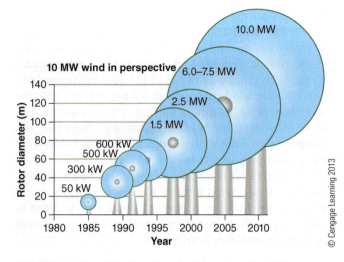

10 MW wind in perspective

FIGURE 1-4 *Growth in size of wind turbines since the 1970s*

resources. To increase farm output capacity and reduce the number of turbines, turbines had to grow in physical size and power rating. Since 2000, the power capacity of many wind turbines has increased to around 3 MW (3,000 kW). **Figure 1-4** shows the size evolution of wind turbines since the 1970s. Increased output has reduced the turbine density of farms but has created other challenges to farm operators and manufacturers. These challenges include blade changes, tower design improvements, power quality, and acceptance of wind farms near communities.

Blade Changes

The power produced by wind turbines is directly related to the swept area of the rotor assembly. This is true for both **vertical axis** and **horizontal axis wind turbines** (**VAWTs** and **HAWTs**, respectively). The relationship is given by

$$P_I = \frac{1}{2}\rho A v^3 C_p$$

where ρ is the density of the flowing air, A is the swept area of the rotor assembly, v is the velocity of the air, and C_p is the coefficient of performance for the rotor assembly. For HAWTs, the relationship of blade length to swept area is given by $A = \pi r^2$, where r is the length of the **blade**. To increase wind turbines' power output requires increasing the length of the blades. C_p is the aerodynamic efficiency of the rotor assembly to extract power from the airstream. This value can vary from 15% to a theoretical maximum of 59.3% (Betz coefficient), depending on the rotor assembly design and velocity of the airstream. Increasing the airstream velocity decreases the rotor assembly C_p value. C_p decreases with increased wind speed because a buildup of air is created upwind of the rotor assembly. This buildup or pillowing effect happens because of airflow resistance created by the tower and rotor assembly. To visualize this effect, consider the flow of water in a stream as it passes over and between stones.

SAMPLE PROBLEM

Determine the rotor swept area and blade length for a 3-MW wind turbine.

SOLUTION

Equation for instantaneous power output:

$$P_I \text{ (watts)} = \frac{1}{2}\rho A v^3 C_p$$

Assume the following conditions:

system-rated wind speed = 15 m/s, air density at standard atmospheric conditions = 1.225 kg/m³ (sea level and 15°C), and C_p = 30%

Rearranging the equation to solve for the unknown value of swept area, we get:

$$A\,(\text{m}^2) = \frac{3,000,000 \text{ watts}}{[\frac{1}{2} \times 1.225 \text{ kg/m}^3 \times (15 \text{ m/s})3 \times 0.30]}$$

$$= 4,837.5 \text{ m}^2$$

$$A = \pi r^2$$

Rearranging to solve for the unknown value of r (blade length);

$$r\,(\text{m}) = (4,837.5 \text{ m}^2/\pi)\,\frac{1}{2} \text{ or the square-root of } 1,540 \text{ m}^2$$

$$= 39.2 \text{ m}$$

The blade length required to develop 3 MW of instantaneous power output is around 40 meters (130 feet). Check out some of the manufacturers' brochures for 3-MW wind turbines on the Internet; they typically have 40- to 50-meter blades. For instance, Vestas manufactures a 3-MW wind turbine with model number V80. The 80 in the model number indicates that the rotor diameter is 80 meters. The blade length will vary, depending on the manufacturer's design estimation of C_p. A decrease in C_p will create a need for longer blades to achieve the same instantaneous power output.

The water builds up over the stones, creating a disrupted flow or turbulent appearance. You cannot see this disrupted airstream, but it is happening. This also shows why the C_p cannot reach 100%. If the rotor assembly were to stop the wind flow completely, then it would act like an air dam. If air was not allowed through the rotor assembly, it would not spin and convert rotational energy into electricity by the **generator**. C_p is approximately 30% for most utility-scale wind turbines. Manufacturers of utility-scale wind turbines typically list their system efficiency at a rated wind speed and power output level.

The sample problem shows the relationship between blade length and power output of a wind turbine. The increase in wind-turbine power output over the past decade has grown from 125 kW to 3 MW, so the blade length has increased from

FIGURE 1-5 *Production area for a wind-turbine blade showing an open mold assembly*

FIGURE 1-6 *Blade assemblies being transferred from a ship to trailers for transportation*

around 10 meters to 50 meters. This fivefold increase in blade length has created technical issues for manufacturing, transportation, and assembly. These issues also include increased material strength, weight, flexing, and tip speed. Some of these topics will be covered in later discussions.

Manufacturing Issues

The manufacturing changes to accommodate these massive blades have created the need to purchase and process larger volumes of resins, glass and carbon structural fibers, and other structural components used within the assemblies. Processing these blades has required larger manufacturing equipment and handling equipment and more facility space. **Figure 1-5** shows a production area for a wind-turbine blade mold.

Transportation Issues

Transporting 50 meter blades is not an insurmountable task, but carrying them across oceans, railway lines, interstate highways, rural roads, and city streets presents challenges. **Figure 1-6** shows blade assemblies being loaded from a ship to trailers for transportation to a wind farm. In many cases, the wind-farm developer has paid to have traffic lights and overhead utility lines moved and road corners upgraded so that trailers can move through communities along the way. This is a disruption to the daily activities of these communities, but it can be managed by proper planning and fostering community support. Once the blades make it to the wind farm, the roads must also provide clearance and support for the long blades, tower components, and crane components.

Assembly Issues

Assembly of the blades to the **hub** is completed at the wind farm as the turbine is being constructed. The three blades may be assembled to the hub and raised to the **main shaft** as a single unit. This method requires two cranes working

together in unison. This author has seen this assembly technique at work. It is amazing to watch experienced crane operators and assembly personnel perform this type of lift. **Figure 1-7** shows a complete rotor assembly being lifted.

Another method is to lift and assemble one blade at a time to the hub. This technique is more time consuming and also requires a crane and assembly personnel who are qualified and experienced. If the process is not followed safely and precisely, personnel and equipment can be quickly lost. **Figure 1-8** shows the blade assembly being lifted to a hub already mounted to the **nacelle** assembly.

A further advance in blade design has been to manufacture these long blade assemblies in two portions that can be bolted together at the wind farm during tower construction. This blade-manufacturing technique makes the blade easier to transport. Another complication to this turbine-construction process is assembly at sea. This requires

FIGURE 1-7 *Complete rotor assembly being lifted to the machine head*

B. Graham Photograph

FIGURE 1-8 *Blade assembly being lifted to hub already mounted on nacelle assembly*

a jack-up barge or other stable floating platform that will support the turbine components and the crane needed for the turbine assembly. **Figure 1-9** shows the assembly of a rotor at sea. Offshore projects are becoming more attractive as turbine size continues to grow. We will discuss offshore wind turbines later in the chapter.

Tower Assembly

Wind-turbine tower assemblies of the 1980s were manufactured as truss or lattice structures like those used for communication towers. **Figure 1-10** shows a lattice-type tower assembly. Some turbine manufacturers still use this tower design for residential and smaller commercial applications. The lattice-type tower is made of many components that can easily be transported and assembled on site. This tower design can also be manufactured as several manageable sections at the manufacturing facility and transported to the wind farm. Depending on the number of preassembled components, this construction method can take considerable time to erect compared to a tube-tower assembly. **Tube towers** are made of rolled steel plate that is welded together in the shape of a tapered cylinder. **Figure 1-11** shows an example of tube-tower sections ready for assembly. The welded-sheet-steel method of fabrication is similar to the construction of

© Cengage Learning 2013 and a W. Kilcollins photograph

FIGURE 1-10 *Lattice-type tower assembly*

REpower—Christian Eiche Photograph

FIGURE 1-9 *Assembly of rotor at sea*

University of Maine at Presque Isle

FIGURE 1-11 *Tube-tower sections ready for assembly*

boilers, large sea vessels, and floating platforms. Three or four of these tube sections can be erected into a tower assembly in a day with a large crane and an experienced crew. Fabricating the tube at a remote manufacturing facility and transporting it to the wind farm on flat-bed or specially designed carriers saves construction time at the site and enables the tower components to be made under strict quality and environmental control conditions. This manufacturing method has become the standard for utility-scale wind turbines over the past decade.

Power-Control Systems

Wind turbines produce electrical power by extracting energy from a moving fluid (air), the same fundamental process used in nuclear, natural gas, coal, biomass, and hydro projects. Nuclear, natural gas, coal, and biomass systems heat water to generate high-pressure steam that flows through a turbine to spin the generator input shaft. Hydro projects use the controlled flow of water through a turbine to spin a generator input shaft for electrical output. Wind turbines use airflow created by the sun's uneven heating of Earth's surface. The drawback created by this natural process is inconsistency in wind speed and direction throughout the day and seasons. This fluctuation can create highs and lows in production power output and alternating current (AC) frequency. Generator power output and frequency are a function of input mechanical power (torque) and rotational speed (revolutions per minute, or RPM) of the prime mover. The prime mover for the wind turbine is the rotor assembly, and its source of kinetic energy is the air flowing past the blades. Generator output power can be calculated using

$$P_I(\text{W}) = \frac{[\text{torque (N-m)}^* \times \text{RPM}]}{9.549}$$

and the output frequency can be calculated using

$$f(\text{Hz}) = \frac{(\text{RPM} \times \text{number of magnetic poles})}{120}$$

Both of these relationships require a consistent rotational speed (RPM) for grid-quality power. With a turbine system that uses the energy from moving high-pressure steam or water, the generator output can be controlled by adjusting the flow of these fluids. Wind cannot be controlled this way, so the wind turbine is at the mercy of nature. Wind-turbine manufacturers have used a few fundamental methods to control the power quality from this inconsistent energy source. One is to only run the turbine when the wind speed is within a specified range. This method uses a **synchronous induction generator** but is limited in production output to times when the wind speed is favorable. The next method would be to produce **wild alternating current** and use a power-conversion system to produce a controlled AC frequency

*Newton meters

SAMPLE PROBLEM

Determine the torque created by the wind on the turbine rotor assembly and the RPM to supply 250-kW to 60-Hz power output to the utility grid using a four-pole generator and a 50:1 gearbox ratio. Assume 100% system efficiency.

SOLUTION

Determine the RPM value of the four-pole generator to produce 60-Hz power output. Rearranging the basic equation $f(\text{Hz}) = (\text{RPM} \times \text{number of magnetic poles})/120$, we get:

$$\text{RPM} = \frac{120(60)}{4 \text{ poles}}$$

$$= 1{,}800 \text{ RPM}$$

We may now use the generator speed (RPM) in the power output equation to determine the torque requirement. Rearrange the variables to isolate torque:

$$P_I(\text{W}) = \frac{[\text{torque (N-m)} \times \text{RPM}]}{9.549}$$

so

$$\text{Torque (N-m)} = \frac{P_I(\text{W})(9.549)}{\text{RPM}}$$

$$\text{Torque (N-m)} = \frac{250{,}000 \text{ (W)}(9.549)}{1{,}800 \text{ RPM}}$$

$$= 1{,}326 \text{ N-m}$$

The gearbox has to supply a constant torque of 1,326 N-m for the generator to maintain full output at 60 Hz. The 1,800 RPM would be much too fast to spin large fiberglass blades on a rotor assembly. So what is the rotor assembly RPM? It is 1,800RPM/50 = 36 RPM; the rotor assembly would have to spin at 36 RPM to produce 1,800 RPM gearbox output. The torque input would be determined the same way using the 50:1 gearbox ratio. Remember, we assume 100% efficiency of the **drive train**. This is never the case, but it will allow us to calculate the input torque without knowing the gearbox efficiency: 1,326 N-m (50) = 66,300 N-m.

The rotor assembly supplies 66,300 N-m of torque and 36 RPM by wind blowing over the blades to maintain full output at the grid frequency.

and voltage output. This power-conversion system would use a **rectifier** electronic circuit to convert the wild AC to a DC output and then use an **inverter** electronic circuit to produce a controlled stepped AC output. This method may use a **permanent magnet** or electromagnet generator to produce a varying AC output signal. **Wild AC** is considered any AC output frequency that varies above or below the local grid value. In the United States, Canada, and portions of South

America, the grid frequency value is 60 hertz (Hz), whereas European and other countries use 50 Hz. Some wind-turbine manufacturers control the output power and frequency by using a **doubly fed asynchronous induction generator** and sophisticated electronic power-conversion equipment. This system will allow the generator RPM to vary within a much wider range than the synchronous generator system mentioned previously and still maintain the proper utility-grid voltage and frequency through system feedback from the power-conversion equipment.

As mentioned previously, the wind is not consistent like other energy sources, so the wind-turbine **control system** has to adjust the output voltage and frequency using power electronics or shut down the wind turbine when these parameters run out of tolerance. Another control method that we will discuss later uses blade pitch controls that are used in many wind-turbine designs today. Pitch control enables the turbine's automatic control system to vary the blade pitch to increase or decrease the torque supplied to the generator. Blade pitch control is also used as a fail-safe primary brake system to stop the turbine in an emergency. Earlier wind-turbine designs incorporated fixed-pitch rotor technology so they could not take advantage of varying pitch to maintain power output.

Public Acceptance of Wind Farms

Gaining public acceptance of larger turbines and increased numbers of wind farms has met with mixed reaction. The general public is in favor of a reduction in greenhouse gases and fossil-fuel use, but how to achieve this goal has been up for debate for the past several years. Wind farms and utility-scale turbines can be collocated with communities given project planning and public education. Community involvement, proper planning, and implementation of local regulations along with consistent implementation of state and federal regulations will allow the United States to achieve its goal of 20% energy production from wind by 2030. Educating the public and fostering community acceptance of wind power as an alternative to fossil-fuel use is the key to going forward with any renewable-energy source.

Project-Development Process
Project planning enables the developer to determine the best location to place wind turbines for maximum output. The old business adage of "location, location, location" is a necessity for wind turbines as well. When the developer is preparing to invest millions of dollars or euros to purchase and install wind turbines, they need to choose the best location for the project. This process requires referring to National Weather Service climate data, National Renewable Energy Laboratory data, or other reputable **wind-resource studies** for a general

project location that is adjacent to a reliable transmission corridor.

Figures 1-12 and 13 show maps of wind-speed classes for the United States and the Canadian maritime province of Prince Edward Island. Maps such as these are useful tools for determining the available wind resources for different geographic areas. Some wind resource maps such as the one for Prince Edward Island superimpose wind resources over electrical transmission corridors to enable a developer to narrow a general area for suitable wind-farm development. Once a general geographic location is determined, the developer needs to contact community officials and landowners for test sites to determine the best location for the project. Some incentives for landowners and the community to work with the developer are land-use royalties and tax payments for operation within the community. Once a specific project location is determined, the developer then works with the state's **land use and regulation commission (LURC)** and **department of environmental protection (DEP)**, the federal **Environmental Protection Agency (EPA)** and **Federal Aviation Administration (FAA)**, along with other regional organizations to prepare environmental and wildlife impact studies that are required for final permitting for a site. Other countries have equivalent governmental regulatory agencies and environmental organizations to promote sustainable use of natural resources. Following environmental protocols alone is not sufficient; it is also important to foster community involvement through town hall meetings and education forums throughout the process to gain public acceptance.

Land-Use Royalties
When wind-project developers find a suitable location for a farm, they work with landowners to determine if the project is acceptable with the landowners' current activities. Many farms have been collocated with agriculture, dairy, forestry, and recreation activities without negative impact. An incentive to landowners for long-term access to their land has been contracted payments in the form of royalties. Land-use royalties or leases are paid to landowners for the placement of turbines, access right of ways, transmission lines, and other ancillary equipment. Collocating wind-project equipment with other land uses allows for added income to landowners or the ability to collect income for land that might not necessarily be used for other activities. Different royalty-payment schemes have been implemented in the United States, Canada, and Europe based on developer philosophy, land cooperative requirements, and local regulations. Royalties in the United States and Canada typically have been paid to the landowner directly affected by the placement of equipment. This has been an income opportunity for specific landowners but does not benefit adjacent landowners. This has created issues for the acceptance of some wind projects. Some European developers and countries such as Germany and France have promoted a philosophy of paying the affected landowner a major portion of the royalty and dividing

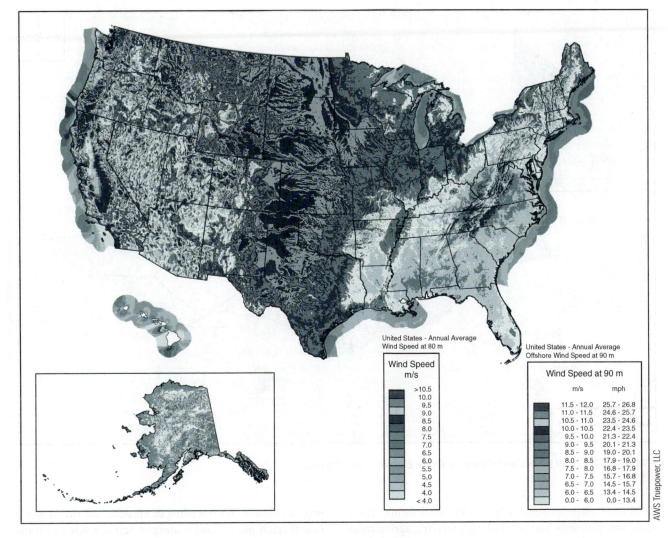

FIGURE 1-12 *Map of continental and offshore wind-speed classes for the United States*

the remainder among adjacent landowners. This philosophy has also been introduced on Prince Edward Island to foster local support for wind projects. Other European countries and Canadian provinces such as Ontario have sponsored local actions to ensure equitable royalties for landowners. Communities with wind farms also get an added benefit of increased tax revenues that come with increased economic development.

Community Taxes

Revenues from wind-farm developers, added employment opportunities, and support opportunities for small business in communities with wind farms have provided tax relief to other landowners and stakeholders. Additional tax revenues have assisted in supporting schools and community infrastructures improvements. These improvements might not have been completed without the added business development brought by wind-farm development. Wind-project development and investment within a community must be

planned to minimize potential environmental impacts for later generations. This author has seen some of the short-term gains and long-term disasters in visits to developing countries. These short-term gains have created eroding hillsides along with polluted rivers and streams that provide recreation for sports enthusiasts. Excessive erosion to hillsides and ridges in Tehachapi, Altamont Pass, and San Gorgonio Pass have been among the complaints about earlier wind projects in California during the early "wind rush" days. **Figure 1-14** shows exposed soil along a wind-project road where the hillside was cut back without reseeding to prevent erosion. Over time, rain will wash the soil into nearby streams. Fortunately, closer project oversight and improved conservation practices have reduce the environmental impact of wind-development projects.

Land Conservation

Oversight by the EPA and DEPs and LURCs has led to improved conservation practices that have limited or

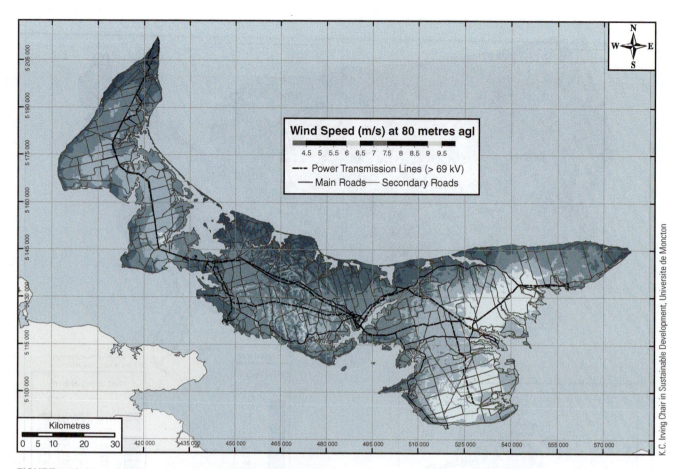

FIGURE 1-13 *Map of wind-speed classes for Prince Edward Island, Canada*

eliminated environmental issues with wind farms. These same organizations have worked with forestry, agricultural, and infrastructural development over the past 50 years to improve land-conservation practices. Practices such as setbacks for cutting of forest areas from streams and rivers,

FIGURE 1-14 *Exposed soil along a wind-project road without reseeding to prevent erosion*

proper contour and reseeding of hillsides and ridges, use of runoff catch basins, and routing of access roads and transmission lines around wetlands have all played a part in improving land conservation. These practices can be seen at more than just wind-farm development sites. Driving by new housing developments, business campuses, and road construction, you can see contours added to the landscape along with runoff catch basins that have been added to allow rainwater to collect and soak into the ground instead of running off and creating erosion problems. These practices also prevent rainwater from carrying sediment and other runoff contaminants into the streams and rivers and creating pollution problems.

Wildlife Conservation

Wind farms can coexist with most wildlife and domestic herds, but they can create problems if placed near migratory bird routes. Lessons learned from the placement of skyscrapers, cell-phone towers, and large electrical transmission links have shown that planning projects to minimize the impact on migratory bird routes is important to an overall wildlife conservation plan. **Figure 1-15** shows various bird mortality rates with human-created obstacles. These data show that wind turbines and wind farms have lower mortality rates than other structures and natural predators. In addition, the

Mortality Cause	Estimated Deaths per Year
Communication tower collisions	4 million to 50 million
Power-line collisions	10 million to 154 million
Vehicle and road collisions	10.7 million to 380 million
Building collisions	100 million to 1 billion
Cats	100+ million
Pesticides and other poisons	72 million
Oil and wastewater pits	2 million
Wind-turbine projects	1 to 7 per wind turbine

© Cengage Learning 2013

FIGURE 1-15 *North American bird mortality rates*

placement of wind turbine towers has required planning for issues other than land and wildlife conservation. Their visual impact has required proper placement to preserve the natural aesthetics of the landscape.

Aesthetics or Visual Impact

Preserving the landscape for wind projects placed near communities and recreation areas has been very important to stakeholders that live in the proximity, use, and regulate these areas. Utilizing the natural contours of the land and placing transmission structures, utility buildings, and transformers so that they cannot be seen from a distance helps to preserve the original landscape. **Figure 1-16** shows the placement of wind-farm equipment to reduce its visual impact. Other practices include the use of colors that blend or contrast with the area to produce the required visual impact of the new project development. **Figure 1-17** shows a color scheme that makes the wind turbines stand out. This practice helps pilots of various aircraft avoid these terrestrial obstructions. Another paint scheme for reducing the visual impact is that made by Enercon. The company paints the base portion of the tube tower with varying shades of green to reduce the visual impact near the ground.

REpower—Jan Oelker Photo

FIGURE 1-17 *Color scheme to improve visual contrast of wind turbines with surroundings*

Another issue with older wind farms is the removal of unused or decommissioned equipment and keeping the area neat. Neighbors and visitors do not want to see piles of damaged turbine components and rusted tower components littered over the landscape. First impressions are important, and someone seeing a poorly kept wind farm with damaged equipment creates a bad impression. Visual impact is not the only issue that neighbors may have with

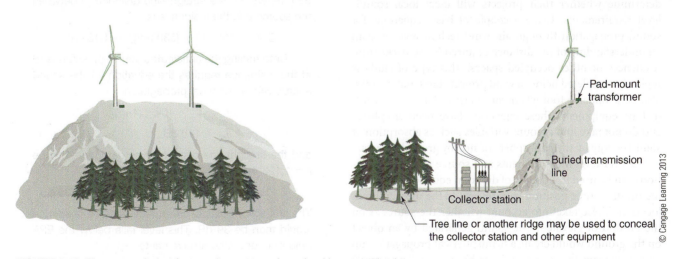

Pad-mount transformer

Buried transmission line

Collector station

Tree line or another ridge may be used to conceal the collector station and other equipment

© Cengage Learning 2013

FIGURE 1-16 *Placement of wind-farm equipment to reduce visual impact on landscape*

wind projects. Noise has also created issues with adjacent landowners.

Noise Impact

Ordinances from the EPA, states, and communities have been established to keep industrial developments and wind projects from creating health issues or nuisance sound problems with adjacent landowners. Sources such as vehicles on freeways, aircraft, and normal conversation create sound waves. The energy propagated from these sources travels through the air and creates vibration within the human ear that is perceived as sound. Sound is perceived as loudness and pitch by the brain. Loudness is perceived as volume or sound-level pressure and pitch as the tone or frequency. Each of these elements can create physiological problems, depending on their level or frequency range. There have been several studies on the effects of sound on humans, including industrial settings with short- and long-term exposure conducted by the Occupational Health and Safety Administration (OSHA), community settings by the EPA and DEPs, and independent studies conducted by the **American Wind Energy Association (AWEA)** and the **Canadian Wind Energy Association (CanWEA)**. Discussions in Chapter 2 will cover information related to OSHA and safe work practices around operating equipment such as wind turbines.

EPA and DEP discussions focus on the sound-level guidelines around a residence, which have set a maximum value of 45 decibels (dB). These agencies have also had discussions around lower sound levels that relate to different wind velocity levels and sound-level changes because of the presence of new wind-turbine installations. Guidelines developed by the Ontario (Canada) Ministry of Environment in 2008 also match those published in the United States with respect to maximum level and changes because of new installations (CanWEA, *Wind Turbine Sound and Health Effects*). These are available through a search of the Internet or by obtaining a copy of local community ordinances that were developed around national and regional standards. Public-policy guidelines allow wind-power developers to determine whether their projects will meet local sound-level requirements. Using a couple of basic equations for sound propagation through air, wind technicians can gain an understanding of the distance required for setbacks from residences or other occupied spaces. This type of study is typically completed before a wind project starts, but it would also be a valuable tool when considering changes to wind-turbine equipment. These equations have been simplified and do not take into account variables such as absorption of sound energy by the atmosphere or nearby ground objects.

Sound-level measurements and source energy calculations can be used in a couple of different equations, depending on the source position with respect to the measurement location. These take into account whether the source is on the ground or elevated. The sound generated by an object on the ground would create a hemispherical propagation up and away from the source and Earth's surface. An example of this would be a train or automobile. This would be represented by

$$L_w = L_p + 10\log(2\pi r^2) \quad \text{or} \quad L_p = L_w - 10\log(2\pi r^2)$$

depending on whether we are calculating the sound-power level at the source or estimating the sound-pressure level, which would be measured at a distance from the source. The variables are:

L_w = sound-power level at the source,

L_p = sound-pressure level measured with a sound meter, and

r = distance from the source in units of meters.

The second set of equations cover a source with spherical propagation such as aircraft at altitude or wind turbines positioned on a ridge or any height that would minimize the ground effects between the sound source and location of the measurement. This would be represented by

$$L_w = L_p + 10\log(4\pi r^2) \quad \text{or} \quad L_p = L_w - 10\log(4\pi r^2)$$

SAMPLE PROBLEM

Determine the straight-line distance between a wind turbine located on a ridge and a nearby residence along with the anticipated sound-pressure–level measurement at the residence. Wind-turbine sound-power level is given as 110 dB at the nacelle. Distance from the residence to the centerline of the wind-turbine tower is 1,000 meters, the ridge height is 450 meters, and the turbine tower height is 80 meters.

SOLUTION

Straight-line distance (r) between the source and residence can be calculated using the Pythagorean theorem, $C^2 = A^2 + B^2$, in which A and B are the sides of the right triangle and C is the hypotenuse. Here C = $(A^2 + B^2)1/2$, A = 1,000 meters, B = 450 m + 80 m = 530 meters, so the straight-line distance (r) between the source and the residence is

$$C = [(1{,}000 \text{ m})^2 + (530 \text{ m})^2]\tfrac{1}{2} = 1{,}132 \text{ m}$$

Determining the anticipated sound-pressure level at the residence requires the equation for the sound source with a spherical propagation:

$$L_p = L_w - 10\log(4\pi r^2)$$
$$L_w = 110 \text{ dB}$$

and the straight-line distance was determined to be r = 1,123 meters.

$$L_p = 110 \text{ dB} - 10\log[4\pi \times (1{,}123 \text{ m})^2]\text{dB} = 39 \text{ dB}$$

The anticipated sound-pressure level at the residence would then be 39 dB. This level falls below the EPA guideline for sound requirements.

Wind-Industry Trends

Wind-turbine manufacturers have made great strides in developing ever larger turbines. A few years ago, manufacturers were working to develop turbines around 5 MW, and recent wind-power journals and science and technology magazines are presenting Enercon and Vestas wind turbines at 7 MW and Clipper Wind Power with a proposed 10-MW offshore turbine. Increasing turbine output also requires an increase in swept rotor area as previously discussed. To make these turbines operate more effectively, they will need higher average wind velocities. Other trends are to produce wind turbines with a **direct-drive** system that eliminates the need for a **gearbox**. This technology eliminates the associated gearbox maintenance but requires even larger generators to accommodate the number of magnetic poles. A previous discussion introduced the equation that relates number of generator magnetic poles, RPM, and output frequency: $f(\text{Hz}) = (\text{RPM} \times \text{number of magnetic poles})/120$. Rearranging the equation to determine the number of magnetic poles would be:

$$\text{Number of magnetic poles} = \frac{120\,[f(\text{Hz})]}{\text{RPM}}$$

These larger turbines have a rotor rotation of approximately 20 RPM, so the number of magnetic poles to produce 60 Hz output would be 360. To continue increasing turbine sizes, farm operators are moving to offshore locations to take advantage of stronger and more consistent wind speeds. European countries have been developing offshore projects for the past two decades, and the United States and Canada have been investigating their feasibility over the past few years. The Canadian province of Ontario and U.S. states bordering the Great Lakes along with states such as Massachusetts, Maine, and New Jersey along the Atlantic coast are in the development stages with state regulations and are working with federal agencies to make offshore wind a reality. Offshore trends and the continued growth of wind farms around the United States and other countries will require larger numbers of skilled wind technicians to construct, commission, and maintain current and future fleets of wind turbines. To ensure wind technicians are trained to a national standard, the AWEA, wind-turbine manufacturers, colleges, and associated professional groups are developing a set of standards to be implemented by training programs.

NATIONAL STANDARD SKILL SET FOR WIND TECHNICIANS

The process to create a basic skill set for wind technicians was initiated at the AWEA Summer Wind Institute at Columbia Gorge Community College in The Dalles, Oregon, in 2008. A subsequent AWEA meeting the following summer in Washington, D.C., formed the AWEA Education Working Group to continue the process of developing a standardized skill set with input from government and industry stakeholders. This author, along with other college instructors, college administrators, industry managers, government officials from the federal Department of Labor (DOL) and the Department of Energy (DOE), and AWEA coordinators defined the DOL job listing for a wind technician and drafted the first industry standard—the "Basic Skill Set"—for entry-level wind technicians. This skill set was further refined throughout 2009 and 2010 during subsequent Education Working Group meetings to include further feedback from industry, government, and academic organizations. States such as Maine and Texas have adopted similar standards for their wind-technician certification and apprenticeship programs through their respective departments of labor. Other states are in the process of adopting similar requirements for their wind-technician certification programs. The standard skill set will also be used by AWEA to develop a board certification exam. Several postsecondary training organizations participated in a pilot test program for the **National Occupational Competency Testing Institute (NOCTI)** during March 2010 (NOCTI). NOCTI has since made the wind technician certification test available on their Web site. Visit its Web site for further details on requirements and testing locations.

Another plan being developed to standardize U.S. training curriculum is the AWEA's seal-of-approval process. This program will enable training organizations to achieve a nationally recognized approval through a committee review process. As of fall 2011, AWEA had granted its seal of approval to seven colleges after reviewing their curricula, facilities, and other criteria as specified in the AWEA's seal-of-approval process. Further details on education resources and the developing wind-energy industry can be viewed at the AWEA Web site (http://awea.org/). Future steps in the process to set national standards for wind technicians will define skill sets necessary for advanced specialties in composites repair, major component inspection and repair, and major component replacement, along with other specialties to be defined by the industry. Most manufacturers have defined specialty tracks for their wind technicians, but the industry presently lacks continuity. The national standardization process coordinated by AWEA and the U.S. DOL is to ensure the continuity of skill sets across the industry.

This author has compiled a summary of wind-technician skills from personal field experience and through discussions with wind-turbine manufacturers, construction groups, industry organizations, and Maine's department of labor officials. These groups encompass wind-energy stakeholders around New England and the United States. Additional information for this summary has been gathered through discussions with European government officials, wind-turbine manufacturers, and training groups in Germany and Spain with the Maine Renewable Energy Mission to Europe during September 2009. **Table 1-1** shows a summary of wind-technician skills. The Maine International Trade Center coordinated the trade delegation made up of former Governor

TABLE 1-1

Summary of Required Technician Skills

Task	Safety[a]	Electronics	Computer[b]	Electrical	Mechanical	Troubleshooting	Repair and Replacement	Maintenance	Communication Skills[c]
Mechanical	X								X
Gearing					X		X	X	
Bearings					X	X	X	X	
Couplings					X		X	X	
Brakes					X	X	X	X	
Fluid power	X								X
Blade pitch		X		X	X	X	X	X	
Braking		X		X	X	X	X	X	
Pumps				X	X	X	X	X	
Valves				X	X	X	X	X	
Actuators				X	X	X	X	X	
Filtration				X	X	X	X	X	
Lubrication					X		X	X	
Oil sampling					X				
Electrical	X								X
Control		X	X	X		X	X	X	
Sensors		X		X	X	X	X	X	
Switching		X		X		X	X	X	
Communication		X	X			X	X	X	
Generator				X		X	X	X	
Power-conversion				X	X	X	X	X	
Surge-protection				X		X			
Environmental	X								X
Heating				X	X	X	X	X	
Cooling				X	X	X	X	X	
Pumps				X	X	X	X	X	
Fans				X	X	X	X		
Filtration							X	X	
Tower	X								X
Foundation					X			X	
Structure					X		X	X	
Access components					X			X	
Grounding and bonding					X	X	X	X	
Blades									X
Exterior					X		X	X	
Interior					X		X	X	
Lightning-protection				X					
Nacelle	X								X
Exterior					X		X	X	
Interior					X		X	X	
Hardware				X	X		X	X	
Safety equipment				X	X		X	X	

© Cengage Learning 2013

[a]Safety practices
Occupational health: OSHA and ANSI requirements
Environmental protection: EPA and DEP requirements
Safety practices: NFPA 70E, ANSI Z359, LOTO, PPE, HRC, working at heights, rigging
Emergency services standard (rescue): NFPA 1983
Weather related issues: hypothermia, hyperthermia, dehydration awareness, prevention, first-aid treatment, CPR, and AED proficiency
Electrical standards: NFPA 70 (NEC)

[b]Computer: Maintenance functions (defragmentation, software updates, and BIOS updates), file transfer, data entry, spreadsheets, word processing, networking, Internet searches

[c]Communication skills: reading, writing, oral, e-mail

John Baldacci and 20 Maine delegates to investigate wind-power programs in Spain and Germany. Some of the delegate's activities in Germany encompassed offshore wind projects and the development of technology focused on deepwater wind applications.

In Spain and Germany, this author spent time with manufacturers, government officials, wind-energy organizations, and training organizations to discuss wind-technician training programs, diversity of technician skills, and the types of jobs available to trained wind technicians. The trip brought insight into training needs within the wind industry for both the United States and Europe. The information gathered in Europe and other conference participation has been beneficial to wind-energy organizations and education groups to which this author belongs. Some of this information will be shared throughout this text to help the reader gain perspective on the wind-energy industry.

CAREER PATHS FOR WIND-ENERGY TECHNICIANS

The emerging wind-power industry has created opportunities for technicians with wind-project developers, site operators, manufacturers, and utility companies. These jobs encompass site planning, construction, commissioning, operations, maintenance, sales, transportation, infrastructure development, and manufacturing to name just some of the opportunities. The only limits to career opportunities are the motivation of the technician. **Table 1-2** lists many of the career opportunities available to wind technicians. Wind-turbine manufacturers and site operators have been providing opportunities to mechanical or electrical technicians for the past several years because there have not been programs in place that combine the technical skills of each discipline. Now several technical and community colleges throughout the United States and Canada are combining these technical skills in wind-technician–related programs. **Figure 1-18** lists some of the programs available for wind-technician training. Many of these programs are set up as two-year degree programs to enable students coming directly from high school to get exposure to a full spectrum of learning experiences. Other certificate programs are set up to fill in experienced students' skill gaps so they can work effectively with wind-turbine technologies. One such international certifying organization is BZEE, which is based in Germany. Its goal is to promote a uniform skill set standard for both North America and Europe. Several state apprenticeship programs are also available to help those in electrical or mechanical careers receive further training to bridge the gap needed to work in the wind industry.

These programs can provide a great start to careers in the wind industry and other related disciplines such as power-system construction, power generation, and transmission. The next section will discuss major sections of wind turbines and their functions. Some of these functions were previously discussed with power generation and the related mathematical analysis for determining input torque, frequency, and sound levels.

TABLE 1-2

Career Opportunities Related to Wind Power	
Opportunity	**Activities**
Site development	Site surveys, wind resource studies, and coordination with landowners and government agencies
Site construction	Earthwork, access roads, foundation, and infrastructure development
Equipment construction	Tower structural assembly, switching, transmission, rigging, and crane work
Transportation	Trucking, railways, cargo ships, coordination, intermodal, and port activities
Installation technician	Cabling, control system connections, troubleshooting, and inspection
Commissioning technician	System functional testing and documentation
Operations technician	Warranty, troubleshooting, component repair, and replacement
Service technician	Troubleshooting, system upgrades, and reliability studies
Maintenance technician	Electrical, mechanical, lubrication, and fluid-power equipment maintenance
Project specialist	Troubleshooting, site evaluation, transportation, and supplier coordination
Composite repair technician	Blade and nacelle inspection and repair
High-voltage systems	System analysis, switching, and utility coordination
Safety technician	Environmental health, safety coordination, and regulatory compliance
Remote monitoring operation	Assessment, troubleshooting, control, and service coordination
Service center operations	Parts, inventory, purchasing, maintenance, and shipping coordination
Training programs	Instructor, lab assistant, planning, and coordination
Manufacturing	Component assembly, quality control, systems development, fabrication, engineering, procurement, production coordination, and management
Field engineering	Field studies, assessment, and system upgrades

Each opportunity also includes moving into supervisory roles and management positions as experience and ambition to succeed warrant.

College/University	State/Province	Degree/Certificate
Delta College	Michigan	Associates
Kalamazoo Valley Community College*†	Michigan	Associates/Certificate
Northern Maine Community College	Maine	Associates
Lakeshore Technical College	Wisconsin	Associates
Lorain County Community College	Ohio	Associates
Mesalands Community College	New Mexico	Associates
Iowa Lakes Community College†	Iowa	Associates
North Iowa Area Community College	Iowa	Associates
Western Iowa Tech Community College	Iowa	Associates
University of Iowa	Iowa	Masters
Northwest Renewable Energy Institute†	Washington	Certificate
Texas State Technical College†	Texas	Associates/Certificate
Texas Tech University	Texas	Doctoral
Cloud County Community College†	Kansas	Associates
Colombia Gorge Community College†	Oregon	Associates
Clinton Community College	New York	Associates
Minnesota West Community and Technical College	Minnesota	Associates/Certificate
Lake Region State College†	North Dakota	Associates
Laramie County Community College†	Wyoming	Associates
Redstone College	Colorado	Certificate
Dabney S. Lancaster Community College	Virginia	Certificate
Holland College*	Prince Edward Island, Canada	Certificate
Northern Lights College*	British Columbia, Canada	Certificate
Lethbridge College*	Alberta, Canada	Certificate
Great Plains College*	Saskatchewan, Canada	Certificate
Groupe Collegia*	Quebec, Canada	Certificate
St. Lawrence College*	Ontario, Canada	Certificate
St. Clair College*	Ontario, Canada	Certificate

© Cengage Learning 2013

FIGURE 1-18 *Education opportunities in North America for energy (partial listing as of October 2011)*

*Bildungszentrum fur Erneuerebare Energien (BZEE) Accredited Schools (www.bzee.de/)
†American Wind Energy Association Seal of Approval Recipient (www.awea.org/)

UTILITY-SCALE WIND TURBINES

Utility-scale wind turbines provide electrical power that is transferred to the utility grid for consumption by the end user. The **utility grid** is the electrical infrastructure consisting of transmission lines, switching, transformers, and substations used to deliver electricity from the power producer to the customer. Manufacturers have developed similar systems with different equipment layouts that meet their operational and manufacturing philosophies. **Figure 1-19** depicts the variations in system layout for utility-scale wind turbines. Wind-turbine systems encompass low-voltage generation in the tower that is stepped up by an external transformer to grid voltage, low-voltage generation that is stepped up to grid voltage within the nacelle using internal transformers, gear-drive systems that turn multiple generators instead of a single large generator, and direct-drive systems that use a single large generator that produces output at relatively low RPM. The wind turbine is divided

into several sections, depending on their function. These sections are foundation, tower, machine head, rotor assembly, and grid connection. Each section will be further subdivided in subsequent chapters during discussions of maintenance activities.

Foundation

The **foundation** provides stability for the wind-turbine assembly. It is designed to match the chosen site's soil conditions and seasonal temperature and moisture variations. Typical tower foundations are made of a large mass of steel-reinforced concrete to provide strength and weight to maintain the tower. This foundation design is considered a spread-footing foundation. Some turbine sites have soil conditions that are not suitable for this style of foundation. These areas require a foundation assembly that uses long threaded rebar to secure the foundation to the sedimentary rock or granite formations in the subsoil. This foundation style enables a more compact foundation than the typical spread-footing

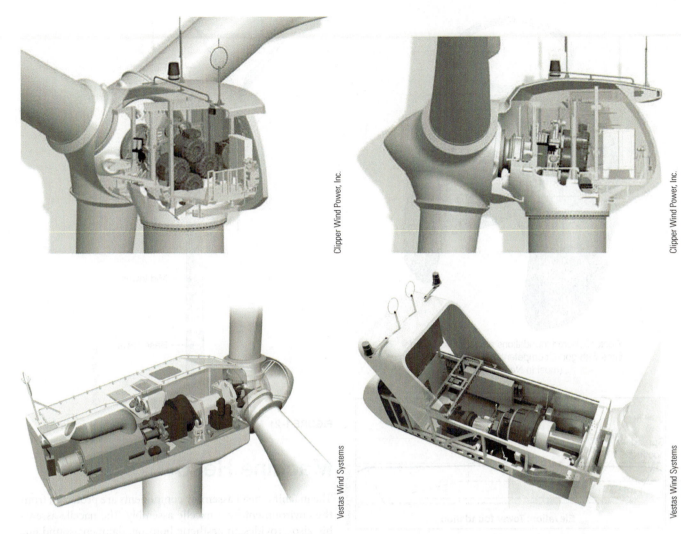

FIGURE 1-19 *Variations in system layout for utility-scale wind turbines*

foundation because it develops stability through the connection with the underlying rock formation and not through its large mass. **Figure 1-20** shows these two foundation assemblies. The next structure in the wind-turbine assembly is the tower. The design of this structure has evolved over time to reduce the number of field-assembly components and time required to construct the wind turbine.

Tower Assembly

The wind-turbine tower design has evolved over time to reduce the time and effort required at the site for construction. Current utility-scale towers may be constructed of tube sections made from prestressed concrete or steel plate that are stacked and bolted together with the use of a large crane and power-bolting tools. One design evolution of tube towers is to make the tube sections in two half shells that can be bolted together on site and then raised into position with a crane and bolted like a solid tube design. As the wind turbines continue to grow in size to the

3- and 4-MW utility scale that have recently been installed, the tower base diameter is ranging around 5 to 6 meters (16 to 20 feet). This diameter becomes difficult to transport over the roadway to the site. Many road and railway overpass structures have a maximum height restriction around 14 feet. This restriction increases the challenge of getting the tower tube structures from the manufacturing facility to the site without traveling along these restrictive routes. Constructing the tubes in two sections is one way to reduce this transportation challenge.

The completed wind-turbine tower assembly is further divided into sections, including the **down tower assembly (DTA)**, the midtower, and the top tower, which includes the yaw deck. **Figure 1-21** shows the major sections of a typical wind-turbine assembly. The DTA houses the control electronics and power-conversion system for many turbine designs. It is also the area designed to provide access to the machine head of the wind turbine. Turbine access from this area is through the use of a ladder, utility hoist, or crew lift equipment to enable transport of personnel, equipment, and tools to the machine head. **Figure 1-22** shows a typical

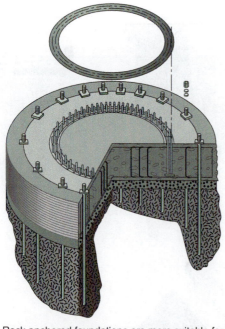

Rock anchored foundations are more suitable for sites with good, competent rock beneath turbines (most in New England)

Cianbro Coporation

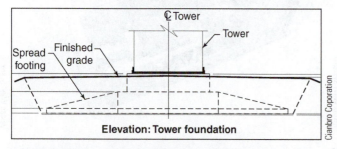

Elevation: Tower foundation

Cianbro Coporation

FIGURE 1-20 *Examples of tower foundation assemblies*

layout of the DTA and ladder-access equipment. The next component of the wind turbine is the machine head. This area contains an extension to the system controls and power production equipment.

First Wind Energy, LLC

FIGURE 1-22 *Typical DTA layout and ladder-access equipment*

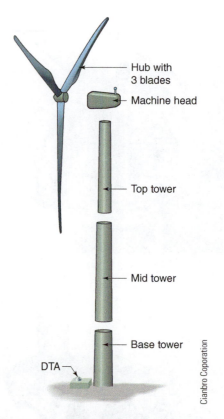

Hub with 3 blades

Machine head

Top tower

Mid tower

Base tower

DTA

Cianbro Coporation

FIGURE 1-21 *Major sections of typical wind-turbine assembly*

Machine Head

The machine-head assembly components are protected from the environment by a nacelle assembly. The nacelle assembly also provides an aesthetic housing, dampens sound energy created during operation, and improves the flow of air around the machine head and tower. Components in the machine head include control and communication electronics, drive assembly, generator(s), power-conversion electronics, heating and cooling equipment, yaw control, pitch control, and a secondary brake. These components and associated

First Wind Energy, LLC

Technical specifications

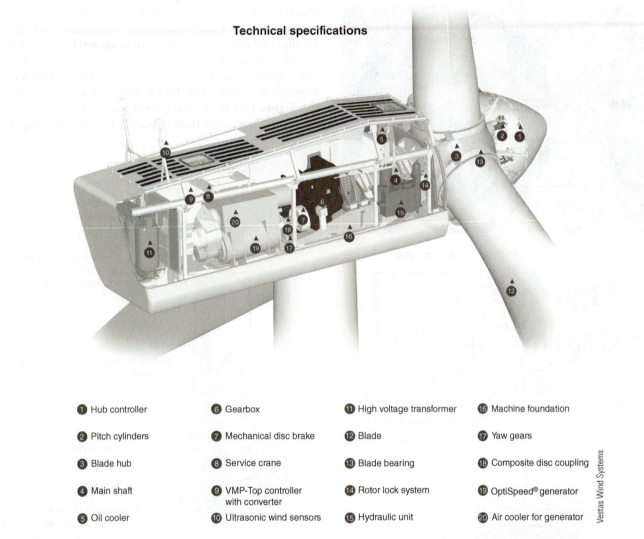

1 Hub controller	**6** Gearbox	**11** High voltage transformer	**16** Machine foundation
2 Pitch cylinders	**7** Mechanical disc brake	**12** Blade	**17** Yaw gears
3 Blade hub	**8** Service crane	**13** Blade bearing	**18** Composite disc coupling
4 Main shaft	**9** VMP-Top controller with converter	**14** Rotor lock system	**19** OptiSpeed® generator
5 Oil cooler	**10** Ultrasonic wind sensors	**15** Hydraulic unit	**20** Air cooler for generator

Vestas Wind Systems

FIGURE 1-23 *Typical machine-head assembly showing major components*

subsystems will be discussed further during maintenance activities in later chapters. **Figure 1-23** shows a typical machine-head assembly listing major components. The rotor assembly is the next component that is connected to the drive assembly in the machine head.

Rotor Assembly

The **rotor assembly** is a structural attachment of the turbine blades to the **drive system**. As mentioned previously, the blades are used to extract the energy from air as it moves by the wind turbine. The hub and blade configuration makes up the theoretical disc area that is used to calculate the wind turbine's power output. Several wind-turbine manufacturers include the rotor diameter as part of the equipment model number, whereas others use the generator rating in the model number. For example, the Vestas V90 has a rotor diameter of 90 meters and consists of three 45-meter blades attached to the hub. An example using the generator rating would be the GE 2.5, which has a 2.5-MW (2,500 kW) rated generator for

electrical production. The rotor assembly also houses blade pitch-control electronics or mechanical linkages, depending on the manufacturer's **pitch system**. **Figure 1-24** shows a couple of blade-pitch drive assemblies. These figures highlight pitch-control systems using either an electric motor drive or a large hydraulic cylinder to actuate blade pitch through mechanical linkage components. The rotor assembly is the prime mover used to spin the generator and produce electrical power, but there needs to be a method of transferring the power to the utility grid. The last component of discussion for the wind-turbine major components is the utility grid connection.

Utility Grid Connection

The utility grid connection for a wind turbine is made through power-switching circuits capable of meeting the voltage and current requirements for the grid collector system. Some manufacturers choose to do this in the DTA, whereas others choose to keep most of their electronics together in the

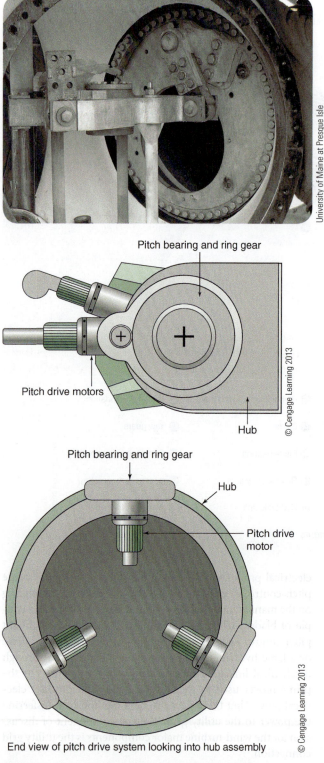

University of Maine at Presque Isle

Pitch bearing and ring gear

Pitch drive motors

Hub

© Cengage Learning 2013

Pitch bearing and ring gear

Hub

Pitch drive motor

End view of pitch drive system looking into hub assembly

© Cengage Learning 2013

FIGURE 1-24 *Examples of blade-pitch drive assemblies*

nacelle. Each style will determine the transformer location, and size and type of power cables between the nacelle and DTA, along with switching components located in the DTA to connect with the farm infrastructure. A **transformer** is used to increase the wind-turbine electrical voltage to match

the utility grid or farm transmission voltage. The transformer assembly may be a can configuration when mounted in the nacelle or a cabinet configuration (**pad-mount transformer**) mounted on the ground by the tower, depending on the system requirements. A typical can configuration can be seen mounted on a utility pole by a residence or an office building. **Figure 1-25** shows examples of electrical infrastructure interconnections with their respective transformer designs.

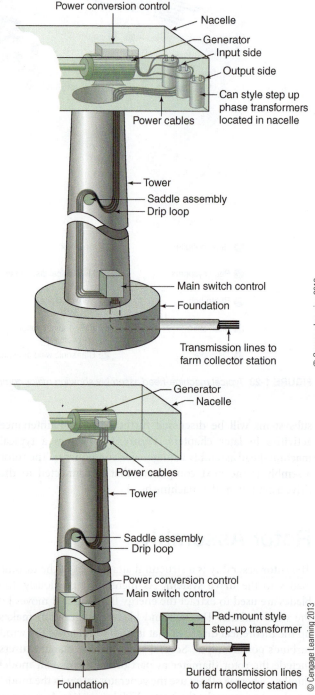

Power conversion control

Nacelle

Generator

Input side

Output side

Can style step up phase transformers located in nacelle

Power cables

Tower

Saddle assembly

Drip loop

Main switch control

Foundation

Transmission lines to farm collector station

Generator

Nacelle

Power cables

Tower

Saddle assembly

Drip loop

Power conversion control

Main switch control

Pad-mount style step-up transformer

Buried transmission lines to farm collector station

Foundation

© Cengage Learning 2013

FIGURE 1-25 *Examples of electrical infrastructure interconnections with respective transformer designs*

These wind-turbine sections will be further discussed in the preventative-maintenance chapters of the text.

Preventative maintenance is a practice that ensures equipment will run efficiently and as designed over its expected life. A wind turbine is a capital investment by a farm operator or developer and should be maintained just as other industries maintain their facilities and equipment. Typically, 10% of the wind-turbine purchase price is spent to maintain the system over its 20- to 25-year life span. The goal of farm operators is to ensure that every dollar spent in preventative maintenance has a **return on investment (ROI)**—that is, a capital gain or increase in revenue from an investment. The majority of farm operators are public companies that have a single focus: make a profit for investors. Making a profit for investors or stakeholders translates into ensuring that they have a quality product (electrical power) and reliable equipment to produce that power. Wind turbines should be ready to run when the wind is blowing and be efficient to generate the most power possible with the available wind resources. Preventative maintenance ensures that wind turbines and farm equipment meet operational and financial goals.

IMPORTANCE OF A PREVENTATIVE-MAINTENANCE PROGRAM

Farm operators and turbine manufacturers refer to available production time as **availability** and measure it as a percentage of production run time versus total time that they could run with available wind resources. This value is determined during contract negotiations between the farm developer and the operations group or manufacturer for warranty contracts. Typically, this value is 95% or higher, with penalties and incentives written into the contract to provide operational goals for both parties. A contract requirement emphasizing equipment availability makes a comprehensive preventative-maintenance program even more important to ensure profitability for all involved. If the equipment is not running, nobody is making money, including customers and shareholders. Maintaining the equipment to the manufacturer's standards ensures that the equipment will operate at peak performance.

ORGANIZATION OF MATERIAL

Safety is our first and most important topic. This topic cannot be stressed enough. Your life and those of your colleagues depends on your situational awareness and understanding of how to prevent injuries. Your own and others' safety attitudes could be the reason why you will be able to return to your family and friends at the end of the day.

The major divisions of the text will include topics related to maintenance fundamentals, preventative-maintenance activities associated with utility-scale wind turbines, and resources that may be used to improve farm operations. Maintenance fundamentals will focus on topics such as fluid power, lubrication, bolting practices, component alignment, and select electrical and mechanical test equipment that may be used to analyze equipment condition and enable a technician to monitor changes with time. The preventative-maintenance section will discuss activities that are performed in the majority of utility-scale wind turbines. This author will note activities that may be seen in several wind-turbine systems and not focus on any particular manufacturer's equipment. A more specific focus will be provided to a technician by the manufacturer's procedures, available technical documentation, and training relevant to each job site. These maintenance activities are presented for reference to typical procedures and may assist in developing a comprehensive plan in the absence of technical documents. Always refer to the manufacturer's documentation and site operation materials for specific equipment.

The final section of the text will discuss information used to develop a comprehensive maintenance plan and what resources are available to monitor and determine equipment performance. Understanding these resources will enable technicians and the farm operations team to determine current equipment condition and develop a strategy to improve overall performance. Improving equipment performance will enable the farm to meet contractual goals of availability and ensure customer satisfaction.

SUMMARY

Wind turbines have been popular since the 1930s for residential uses such as lighting and powering electrical appliances. Rural electrification of North America and Europe brought on a decline in wind-turbine systems for all but the most remote locations until political unrest in the Middle East in the early 1970s brought an end to inexpensive electricity produced by foreign petroleum. Many governments during the 1970s and 1980s promoted renewable-energy sources such as wind, solar, biofuels, and hydro to supplement their national energy needs. The global resurgence in wind power was brought on when several governments provided monetary incentives to investors in the form of tax credits and feed-in tariffs. These incentives reduced the uncertainty for investors and made it possible to profit with manufacturing and development of wind projects. Wide-scale use of wind-turbine systems has made it possible to develop larger, efficient, and more reliable designs that can last 20 years or more. The longevity of these systems has brought on new career opportunities for individuals with sound backgrounds in electrical and mechanical systems. Career opportunities in wind can be as diverse as site development, construction, operation, maintenance, and manufacturing. The goal of this text is to provide insight into the required skills to be a safe and proficient maintenance technician in the fast-growing wind industry.

REVIEW QUESTIONS AND EXERCISES

1. Describe the purpose of the production tax credit (PTC).
2. Contrast U.S. energy-policy changes with those of European countries such as Denmark and Germany after the 1973 OPEC oil embargo and subsequent oil-supply uncertainties.
3. List several wind-turbine manufacturers that have a manufacturing presence in the United States.
4. Which U.S. state has the most installed wind power?
5. What is meant by the term *wild AC*?
6. Name a few U.S. wind-data resource organizations.
7. What are land-use royalties?
8. Name two U.S. organizations that are developing wind-technician standardized competency tests.
9. What is the purpose of a preventative-maintenance program?
10. List some of the required skills a wind-turbine technician should possess.
11. Estimate the coefficient of performance (C_P) for a 4.0-MW–110 wind-turbine generator. Turbine information: rotor diameter (D) = 110 m, rated nameplate capacity = 4.0 MW, and rated wind speed = 14 m/s. Use standard atmospheric conditions (sea level and 15°C) for your calculations.
12. Determine the sound-power level of a wind turbine located on a ridge 5,000 feet away from a residence with a sound-pressure reading of 42 dB measured outside a building.
13. A direct-drive wind turbine operates with a rotor rotational speed of 19 RPM at full power output. Determine the number of magnetic poles required per phase to match a grid frequency of 50 Hz.
14. Determine the input torque provided by a 100-m diameter rotor assembly to produce an instantaneous power output of 3.2 MW at 20 RPM for a direct-drive wind-turbine generator system.

REFERENCES

AWEA. 2011 Statistics (www.awea.org/learnabout/industry_stats/index.cfm).

CanWEA. *Wind Turbine Sound and Health Effects*. Sound White Paper December 11, 2009. (http://www.canwea.ca)

National Occupational Competency Testing Institute (NOCTI). Pilot Testing. (http://www.nocti.org)

Wind technician certification test link. (http://www.nocti.org)

CHAPTER

2

Tower Safety

KEY TERMS

American National Standards Institute (ANSI)
body belt
body support
Canadian Centre for Occupational Health and Safety
  (CCOHS)
Canadian Standards Association (CSA)
carabiner
clearance distance
competent inspector
connector
D-ring
free fall
horizontal lifeline
job briefing
ladder safety device

ladder safety systems
lanyard keepers
National Fire Protection Association (NFPA)
Occupational Safety and Health Administration
  (OSHA)
personal protective equipment (PPE)
self-retracting lifeline (SRL)
shock absorber
shock-absorbing lanyard
tailgate meeting
tower-rescue systems
trauma straps
vertical lifeline
work-positioning devices
work-positioning lanyard

OBJECTIVES

After reading this chapter and completing the review questions, you should be able to:

- Understand the need for safe work practices.
- Discuss reasons for a safe work attitude.
- List organizations that develop and propose safety standards.
- List organizations that regulate and enforce safety regulations.
- Discuss how an employee's attitude can play a role in an injury.
- Discuss three methods of preventing employee injuries.
- Understand the need for a documented emergency rescue plan.
- Understand the need for coordination with local emergency-response groups.

- Understand the need for job-safety briefings or tailgate meetings.
- Discuss reasons for practicing tower-rescue activities.
- Understand the need for daily safety equipment inspections.
- Understand documentation requirements for safety equipment inspections.
- Discuss hazards associated with working at heights.
- Discuss the importance of vertical clearance for a fall-arrest system.
- Discuss requirements of a tower-rescue plan.
- Understand the need for prompt action in executing a tower rescue.
- Discuss the basic steps in a tower rescue.

INTRODUCTION

Safety is part of our everyday life. Most people think of safety as an inconvenience we are required to observe at work in order to keep the boss happy or to prevent us from losing our jobs. The truth is that we develop basic safety skills from the time we are very young. Our parents or grandparents were

always telling us not to play with matches, not to run with sharp objects, and to look both ways before crossing the road. The list goes on and on. We probably learned many safety lessons on our own by witnessing the horror of injury or death to someone we knew or saw in the news. These injuries or deaths made most of us realize we are not immortal and that the situations that caused these events could have been avoided.

Yes, in most cases they can be avoided. Saying something was just an accident and could not have been avoided is the wrong approach. With any accident, there is a definite cause-and-effect relationship that could have alerted someone to avoid the situation. If an activity has known risks involved, then knowing what protective equipment to use can minimize injuries and protect a person. Knowledge is a key factor in accident avoidance, and proper training is the way to acquire knowledge.

Another key factor is attitude. A mind-set of "It will never happen to me" or "It's no use—I can't do anything about it" will blind you to the possibilities of protecting yourself. I have always told colleagues, "If it doesn't look safe, don't do it." If an activity doesn't look safe, then ask your supervisor or safety coordinator about alternate methods or safety gear that will improve the activity's safety. Ask before you act. Take the time to review safety procedures and equipment manuals, don appropriate safety gear, and be prepared. To ensure the focus is on safety, each day should start with a **job briefing** or **tailgate meeting** to review planned activities. These meetings should include discussions on pertinent safety procedures, technical documents, review of safety gear and tools, and specify the status of personnel training required to complete activities. Make sure you and your colleagues are prepared so that you will return home to your families and friends at the end of the day. This should be automatic, right? Many companies and individuals figure it takes too long or it may cost too much to play it safe. We just need to get it done, and no one has been hurt in the past doing it this way. It only takes one death or serious injury to change minds—but this option is rarely available to the victim of an accident. A safe workplace is a requirement, not a privilege. Industry groups and government agencies have been working to ensure safe work practices are a requirement.

Government Safety Regulations

Many governments throughout the developed world have regulated safe work practices to ensure the protection of company employees. The U.S. Congress enacted the Occupational Safety and Health Act in 1970, and President Richard Nixon signed it into law in December 1970. Since then, the **Occupational Safety and Health Administration (OSHA)** has been directed to ensure that safe work practices are followed in both private-sector and federal workplaces. The **Canadian Centre for Occupational Health and Safety (CCOHS)** and European safety groups have been tasked by their respective governments to regulate and enforce workplace-safety practices. OSHA has published workplace-safety regulations such as 29 CFR 1910 (General Industrial) and 29 CFR 1926 (Construction Industry) to ensure that companies have up-to-date reference materials to implement

workplace-safety programs. Several standards organizations such as the **American National Standards Institute (ANSI)**, the **National Fire Protection Association (NFPA)**, and the **Canadian Standards Association (CSA)** work with industry leaders to develop uniform safety guidelines. These guidelines reflect best industry practices for activities such as fall-prevention, electrical work, and chemical-related activities to name a few. These guidelines are then proposed to government regulatory agencies for acceptance as safety regulations.

This chapter will focus on regulations, standards, and best work practices that relate to working at heights in the wind-power industry. Other work place hazards such as those associated with electrical, mechanical, and outside work activities will be discussed as part of the safety program development process in the next chapter. Many think the first wind-technician task of the day is to climb the tower. Actually, the first task is a job briefing. Before any work activity, technicians should receive available safety information, work instructions, and a list of required tools and safety gear. This is the time to discuss what needs to be accomplished and determine that everything is in place, including required training. Safety training includes more than the directive "Here is a harness, I'll meet you at the top." A technician's working knowledge of occupational hazards such as potential fall hazards and preventative measures will protect them and their colleagues. Once the job briefing is completed, the team is ready to climb and get to work.

This author has heard many new wind technicians say "What a view!" the first time they step onto the nacelle. **Figure 2-1** shows some of these views from the top of wind turbines. Working on wind turbines has the added benefit of an office with the best possible view.

WIND-POWER INDUSTRY SAFETY HAZARDS

It is a priority for businesses to protect their employees from work-related safety hazards. These hazards come in many forms, depending on the industry or work activity. The top three working hazards associated with the wind-power industry are:

1. falling from the tower or nacelle,
2. being struck by a falling object, and
3. shocks and electrocution.

Determining work-related safety hazards is an important step in eliminating or reducing their potential to cause an injury. The top hazard associated with the wind industry is falling, so this is a great place to start our discussion. There are three fundamental practices to eliminating or reducing hazards in the workplace: engineering practices, management practices, and use of personal protective equipment.

Cianbro Coporation

Cianbro Coporation

FIGURE 2-1 *Two views from tops of wind turbines*

TABLE 2-1

Examples of Engineering Changes: General Industry and Construction Applications	
Hazard	**Equipment Design Change or Addition**
Falls	Handrails
	Guardrails
	Warning lines
	Restraint systems
	Ladder cages
	Platform
	Barrier
	Door
	Hatch
	Move activity to lower elevation
Falling object	Overhead barrier
	Guardrails
Pinch points	Guard
	Enclosure and interlock switch
Rotating shaft	Guard
	Barrier
	Enclosure and interlock switch
Electrocution	Guard
	Barrier
	Door
	Full enclosure
	Reduce voltage
	Ground fault circuit interrupter (GFCI)
Thermal burn	Guard
	Barrier
	Enclosure
Chemical burn	Guard
	Barrier
	Enclosure
	Ventilation

© Cengage Learning 2013

The first step to eliminate a workplace hazard is to change the situation that presents the hazard. Engineering practices is one way to make this change.

Engineering

Eliminating a hazard through engineering practices may include changing the design of the product, process, or equipment, adding safety features, or developing tools that ensure safe operation. Each step, whether separately or in conjunction with another, can prevent or substantially reduce the probability of a workplace hazard. **Table 2-1** lists several engineering practices that could eliminate a safety hazard. These changes can be as simple as adding guards, handrails, or locking devices or changing the lighting around a safety hazard. They can be as complex as redesigning a system or even installing new equipment that does not have the associated hazard. If changing the equipment is not a practical option, the next step is to implement management practices

so operators are aware of the hazard and are trained to work safely to prevent injury.

Management

Management practices to eliminate or reduce work hazards include policies, procedures, and training to ensure activities are performed safely. Training is the key to effectively implementing management practices. Developing policies, procedures, and work instructions alone do not ensure a safe work environment. Employees need to be involved with the development of management practices such as policies, procedures, and any other requirement they may be affected by in the process of their activities. Experienced employees have practical knowledge of these hazards and can be valuable resources in developing solutions to problems. Their

TABLE 2-2

Examples of Management Practices	
Hazard	**Example of Policy or Procedure Implementation**
Fall	Only climb when winds are below 20 m/s.
	Do not enter the hub unless winds are below 15 m/s.
	Only qualified employees approved to climb.
	Develop training standards for climb-qualified employees.
	Only ANSI- or CSA-approved equipment can be used.
Overhead hazard	Do not get within 1,000 feet of a turbine if conditions can create ice falling from tower, nacelle, or blades.
	Do not leave tools and components on the top of the nacelle without being secured.
	Signs and barriers must be posted in hazard areas to prevent entry by unauthorized employees.
Heat exhaustion	Stay hydrated and limit time exposed to temperatures at or above 90 °F to 30 minutes.
Electrocution	Only qualified employees shall work on live electrical circuits.
	Set training standards for qualified electrical technicians.
	Establish procedure to define approach boundaries around live electrical activities.
	Do not enter a wind-turbine tower when lightning has been detected within 30 miles of the site by the National Weather Service or other weather-monitoring service.
Cold water	Exposure suits must be worn during transportation from shore to offshore wind turbines.
	Only water-survival–qualified employees may work on offshore projects.
Surface-vessel emergency	Only qualified personnel shall be allowed to transit by surface vessels to off shore wind projects.

© Cengage Learning 2013

participation will aid the implementation process and assist in enforcing policies. Once the management practices are developed and documented, all at risk employees need to be trained to these procedures. **Table 2-2** lists examples of management practices to reduce exposure to workplace hazards. Training is not the final step in the implementation process. There needs to be periodic assessment and adjustments made to procedures to ensure they continue to be effective. Safety policies and practices should never be thought of as "etched in stone." These documents should be fluid in nature to meet the needs of new regulations, equipment, processes, technology, and employees. These policies must also be applied consistently with appropriate incentives and penalties for appropriate and inappropriate safety behavior. When engineering and management activities cannot eliminate or reduce employees exposure to a hazard, personal protective equipment must be implemented to ensure safety.

Personal Protective Equipment

A variety of **personal protective equipment (PPE)** items are available in the marketplace for almost any hazard that may be encountered. **Table 2-3** shows examples of hazards and available PPE to reduce or eliminate injuries associated with these hazards. Knowing the PPE that is appropriate for any particular situation is virtually impossible without proper training or guidance from qualified resources. PPE may be the last line of defense, but it must be planned and

implemented through the engineering and management activities previously presented. Regulatory agencies (OSHA or CCOHS) along with standards organizations (ANSI, NFPA, or CSA) have documented the requirements for appropriate PPE and guidelines for proper use. This chapter discusses safety practices associated with tower-climbing activities. These activities will be presented along with pertinent regulations and standards that can be used for guidance with preparing a job briefing and assessing appropriate PPE, suggestions on equipment inspection, and documentation requirements. This information is not intended to cover all situations and should be used as a reference with company-specific documents, safety regulations, and appropriate standards. Engineers and safety specialists have devoted years to understanding industry standards and regulations so follow their recommendations when preparing for a climbing activity. Remember, it is an employer's responsibility to establish and implement an effective safety program, but it is the employee's responsibility to adhere to those policies, procedures, and practices that promote a safe work environment.

Working at Heights

Activities that require an employee to work on an elevated platform can be considered working at heights. These platforms could be temporary scaffolding used during a construction project, permanent ladder assemblies on elevated equipment, lattice structures used for communication towers or transmission-line supports, and ladders used for access to wind turbine upper levels. Each activity is covered under

TABLE 2-3

Examples of Personal Protective Equipment (PPE)	
Hazard	**PPE**
Sound level: above 85 decibels (determined by time-weighted average)	Ear plugs or muffs
Cold temperatures	Thermal-barrier jackets or parkas
Cold water	Exposure suit
Overhead falling objects	Hardhat
Fall	Full-body harness, shock-absorbing lanyard, and anchor
Foot punctures or crushes	Steel-toed or composite-toed safety shoes
Hand cuts and abrasions	Cut-resistant safety gloves
Poisonous atmosphere	Self-contained breathing apparatus (SCBA) and chemical-exposure suite
Underwater	SCUBA, wet or dry suit, and associated gear
Arc-flash exposure	Arc-flash suit with appropriate calorie rating: 8 cal/cm², 25 cal/cm², etc.
Extreme elevated temperature	Aluminized exposure suite such as used for firefighting and around molten-metal processing
Sunlight	Sunglasses
Flying debris	Safety glasses or face shield
Chemical exposure	Self-contained exposure suit

© Cengage Learning 2013

different industry classifications. The classification determines government regulations that must be followed to meet safe workplace practices.

General industrial standards cover safety practices related to processing facilities, logging, transportation, and agricultural activities, along with the operation and maintenance activities at power-generation, transmission, and distribution installations. Construction standards cover safety practices related to steel erection, bricklaying, bridge work, crane activities, and some power-generation and transmission activities. Industrial settings typically have fixed elevated working surfaces such as catwalks, installed electrical generation and distribution systems, and other installed systems. The OSHA General Industry Standard 29 CFR 1910.269 sets 4 feet (1.2 m) and above as the level requiring fall-arrest equipment if no other protection is provided. At this level, employers are required to develop, document, and implement a safety policy for employees to use precautions necessary to protect themselves from a fall. These policies may specify employee training, access equipment standards, and PPE, along with inspection and use requirements. The OSHA Construction Standard 29 CFR 1926.501 sets 6 feet (1.8 m) and above as the level requiring fall-arrest equipment if no other protection is available. This level could be above the ground or an adjacent work surface that employees may use to work from during an activity. Employers regulated by this standard are also required to develop, document, and implement a safety policy to protect employees from a fall hazard. This policy covers temporary working structures, so it provides guidance on the use of scaffolding strength, safety nets, and horizontal and **vertical lifelines**, along with other fall-prevention and fall-arrest systems. An exception to this regulation is the erection of steel framing for large structures.

OSHA sets 15 feet (4.6 m) for a maximum unprotected level before fall-arrest equipment is necessary. If employees have completed a specific training program in accordance with OSHA requirements, they can work in controlled access zones up to 30 feet (9.1 m). Refer to your organizations site specific policies for more details.

Should a fall hazard still remain after engineering and management measures have been implemented, PPE will need to be determined for each climb activity. Wind-tower activities require fall-arrest PPE because technicians are required to climb 70 foot (22 m) plus ladder sections, traverse the top of the nacelle to work on metrological equipment, or enter the hub from outside the nacelle. Each activity exposes technicians to unprotected walking and working surfaces greater than the 6-foot requirement of OSHA 29 CFR 1926. The fall-arrest PPE requirement for these activities would comprise a full-body harness, a twin-leg **shock-absorbing lanyard**, and a **ladder safety device**. **Figure 2-2** shows examples of these PPE that should be used to meet regulatory requirements.

Fall-Arrest Systems

Fall-arrest systems include anchorage, connectors, and body-support, as well as rescue systems. Each system is employed as necessary, depending on the application. Several employees working on a structure could have a fall-arrest net setup below them to prevent an injury from a fall. The fall-arrest system may be used to protect an individual employee working on a steel structure. This system would require a horizontal lifeline to enable the worker to move freely while also providing a secure anchor system should a fall occur. A **horizontal lifeline** is a cable system connected between two

FIGURE 2-2 *Examples of personal protective equipment that meet regulatory requirements*

FIGURE 2-3 *Example of horizontal lifeline setup*

anchor points that can be used as an anchor for a fall-arrest system. **Figure 2-3** shows an example of a horizontal lifeline setup. The employee would also need an appropriate body support and connectors to ensure he or she were attached to the lifeline system. A properly rated horizontal lifeline could be used to arrest the fall of two individuals working on a structure. Should one employee fall, it is likely that the second employee will also fall. When they come to rest on their lanyards, they better be friends because they will be up close and personal.

OSHA 29 CFR 1926.502 regulations, ANSI Z359.1, and CSA Z259.11 standards are good references to equipment requirements used for fall-arrest systems. The purpose of the fall-arrest system is to ensure the employee does not get injured during the deceleration process after the free fall. **Free fall** is the total distance a person can fall before he or she hits an object or begins the deceleration process created by a fall-arrest system shock-absorbing device. In short, it is not the fall that will cause the injury but the sudden stop at the end. A 200-pound (91-kg) person falling 6 feet (1.8 m) will create around 5,000 pounds (22.2 kN) of impact force on the body as it decelerates against another object. Solid structures such as concrete, steel, stone, or soil will not move, so the impact energy will be absorbed within the human body. This energy absorption process creates dislocated joints, broken bones, and soft-tissue damage. The purpose of a fall-arrest system is to reduce the impact on the body through decreasing the fall distance and absorbing some of the kinetic energy acquired during the fall.

Forces

Early fall-arrest systems were little more than ropes tied around an employee's waist or a rope connected to the employee's belt. Still, employees were sustaining debilitating injuries. The fall mechanics remained the same. A 200-pound (91 kg) person using a simple **body-belt** support and rope lanyard still creates a 5,000-pound (22.2 kN) impact force with a 6-foot free fall. **Figure 2-4** shows an image of a person hanging in a body belt. Injuries with this fall-arrest system

FIGURE 2-4 *Worker hanging in body belt*

could include a broken back or death from internal crush injuries. The force of the fall is spread over a small region of the abdomen but also bends the person in half at the waist, creating stress on the spine during deceleration. If the belt was loose enough to slide to the side during the fall, then the person would suffer an even more serious injury because of side impact on the spine. There had to be a better way.

As of January 1998, body belts were no longer accepted by OSHA as a body support for fall-arrest system. Since that date, body belts should only be used as positioning devices to prevent falls. The goal of newer body-support equipment is to minimize the fall impact on the body by spreading the forces over the entire torsos. Correctly designed fall-arrest systems are intended to distribute forces over the torso during deceleration, but what if the person hits an object below his or her work area during free fall before the system is engaged? Ensuring that the proper **clearance distance** exists below the work area is as important as choosing the correct fall-arrest system.

Fall Clearance

Considerations for fall-clearance distance below the work area take into account the length of the connecting devices, the deployment of shock-absorbing devices, the location of the anchor with respect to the dorsal D-ring on the harness, and the ability of the harness to slide up the back during deceleration. These factors also consider that the anchor point is located at the technicians head. If the anchor point was at the technician's feet, then his or her height would have to be added to the total fall distance. Typical fall-arrest systems require 12.5 feet of clearance below the work area to ensure proper function. Objects connected to a lanyard do not always drop straight down, so consider any objects to the side of the fall zone that may be hit because of a swing fall. Swinging into an object during a fall can create a severe injury or death just as falling onto an object. The following table shows how the 12.5-foot fall distance is determined.

Total Fall-Clearance Calculation Example

Distance	Factor
6.0 feet (1.8 m)	Lanyard free fall distance
3.5 feet (1.1 m)	Shock-absorber deceleration distance
1.0 feet (0.3 m)	Slide of the D-ring from the center of back to just behind head*
2.0 feet (0.6 m)	Safety factor to ensure feet do not hit lower surface
12.5 feet (3.8 m)	Total drop distance

*Properly adjusted harness. A loose harness may allow occupant to fall out of the back if the D-ring slides above his or her head.

What about the OSHA 29 CFR 1926 requirement to wear a fall-arrest system above 6 feet with no other protection measures? If a 6-foot lanyard was used with a working surface at 6 feet, then the person would hit the ground or lower work surface during free fall. How could this be prevented? This may be prevented by the use of a self-retracting lifeline (SRL) that can decelerate a fall in less than 3.5 feet provided it is anchored at a location well above the technician's head. This scenario would still require 1 foot for the slide of the harness and a safety factor of 2 feet (3.5 ft + 1 ft + 2 ft = 6.5 ft). This setup pushes the drop distance into the safety-factor zone by 0.5 feet (6 inches or 15.24 cm), so it may be possible for contact with the lower level. If the SLR used for the application has a deceleration distance less than the standard 3.5 feet, contact with the lower surface would be avoided. These calculations show how important it is to understand the safety gear used for each work activity.

Body Support

Body-support equipment such as a full-body harness distributes the load over the entire torso during fall deceleration. This eliminates the spinal and crush injuries associated with the body-belt support system. **Figure 2-5** shows examples of full-body–support systems. Refer to ANSI Z359.1 or CSA Z259.10 standards for more details on body-harness requirements. A variety of full-body harness designs are available, depending on the PPE requirements of job. The location of the D-ring(s) determines the body-harness application. A **D-ring** is a metal ring-shaped component that is attached to web straps on a body harness. These rings are used to secure **carabiners** or snap hooks of lanyards and other safety devices to the body harness. A D-ring located on the dorsal

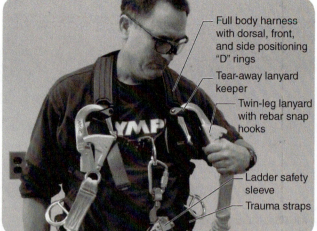

Full body harness with dorsal, front, and side positioning "D" rings

Tear-away lanyard keeper

Twin-leg lanyard with rebar snap hooks

Ladder safety sleeve

Trauma straps

© Cengage Learning 2013 and a W. Kilcollins photograph

FIGURE 2-5 *Examples of full-body–support systems*

surface (back) of the harness would be used to attach a fall-arrest lanyard. The dorsal position enables a technician to work without the lanyard interfering with activities but to deploy during a fall without creating facial injury. A D-ring located at chest height would be used for descent control as part of a ladder safety system or with a rescue descent system during a nacelle emergency evacuation. Typical wind-tower work activities require a body support with D-rings located on the dorsal surface, upper chest, and hips. D-rings located on the hips are used to attach a **work-positioning lanyard** and allow a hands-free posture.

Choosing the correct body support for the job is only part of the safety process. One must inspect the harness before each use and don the harness correctly. A damaged or inadequately tightened harness may not arrest a fall. There have been situations that workers have fallen out of their harness during a fall because it was worn incorrectly. Harness manufacturers include safety booklets with all safety equipment. Take the time to become familiar with your equipment brand and model information before first using it. Follow the manufacturer's care and handling recommendations to ensure safety and reliability of the equipment over its intended life.

Each day a technician should start his or her activities by inspecting safety gear needed for planned activities. Think of your safety gear as an insurance policy. This attitude will ensure that you return home to family and friends at the end of each day. Follow these simple guidelines to make sure your insurance policy will be there when you need it most. It is possible that damage to your safety equipment could have occurred during activities at the end of the previous day or from storage over a period of time. If something does not look right during the inspection, ask your supervisor or safety coordinator. Resolve the question or replace the component before heading to the job site.

Daily Inspection. Inspection of the hardware components on the harness is a good place to start. Look at the D-rings,

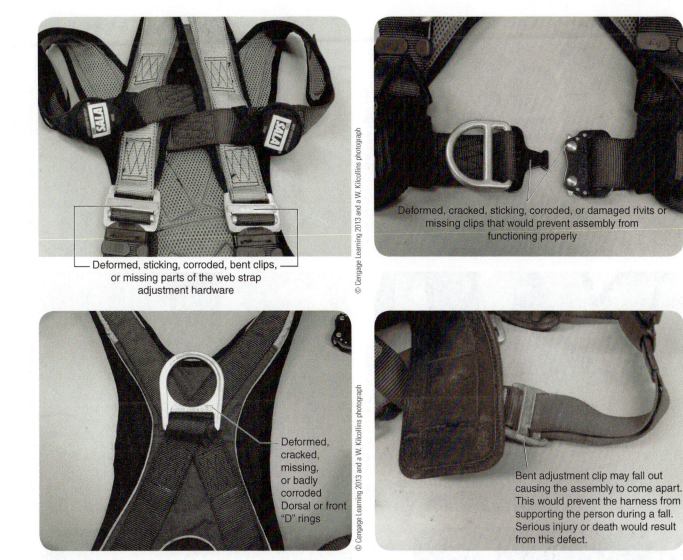

FIGURE 2-6 *Defects that may be encountered during daily inspection*

pass-through buckles, quick-connect clips, parachute buckles, and buckle-and-tongue assembly for belt-style harnesses. Many new harness designs have quick-connect clips to save time when donning the harness. Inspection items for metal components include cracks, rough edges, distorted shapes, corrosion, and missing parts. **Figure 2-6** shows examples of defects that may be encountered during a daily inspection. Buckles and clips should work smoothly without any binding. For workers in climates below freezing, take care to ensure that water or snow do not get into buckles during inspection and donning. The buckles will freeze and require time in a heated vehicle or shop to melt and dry out. Do not make the mistake of thinking it will melt and work properly later. If it is cold enough to freeze outside the tower, then it will still be below freezing in the tower during the climb. A clip or buckle that comes apart during normal activities could be fatal should the technician come out of the harness during a fall. After connecting a web strap buckle, like a pass-through clip or quick-connect buckle, give a tug to ensure it is locked.

Harness webbing and straps need to be inspected for defects such as cuts, tears, broken fibers, frays, abrasions, mold growths, discolorations from chemical exposure, melted or burned sections, and brittle sections because of ultraviolet (UV) exposure from sunlight. Check sewn connections for missing threads, frays, and breaks in the thread pattern. Refer to the manufacturer's product documentation for details on inspecting sewn connections. **Figure 2-7** shows examples of areas on the harness webbing that should be inspected for damage. The body harness should always be stored in a cool, dry, and clean environment away from sunlight. Do not leave your harness hanging in the back window of the service truck as a badge of courage. The webbing will be damaged by extended periods of excessive heat and sunlight exposure. Do not wear a body harness after the webbing becomes saturated with water. Water-saturated webbing degrades in strength and may fail under the impact of a fall. Wearing a body harness for short periods in the rain does not allow it to become saturated, so the strength should not become compromised.

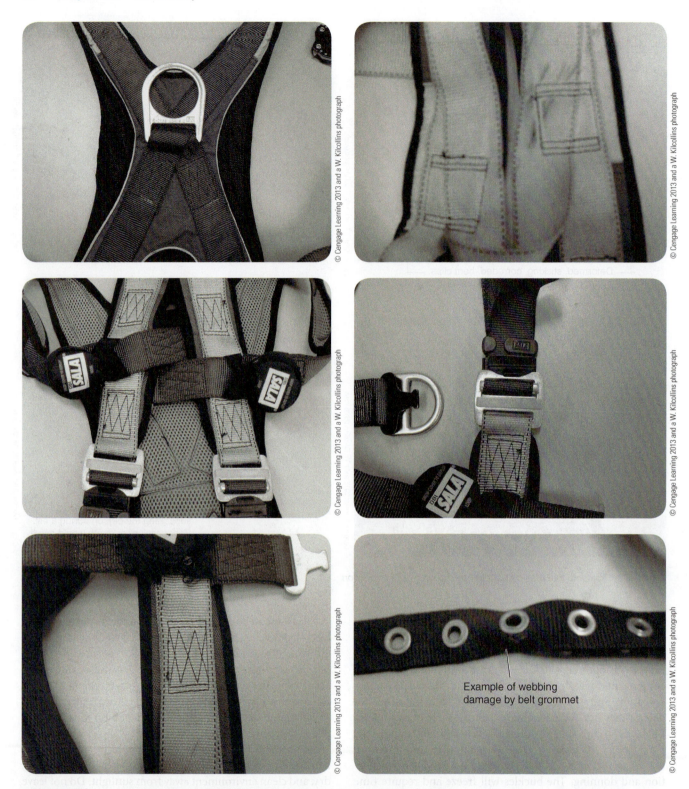

FIGURE 2-7 *Examples of harness webbing and Stitching that should be inspected for damage*

Once the harness is allowed to air dry, the webbing will return to full strength.

Labels and manufacturers tags should be present and legible for a harness to meet OSHA regulations. The manufacturer's name, contact information, model number, production

lot information, and any inspection checks should be available. If any of this information is not present, then give the harness to your supervisor or safety coordinator for proper disposal. **Figure 2-8** shows examples of manufacturer's labeling requirements.

Example of webbing damage by belt grommet

FIGURE 2-8 *Examples of manufacturer's labeling requirements*

Month	Color
January	White
February	Gray
March	Light blue
April	Dark blue
May	Violet
June	Light green
July	Dark green
August	Yellow
September	Orange
October	Light brown
November	Dark brown
December	Black

FIGURE 2-9 *Possible matrix for color coding monthly safety-inspection tags*

These colors correspond to available vinyl tapes. Consult site policy on application of a colored tape to a body harness. Tape should never be applied over any portion of the harness that requires frequent inspection because it could mask a defect. Always remove previous color code to eliminate confusion. Colored tags secured with plastic wire ties may also be used to provide inspection status.

Monthly and Annual Inspections. Monthly and annual inspections include the same daily inspection criteria but will be completed by a competent inspector who has been designated by the employer. A **competent inspector** should be trained to the ANSI Z359.2 or CSA Z259 standard as required by local regulations. Refer to national, state, or provincial standards that cover the location of the wind farm for details on training requirements. National regulations such as OSHA 29 CFR 1910.66 General Industry, OSHA 29 CFR 1926.502 Construction, or CCOHS regulations indicate specific industry-related inspection information along with associated documentation requirements. Forms developed on site or supplied in a manufacturer's product pamphlets may be used by a competent inspector or safety coordinator to document monthly findings. Monthly inspection forms such as the examples shown in Appendix A are set up for individual components. A good visual clue that a harness assembly has passed the monthly inspection process is to affix a colored tag to the harness. **Figure 2-9** shows a possible matrix for color coding monthly safety-inspection tags. Annual inspection forms may provide easier access to safety records if they are set up as a spreadsheet to supply data on all active service and disposed safety gear. This information is required for proof of regulatory compliance should an injury result from a safety-device failure. Refer to government regulations to determine how long these documents have to be maintained.

Donning a Full-Body Harness. Donning a safety harness is like putting on a jacket. Pick up the harness assembly by the dorsal D-ring and ensure that the leg straps and chest straps are not tangled. These straps should dangle down away from the dorsal D-ring as it is being held in front of you. **Figure 2-10** shows an example of a harness being donned. Rotate the harness so the dorsal D-ring is away from you. Place your right arm into the loop formed by the front and shoulder web straps on the left. Slide the back web strap crossover onto your back and place your left arm into the opposite loop formed by the front and shoulder web straps. Bring the front web straps together and latch the quick-connect clip. If the harness is equipped with a belt-strap assembly, place the belt's free end into the buckle and pull it up snugly to your chest and then position the tongue into a belt grommet that will allow a couple of inches (5.5 cm) of space between your chest and the buckle. Spacing is the same between the different styles of connector assemblies and the chest. Adjust the web strap in the buckles as necessary to get the correct fit.

It may be easier to take the harness off and make these adjustments. Reach down between your legs and find the web strap or belt for the left side and bring it up in front of you. Latch the leg strap clip into the lower left side web strap of the harness. For the belt style, follow the previous procedure to connect the belt-buckle assembly. Follow this same

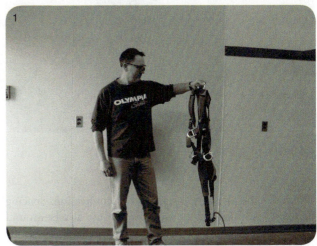

FIGURE 2-10 *Donning a full-body harness (Continued)*

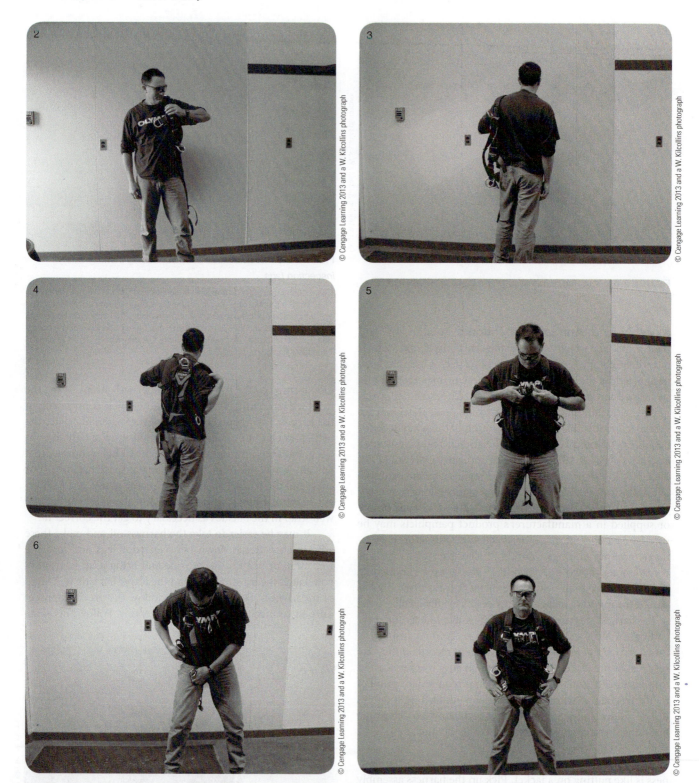

FIGURE 2-10 *(Continued)*

procedure for the right leg web strap connection. Spacing between your hips and the web strap should also be around 2 inches when you are standing up straight. You should be able to slide your hand, with fingers extended, between the web straps and your body to gauge the correct spacing. If the spacing is smaller, it will require considerable effort to move in the harness, and skin chaffing may occur under the web straps. If the spacing is too large, you may come out of the harness when the harness dorsal D-ring slides up over your head during a fall. There has also been a documented severe

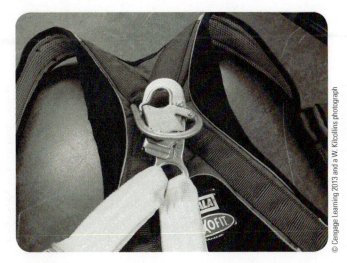

FIGURE 2-11 *Example of a fall-arrest lanyard false-connection*

groin injury caused by a person falling in a loose harness. Take the time to properly adjust the harness each time it is worn to ensure a proper fit.

When the full-body harness is adjusted for a correct fit, attach the safety devices necessary for the work activities planned. Climbing an internal ladder or a ladder attached to a lattice tower would require a ladder safety device provided the ladder has a vertical safety line or a safety rail.

A ladder system without a vertical safety line or a rail will require a twin-leg shock-absorbing lanyard for the climb. This twin-leg lanyard system would also be used for elevated work away from the ladder or when traversing the top of the nacelle to the hub or to work on the metrological mast.

CAUTION: **After your full-body harness is set and safety devices are attached, have a colleague inspect the devices for proper attachment. Attaching the lanyard assembly over your shoulder may give you a false sense of safety. It may have snapped into position or may not move when pulled on, but it still may not be right. A second set of eyes will ensure the devices are attached properly. Figure 2-11 shows a lanyard that may feel attached but is not.**

Care and Cleaning. Care should be taken to ensure your harness is kept in good condition. Previously presented inspection guidelines mentioned damage from burns, elevated temperatures, sunlight, and chemical exposure. Burns can be caused by a harness being exposed to high-temperature work activities such as welding, torch cutting, and grinding. Direct contact with hot surfaces or sparks generated during hot work activities will melt synthetic web straps and leave damaged fibers or holes through the straps. This type of damage can be avoided by choosing a harness material such as Nomex that is designed for activities such as hot work. Long periods of elevated temperature, sunlight, or chemical vapor exposure may also degrade the synthetic fibers of the web straps. Each of these environments can oxidize the synthetic fibers, causing them to become brittle, rough, and discolored. If an inspection of the web straps shows the fibers

are no longer flexible or even break during flexing, alert the site supervisor or safety coordinator.

Harness web straps soiled by oil, grease, paints, or perspiration will degrade in strength over time. Petroleum products should be wiped off with a clean cloth to minimize penetration into the fibers. Wipe paint from the web straps before it dries to prevent hardening of the straps. Wash harnesses in water and a mild soap solution to remove perspiration and other organic contaminants that promote the growth of bacteria and mold. Bacteria and mold will degrade the web straps with time and reduce the usable life of the harness. Hang the harness assembly to dry in a clean environment away from direct sunlight. Do not use a hot air source to dry the harness assembly. A competent inspector should verify the condition of the harness after the cleaning process before it is returned to service.

Connectors

Body support is only one part of a fall-arrest system. Another vital component in this system is a **connector** that is used to attach the body support to an anchorage point. The different anchorage types will be discussed later. Fall-arrest system connectors may be carabiners, snap hooks, lanyards, self-retracting lifelines, and ladder safety systems. **Figure 2-12** shows examples of fall-arrest system connectors. Snap hooks or carabiners are used to attach fall-arrest components to the body harness or an anchorage system. A larger snap hook known as a *rebar snap hook* is useful in attaching a lanyard to a ladder rung, horizontal lifeline, or rail. Always follow the manufacturer's recommendations with snap hooks and carabiners to ensure proper use. The fall-arrest lanyard is made of a strong flexible material used to tie the body-support harness to an anchorage. **Figure 2-13** shows examples of twin-leg fall-arrest lanyards used in the wind-power industry. Use of a twin-leg lanyard for work on wind turbines ensures that there is always a lanyard connector available for 100% tie-off when moving from one anchor to the next. The flexible material of a lanyard may be constructed of a synthetic web strap or a synthetic rope that meets the 5,000-pound (22.2-kN) minimum impact strength requirement. A previous discussion stated that a 200-pound (91-kg) person falling 6 feet would create a 5,000-pound impact force. OSHA and CCOHS set the minimum strength requirement for the lanyard assembly at this level. When a lanyard stops a free fall, 5,000 pounds of impact force is a considerable amount of force for the human torso to absorb. Recent regulations have taken this 5,000-pound impact force standard another step further by reducing the body impact to 1,800 pounds (8 kN) or less. This is accomplished through the use of a shock-absorbing system. The goal of the shock absorber is to deform or tear away sewn strips that will absorb the kinetic energy of the fall and decelerate a body at a lower rate over a 3-foot (0.9-m) distance. This improvement increases the total drop distance of the fall-arrest system but has decreased injuries by the reduced impact force applied to the body. Shock-absorber assemblies can withstand as much as

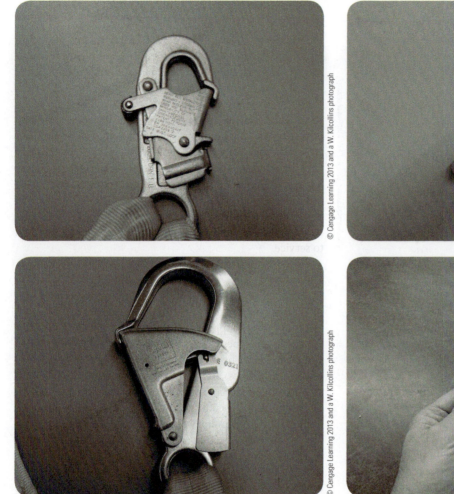

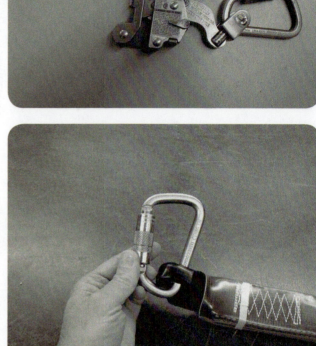

FIGURE 2-12 *Fall-arrest system connectors*

a 400-pound (1.78-kN) force before deployment, so they can be used during activities that require limited positioning tension on the lanyard. A work-positioning lanyard is preferred for activities that require a hands-free posture instead of applying a tension force to a fall-arrest lanyard.

Self-retracting life line (SRL) assemblies are designed with a web strap that is coiled into a housing and spring loaded to allow a variable length of lanyard to be deployed as a person travels along an elevated work surface. **Figure 2-14** shows an SLR assembly. The SRL is designed so that lanyard will extend and retract easily, but when the strap is accelerated because of a sudden movement or fall it will lock and deploy an internal shock absorber at forces above 400 pounds. The locking action is the same as that employed by an automobile seat belt. Replacing a conventional lanyard assembly with an SRL will decrease the overall drop during a fall by the free length of a lanyard (approximately 6 feet). A properly functioning SRL can arrest a fall within the 3.5-foot length of its internal shock absorber.

Ladder safety systems are used to prevent falls during climbing activities. Several systems are available for use in tower-climbing applications. **Figure 2-15** shows examples of ladder safety connectors. Basic systems employ a wire rope or rail system that enables an adapter attached to the front D-ring of the harness to be connected to a vertical safety line. The safety sleeve or adapter is attached to the safety line with freedom of motion up or down the ladder, but it locks in place during a sudden increase in motion. The locking action of a properly functioning system takes place within a few inches (8–9 cm). The total drop of a person falling with this type of safety device is within 2 feet (0.6 m), which includes the length of the carabiner attached to the D-ring and the upward shift of the harness as the person comes to rest after the fall.

CAUTION: Some technicians hold the lock up on their ladder safety connector during descent to prevent it from locking. This action prevents the connector from locking during a controlled rapid descent, but it also prevents the ladder safety system from activating if the technician slips on a rung. Should a slip occur, it may turn into a free fall of several feet before entangling the ladder or striking a deck. This may be enough distance to break an arm or leg or suffer a facial injury. Do not routinely bypass safety devices. Sooner or later, the odds will be stacked against you and

© Cengage Learning 2013 and a W. Kilcollins photograph

FIGURE 2-14 *Example of an SRL assembly*

© Cengage Learning 2013 and a W. Kilcollins photograph

Slider assembly

B. Dutil Photograh

© Cengage Learning 2013 and a W. Kilcollins photograph

FIGURE 2-13 *Twin-leg fall-arrest lanyards*

Sleeve assembly

© Cengage Learning 2013 and a W. Kilcollins photograph

FIGURE 2-15 *Examples of ladder-safety connectors*

there will be an injury. Accidents may happen, but planning and prevention decrease the probability of occurrence and the severity of injury.

Daily Inspection. The daily inspection process for fall-arrest system connectors includes visual and functional

© Cengage Learning 2013 and a W. Kilcollins photograph

© Cengage Learning 2013 and a W. Kilcollins photograph

FIGURE 2-16 *Opening snap hook or carabiner*

testing. Visually inspect the metal components for damaged gates, locks, barrels, and springs. Ensure that the metal components are not corroded, bent, or cut in a way that would compromise their strength or prevent gates or barrels from closing automatically. Gates and barrels should close each time without assistance, and they should only open when the lock assembly is depressed. **Figure 2-16** shows the process for opening a snap hook or carabiner. If assemblies do not close automatically, clean and lubricate them with a minimal amount of light oil. Do not allow the lubricant to contaminate the webbing or rope portions of the lanyard. If cleaning does not correct the problem, then turn in the lanyard for a replacement. A damaged carabiner or snap hook gate may prevent it from staying closed during use. **Figure 2-17** shows inspection points of the snap hooks and carabiners.

Visually inspect the webbing strap or rope material for defects such as burns, holes, abrasion, cuts, contaminants, knots, or rope strands that do not lay correctly. Knots in the rope or web strap may degrade the material strength by 50% or more under an impact load. If the knot cannot be removed without damaging the webbing or rope, then request another lanyard from your supervisor. The rope lay should be continuous and without any openings. Ask a competent inspector to verify the condition of a questionable rope before returning it to service. **Figure 2-18** shows defects of lanyard web straps and ropes. Examine the doubled and sewn ends of the lanyard web strap for abrasions, damaged, or missing stitches. If stitches are missing or are damaged, then have the connection verified by a competent inspector for usability. Recent changes to ANSI and CSA standards have required an impact indicator to be incorporated into lanyards at the doubled and sewn connection with the snap hook assembly. The impact indicator is a small strip of red material that is sewn into the end of the fold. If the lanyard has been impacted, the red strip will extend out past the fold. Remove the lanyard from service if impacted or suspected of being impacted and request another lanyard.

A manufacturer's labels and tags should be present and legible for a lanyard to meet OSHA regulations. The manufacturer's name, contact information, model number, production lot information, and any inspection checks should be available. If any of this information is not present, then the lanyard should be given to your supervisor or safety coordinator for proper disposal. **Figure 2-19** shows examples of manufacturers' labeling requirements.

Lanyard **shock-absorber** assemblies come in a couple of different configurations such as a web strap section sewn into an accordion that is enclosed in a fabric bag or a web tube that is connected on each end to snap hooks and collapsed in length by the use of multiple bungee cords running through the center of the assembly. The purpose of the accordion section or the bungee cords is to absorb energy during the deceleration portion of the fall to reduce the impact force transferred to the human body. Some fall-arrest system manufacturers demonstrate equipment during on-site visits. If a manufacturer's representative is near your site, request a visit and demonstration. One demonstration may include dropping a 200-pound weight on the end of a shock-absorbing lanyard from a test stand. After this test, grab the deployed shock absorber: it will be hot to touch. Why? Kinetic energy from the falling weight was absorbed by the shock-absorbing pack as it was stretched or deformed. This absorbed energy is then released in the form of heat. This is the same energy that, if absorbed by the human body, would break bones, dislocate joints, and cause soft-tissue damage. Another example of this energy transformation is seen in heating the disc and drum brakes during the process of stopping a car.

Shock-absorber assemblies should be inspected before each use for signs of damage. **Figure 2-20** shows examples of damage to shock-absorber packs. Damage may include chafing of the web strap connecting loops, torn or abraded covering over the accordion pack, holes or tears in the web

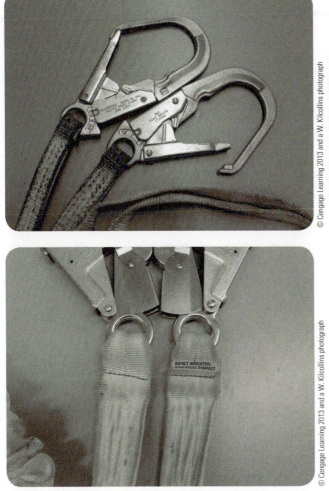

FIGURE 2-17 *Inspection items of snap hooks and carabiners*

tubing over the bungee cords, and chemical or sunlight damage to the fabric. Remember: the web tubing over the bungee cords protects the structural component of the lanyard. The bungees cords are the shock absorber for this type of lanyard.

Monthly and Annual Inspection. Monthly inspections follow the same criteria as discussed for daily connector-inspection activities but should be conducted by a competent inspector as outlined previously. Documentation of monthly

FIGURE 2-18 *Defects in lanyard web straps and ropes*

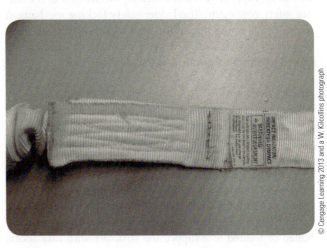

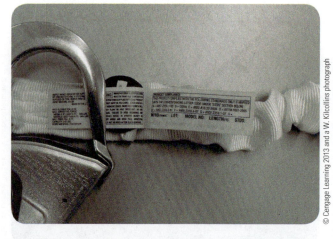

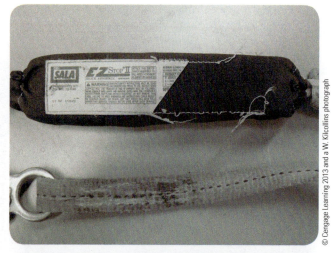

FIGURE 2-19 *Examples of manufacturers' labeling requirements*

or annual inspection information may be required to show compliance with national regulations. Use forms supplied by the component manufacturer, those in Appendix B of this text, or those developed by your company safety department. Compile all annual inspection information in the same location such as a computer spreadsheet to assist in tracking safety-equipment status.

Care and Cleaning. Connector assemblies should be stored in a cool, dry, and clean environment to prevent corrosion. Web strap and rope components should also be stored away from direct sunlight, chemical vapors, and extreme heat exposure to prevent a breakdown of the synthetic fibers.

Connection devices should be cleaned on a regular basis to ensure proper function. Clean soiled carabiners, snap hooks, and other metal components with a clean dry cloth to remove grease, oil, paint, and other contaminants that would cause the component to stick or corrode. If the spring action is not smooth after cleaning, then lubricate with light oil and activate the lock or gate to work in the oil. Apply only a minimal amount of oil to the snap hook gate, lock, or carabiner barrel. Wipe off excess oil so that it does not contaminate the web strap, rope, or get on the locking mechanism of the ladder safety connector. The ladder safety locking mechanism works by clamping the rail or wire rope, and any lubricant present in this area will reduce or eliminate the friction required to prevent slippage and make it inoperable.

Anchorage

Anchorage components are used to connect lanyards and lifelines to ridged structures. These components can be permanent or portable and are classified as either certified or noncertified. Certified anchorage hardware must be designed by a professional engineer who is knowledgeable of fall-arrest system requirements. Certified anchorages are required to be tested to verify that they can withstand a minimum impact load of 5,000 pounds. Permanent anchorage points may be bolted or welded to a supporting structure with easy access to facilitate connecting a fall-arrest system. These anchorage points need to be painted with a distinctive color, periodically

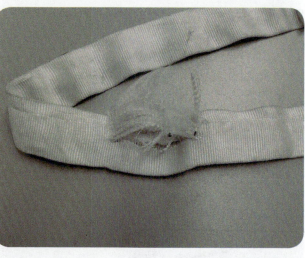

FIGURE 2-20 *Examples of damage to shock-absorber packs*

inspected, inventoried in the safety program, and their status must be recorded with other facility safety information.

Noncertified anchorage points do not require the design and testing process of the certified anchorage points, but they still are required to meet the same standard as other fall-arrest components. Noncertified portable anchorage points may be connected to structural beams, truss assemblies, or any other component that can withstand an impact load of 5,000 pounds. **CAUTION: Electrical conduits, water pipes, and air lines, along with other utility service components, are not acceptable anchorage points.** These typically will not support the weight of a person and will break under the stress of an impact load created by a fall. Does the structure look like it can support the weight of an automobile? If it does not, then it will not support the impact force of a fall. Portable anchorage systems include clamping mechanisms and tie-off adapters that can be purchased from safety-supply dealers. **Figure 2-21** shows an example of a temporary anchorage adapter. Care should be exercised when using web strap tie-off adapters to ensure the supporting structure does not have sharp edges that will chaff and degrade the strength of the strap. Always follow the manufacturers' recommendations for assembling and verifying a portable anchorage setup. Do not leave portable anchorage assemblies exposed to the environment for long periods of time. As the name implies, they are not permanent and should be stored, cleaned, and inspected like any other safety device to ensure reliability.

Other Equipment

Other components that may be used with a fall-arrest system include trauma straps, lanyard keepers, and work-positioning devices. When a person is suspended in a full-body harness, the leg straps of the harness apply pressure to the inside of the legs. This pressure is uncomfortable and can be deadly if the person is suspended for a long period of time. This

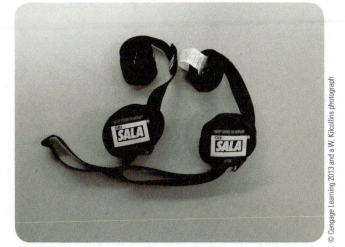

FIGURE 2-22 *Example of a trauma strap*

pressure point is the same location as a major artery (femoral artery) that carries blood down the inside of each leg. Applying pressure to this artery is like placing a tourniquet on each leg to stop blood flow. Without circulation, blood pools in the legs and body tissues are depleted of oxygen and accumulate waste products, and clots may form. After the rescue is performed and pressure points are relieved, this blood flows back into the lungs and heart, creating issues such as heart attack or pulmonary embolism. Adding a set of **trauma straps** to the body-support harness will enable a suspended person to stand in the straps and release the pressure from the legs. **Figure 2-22** shows the use of a trauma strap. This will enable a conscious victim to safely wait for the rescue team to retrieve him or her from the side of the nacelle or rotor. Without trauma straps, the rescue team will be racing against time to prevent a secondary injury or death from suspension trauma created by the loss of blood flow in the legs. Serious injury or death may occur in as little as 20 minutes without the use of trauma straps.

Lanyard keepers are clips or plastic D-rings that are used to secure fall-arrest lanyards when they are not needed. **Figure 2-23** shows examples of lanyard keepers mounted on a full-body harness. A wind-tower lanyard assembly is typically a twin-leg system with large rebar snap hooks located on each free end. With the 100% tie-off rule for work on the nacelle and hub, there is a need to have the spare leg connected to the harness during work activities. Mounting the spare leg of fall-arrest lanyard high on a chest web strap prevents the lanyard from becoming a trip hazard. The use of the hook and loop strips to mount the keepers will allow the spare leg to be torn away from the harness in the event of a fall. If the keeper did not tear away during the fall, the fall-arrest lanyard shock absorber would not deploy properly. This would increase the risk of an injury from higher-impact forces felt by the body due to improper deployment of the shock absorber.

A full-body harness with D-rings located on the hips is used to attach **work-positioning devices** such as chain, web strap, or belt assembly. These D-rings should never be used to

FIGURE 2-21 *A Temporary anchorage adapter*

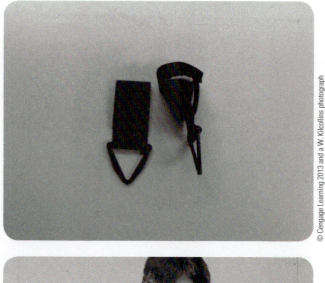

© Cengage Learning 2013 and a W. Kilcollins photograph

© Cengage Learning 2013 and a W. Kilcollins photograph

FIGURE 2-24 *Example of positioning devices*

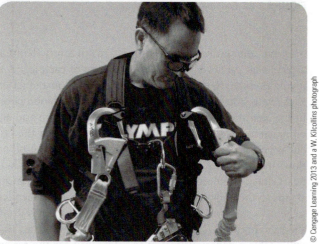

© Cengage Learning 2013 and a W. Kilcollins photograph

FIGURE 2-23 *Lanyard keepers mounted on a full-body harness*

park your twin-leg fall-arrest lanyards. These devices may be wrapped around a ladder, pole, or other structure to prevent a technician from falling while allowing them to work with their hands free. Some positioning devices use snap hooks for attachment to permanent anchorage points to create a hands free posture. Position anchors and belt assemblies are fall-prevention devices and not fall-arrest devices.

CAUTION: Work position anchorage points are only required to meet a 3,000-pound force rating. Always use a fall-arrest system mounted to an appropriate anchorage point when using positioning devices. A technician who slips while moving between positioning anchors will be protected by the fall-arrest system. Figure 2-24 shows examples of positioning devices.

Case Study

Background

A group of contractors was hired to replace a major component of a wind turbine. Contractors included crane services,

riggers, and a major component-replacement team. Each contractor was required to provide proof of related technical and safety training for each employee. This training documentation requirement is a standard throughout general industry. OSHA requires the coordination of subcontractor employees to follow an organization's safety plan while working on site. This includes participation in planning meetings, job briefings, job-safety procedures, communication plans, and confined-space procedures, among others. The contractors agreed to the conditions and submitted appropriate training records, equipment-inspection certificates, and licensing credentials as required for the project. Project team leaders also participated in planning sessions to cover job scheduling and training with site safety procedures.

Issue. Day two of the project involved initial setup and inspection of major equipment. Other activities included the inspection of safety gear before the component-removal breakdown phase started. The site safety coordinator requested the fall-arrest equipment from the climb team for inspection. Six sets of climb gear were presented for inspection. Inspection determined that three sets did not meet requirements. Defects included a couple of bent carabiners and a sticking lanyard rebar snap hook. The safety coordinator presented the defective equipment to the subcontractor team leader and asked if the climb gear had been inspected earlier that day. The team leader said the gear had been inspected by the team members. A meeting was held with the team members to confirm the safety gear had been inspected. Team members said they inspected their gear before they left the shop and headed to the job site.

Did the employees miss the defects during their inspection? The climb team was presented with the gear to determine if they could see any issues. This technique is a good learning tool. Each employee was able to point out the defective gear and state why it was not usable. Results of the meeting indicated that training did not seem to be the issue.

How was the gear damaged during the trip to the site? The climb team was asked how it stored and transported the gear to the job site. Some team members had stored their gear in a job box on the back of the service truck. Their gear was not found to be in the defective group. Three employees had tossed their gear bags on the back of the service truck with the tools and large fixtures used for the change-out procedure. This group of safety gear was found to be defective. One of the climb team members recalled having to move a large fixture that was laying on the safety gear bags when they were unloading the equipment. The load had not been secured properly, so the fixture was able to roll onto the gear bags during the drive up the ridge.

Conclusion. Poor storage practice was determined to be the root cause of the damaged safety gear. The safety gear was tossed in the back of a service truck with loose tools, structural components, and fixtures for the drive to the site. A large fixture rolled onto the gear, bending the ladder sleeve carabiners and a lanyard rebar snap hook. This shows the importance of proper care and handling of all safety gear. These defects may have been discovered during the process of donning safety gear, so the team members would have had a chance to exchange the components for acceptable ones. But what if these defects were not detected and failed when they were needed?

TOWER RESCUE

Discussions to this point have been about fall-arrest systems, but how does a person get him- or herself back onto the working surface or down to a lower level? A safe rescue from a fall is as important as not sustaining an injury during a fall. Rescue from a fall can be through self-rescue or rescue provided by co-workers. Rescue locations that one might encounter during work in and around wind-turbine equipment include ladders, nacelles, and rotor hubs. Each location provides its own set of challenges. To ensure a successful rescue from any emergency situation, there must be a plan. OSHA 29 CFR 1910.38 and 29 CFR 1926.35 require employers to have an emergency action plan (other standards may also apply). The purpose of the plan is to determine how the emergency situation will be handled before it happens. This prevents colleagues from losing time by debating who will run to get the gear and who will perform the rescue. The emergency action plan should include items such as procedures, responsibilities, training, and a review process. It should also define items such as communication, support organizations, equipment, and practice activities. Procedures should include approved methods, personnel, and equipment to perform an emergency rescue efficiently and safely. A universal rule in any rescue situation is that the personnel performing the rescue should never place themselves in harm's way. If this is not followed, then the rescuers themselves may need rescue.

Responsibilities should be listed by type of emergency situation or by the personnel available at the time of an emergency. An emergency rescue required during hours when there may be a fully staffed operations building would be handled differently than at night when two technicians are on site for a service call. Have a plan and train everyone on their responsibilities. Operations in remote locations should follow these guidelines: any tower or internal nacelle activity should use a minimum of two rescue-qualified technicians, and any activity that requires climbing on top of the nacelle or entering the hub should have three rescue-qualified technicians. Two qualified technicians for internal activities will enable one technician to rescue the other if a fall should occur. Three technicians for work on the nacelle or in the hub will allow the third technician to be available at ground level should an unconscious colleague be lowered down after a fall or medical emergency. These suggested numbers should enable a successful rescue in most emergency situations.

Qualified rescue technicians should have training in intermediate first aid, cardiopulmonary resuscitation (CPR), and use of an automatic external defibrillator (AED), along with hands-on training with the rescue gear used at the site. A technician who falls from the nacelle because of lost balance will be assisting with the rescue process. If the worker fell because he or she had an underlying medical issue, then the rescuer will be working alone, so knowing emergency care for these situations may mean the difference between life and death for the colleague.

An annual review process enables a look into the emergency rescue plan and determines if it is effective or needs to be modified to improve performance. The best way to determine if a plan works is to develop drills that allow practice with different emergency scenarios. Give everyone a chance to practice their roles under controlled conditions. If your organization uses outside support such as a local fire department or ambulance service, then practice with their employees. These groups typically will not climb, but they will be waiting at the base of the tower when your group lowers a victim. This practice will enable the groups to learn necessary communication practices vital to a successful rescue. The basic call out for these support services is a 911 phone call, but when they get on site the communication method has to be defined. This is where practice and a coordinated plan become effective. Purchase compatible two-way radio systems that work for both wind-farm communication and emergency organizations as necessary.

Rescue Systems

There are many **tower-rescue systems** available from reputable suppliers, so choose the best one for your needs. The best advice is to keep the system simple. Some systems are as elaborate as rock-climbing gear with plenty of widgets to assemble, whereas others are as simple as a block-and-tackle pulley set with some rope. There will be considerable stress

FIGURE 2-25 *Examples of tower-rescue systems*

during an actual rescue, so if the system is difficult to use during a mock rescue, it may not be the right system for your group. **Figure 2-25** shows examples of tower-rescue systems. Wind-turbine manufacturers typically have existing rescue plans and designated equipment for their employees. They also have experienced staff to train new technicians on the use of their system so the work of deciding what system fits best has been done. If you are deciding on a new system, speak with a reputable supplier and request a demonstration

of the system and allow your personnel to practice with it. Once you choose a system, practice as often as possible to improve skills and maintain proficiency.

Rescue or Evacuation

Two types of rescue can be used for a fallen individual: self-rescue or assisted rescue. A self-rescue may be as easy as grabbing the structure and climbing back up to the working area. If there is nothing that the worker can grab to help climb back up to the surface, a colleague can drop a rope for the worker to climb on or use a rescue system to capture a harness D-ring and raise the worker back up to the deck. A tower evacuation is another type of self-rescue. This may be necessary if a fire starts in the tower, nacelle, or hub. A fire in the tower prevents a normal egress down the ladder because of smoke and possibly flames, and a fire in the hub or nacelle may block exiting down the tower, so another egress method is required. Portable or permanently mounted evacuation equipment will enable a safe egress around the fire if the primary evacuation route is blocked.

The challenge with an assisted rescue is handling an unconscious person who cannot assist in the rescue process. This person may be suspended in his or her ladder safety device on the ladder, suspended in a fall-arrest lanyard outside of the nacelle or hub, or lying in the hub unconscious. The purpose of tower-rescue equipment is to assist with lifting the person back onto the nacelle or lowering him or her down the tower. This type of rescue is time sensitive, so prompt action must be taken to get the person to medical attention. Having an emergency action plan and following it is great, but you also need to know how to use the rescue gear. There typically will not be time for another person to climb up and perform the rescue.

Basic Rescue Steps

The steps required to perform a rescue will depend on the rescue equipment used and the medical emergency encountered, but the steps to initiate a rescue should be the same. These steps should include the following:

- Call for help (two-way radio or cell-phone contact with the office or 911 dispatcher).
- Assess the situation.
  - Is the person conscious?
  - Is the person unconscious?
  - Is first aid necessary?
  - Is there a medical emergency?
  - Is assistance required?
  - Will the rescue put you in danger?
- Assist with self-rescue if the colleague is conscious.
- Pass along initial assessment if the colleague is unconscious or unable to assist.
- Set up rescue equipment.
- Perform rescue.
- Perform first aid as required to stabilize the colleague and wait for the emergency medical team for assistance and transportation to the hospital.

© Cengage Learning 2013 and a W. Kilcollins photograph

These steps provide an outline for a basic rescue procedure. Practice with the rescue system at your site—this cannot be emphasized enough. Rescue practice should be included in emergency plans.

Equipment Inspection

Portable tower-rescue equipment should be inspected before each use to ensure required components are present and functional. The inspection should be completed before leaving the operations building. If there is a problem, it can be corrected without a return trip. The job for the day may be 100 miles (160 km) across the county or in another state. Save time and money by planning ahead.

Inspection points must ensure the following:

- No visible damage to components.
- Snap hooks and carabiners function.
- Rodents have not made the rope into a nest.
- Rope is not tangled.
- Components are clean and dry.
- Required safety tags are present.

CAUTION: If your site has different rescue-system rope lengths available, make sure you have the correct length for the intended tower. This inspection mistake could have grave consequences.

Rescue or evacuation systems that are permanently mounted in a nacelle should be inspected before beginning work or while equipment is being hoisted up the tower. Follow the same basic process as with the portable system. One exception may be that a permanent mounted system uses a wire rope and not a synthetic rope. If the system uses a wire rope, make sure the metal components are not corroded or inoperable. Portable systems should require a competent inspector review them each month, but a permanent system may not be looked at for several months between service and maintenance intervals. Exposure to the environment will deteriorate metal components and may make the rescue system inoperable.

SUMMARY

Working in the wind-power industry requires climbing and working on towers, so it is important to understand the hazards associated with these activities. National standards organizations such as ANSI or CSA in North America and those in other countries have developed recommended safe work practices and standards for safety equipment to improve industry safety. These organizations have also worked with government regulatory agencies such as OSHA or CCOHS to promote uniform standards that can be implemented equitably no matter the size of the organization.

This chapter has presented information on safety planning related to working at heights along with the necessary practices and personal protective equipment that should be used as part of the plan. Having a plan is only the first step in the process of ensuring a safe workplace. Making sure that employees are trained to the plan, supplied with proper safety equipment, and understand the use and care of the equipment will prevent injuries. Should an emergency occur, an implemented emergency rescue plan as part of the overall safety plan will ensure that employees can respond effectively as a team or with outside emergency-response groups for the best outcome.

REVIEW QUESTIONS AND EXERCISES

1. Name some North American organizations that publish safety standards.
2. What is the role of a national standards organization in implementing safety regulations?
3. What are the top three safety hazards in the wind-power industry?
4. List three fundamental practices used to reduce or eliminate a hazard.
5. What is the importance of a job-safety briefing or a tailgate meeting?
6. List PPE required for a tower-climbing activity.
7. What is the minimum impact force a fall-arrest system must endure without failure?
8. Why is it important to determine the fall-clearance distance for an elevated work area?
9. Why is it important to inspect safety equipment before each use?
10. List inspection points for a full-body harness.
11. List environmental factors that can deteriorate synthetic web-material strength.
12. Is it necessary to maintain documentation on fall-arrest equipment inspections? Why?
13. How many employees should be available for work on top of a wind-turbine nacelle? Why?
14. What injury is associated with a person hanging in a full-body harness for an extended period of time without trauma straps?
15. List two local organizations that may assist with an emergency rescue.
16. List the basic steps for a tower rescue.

CHAPTER 3

Workplace Safety

KEY TERMS

affected employee
approach boundaries
arc flash
attendant
confined space
dehydration
department of environmental protection (DEP)
ditching
electrically safe work condition
electrical work permit
entrant
entry supervisor
Environmental Protection Agency (EPA)
fire-resistant (FR)
frostbite
hazard communication (HAZCOM)
hazardous materials (HAZMAT) team

hazard risk category (HRC)
hot work
hyperthermia
hypothermia
hypoxia
job-safety analysis (JSA)
lockout–tagout (LOTO)
material safety data sheet (MSDS)
National Fire Protection Association (NFPA)
neurotoxin
permit-required confined space
personal flotation device (PFD)
pull, aim, squeeze, sweep (PASS)
qualified person
self-contained breathing apparatus (SCBA)
United States Coast Guard (USCG)
zero-energy condition

OBJECTIVES

After reading this chapter and completing the review questions, you should be able to:

- List steps to implement an electrical safety plan.
- List organizations that develop and propose electrical safety standards.
- List organizations that regulate and enforce industry safety standards.
- Discuss some requirements for an annual safety program review.
- Discuss hazards associated with electrical systems.
- Discuss the approach boundaries associated with electrical work.
- Understand the relationship between personal protective equipment and hazard risk categories.
- Discuss steps to execute and release a LOTO procedure.
- Discuss the need for a material safety data sheet.
- Discus the need for proper chemical handling and labeling practices.

- Discuss some basic requirements for emergency action plans.
- Discuss hazards associated with confined-space entry.
- Discuss personnel requirements for confined-space entry.
- Discuss hazards associated with hot-work activities.
- Discuss hazards associated with mechanical systems.
- Discuss hazards associated with rigging operations.
- Discuss hazards associated with fluid power systems.
- Discuss hazards associated with working in hot and cold environments.
- Discuss hazards associated with offshore wind projects and water survival.

46

INTRODUCTION

Working in the wind-power industry is no different than in any other industry. It has its challenges in construction projects, troubleshooting, and maintenance activities. There are also hazards associated with these and other activities that technicians may be involved with every day. The goal of any industrial safety group is to determine the hazards and create an action plan to reduce or eliminate hazards. Company safety groups are not alone in this challenge. They can draw from national standards organizations as well as federal regulatory groups that have published standards that can be used to develop and implement safe work practices. The previous chapter discussed the hazards associated with working at heights and the actions necessary to reduce or eliminate the hazards associated with falling. The Occupational Health and Safety Administration (OSHA), the American National Standards Institute (ANSI), and other organizations have extensive information to assist with hazards that may be encountered in non–tower-related work activities. Non–tower-related activities include working with electrical, mechanical, fluid-power, and environmental control systems, along with chemical operations, hot work, confined space, and working in outside environmental conditions. Some of the hazards associated with each of these are as follows.

Examples of Industrial Work Hazards	
Work Activity	**Associated Hazards**
Electrical	Shock, electrocution, arc-flash burns
Mechanical	Pinch, crush, entanglement, laceration, amputation
Rigging	Pinch, crush, entanglement, head injury
Fluid power	Injection, pinch, crush, burns, poisoning
Environmental controls	Pinch, laceration, amputation, burns, poisoning
Chemical handling	Burns, asphyxia, poisoning
Hot work	Burns, asphyxia, eye injury, shock, electrocution, laceration, lung damage
Confined space	Asphyxia, burns, entanglement, crush
Outside environment	Bites, stings, poisoning, hypothermia, hyperthermia, dehydration, sunburn, lightning strike, frostbite
Offshore	Bites, stings, drowning, hypothermia, frostbite, crushes

Each of the associated work hazards may end in death for a worker that has not properly planned and followed safe work practices. Many established companies have developed and documented safety programs, but others are still in the beginning stages of this process. The goal of this chapter is to present information that may be used by workers, supervisors, or managers to determine work-site hazards, develop safety plans, and implement safe work practices. The first example will briefly discuss hazards associated with electrical systems and then develop into general topics to consider with safety program development and implementation.

ELECTRICAL SYSTEMS

Working with electrical systems not only can be interesting and challenging but also hazardous if appropriate safeguards are not observed to protect the electrical worker or other personnel around the activity. Government regulations and national standards organizations require any electrical activity that is carried out on or around a circuit voltage above 50 V must have an established safety plan with documented safeguards. Wearing appropriate personal protective equipment (PPE), isolating energy sources, preventing unauthorized access, and understanding the risks associated with the electrical activity are some of the necessary safeguards. **National Fire Protection Association (NFPA)** (70E) and OSHA regulations (OSHA 29 CFR 1910 subpart S, 29 CFR 1926 subpart K) describe in detail the safety practices necessary to work with electrical systems to minimize hazards. These resources, suggestions presented in this section, and those documented in your company's safety program should assist in safely working around electrical hazards whether they are 300 feet up a wind-turbine tower or working on the floor of the operations building.

Electrically Safe Work Environment

Federal regulations, standards, and company safety policies all have one common theme: to make the work environment as safe as possible no matter what activities are being preformed. To ensure a safe work environment, a program should be in place to determine each workplace hazard and create a plan to eliminate or reduce them. Whether it is protection from a fall or electrocution, the process should include the same steps. Working on a wind farm includes activities with electrical systems, so the safety plan must include all necessary steps to provide an **electrically safe work condition.** This means that all electrical hazards must be determined and eliminated so a person can work safely. Having a plan is not enough to make a safe work environment. The plan has to be implemented through company policy, employee training, written procedures, tools, equipment, and documentation that can be used to track the effectiveness of the plan.

Planning

One of the first steps in the planning process is to determine the hazards associated with the work environment. This part of the process should be developed by a qualified team that is familiar with each activity performed at the facility. Each activity should be evaluated for type of hazard, severity of the hazard, how often the activity is performed (frequency), and what measures can be taken to prevent the hazard.

Here is an example of a work-activity evaluation. A 15-amp fuse needs to be replaced in a 240-V three-phase disconnect panel. What are the associated hazards? For this process, these would be electrical shock, electrocution, and arc flash, not to mention any pinch or crush hazards associated with having a body part in the equipment during an unannounced start-up. How severe are potential injuries? Injuries from the fuse-change process could be anything from an electrical shock to an electrocution death. How often would this fuse be replaced? The frequency of this activity may be once a quarter or less. How could an injury be prevented? Measures to prevent an injury may include **lockout–tagout (LOTO)** of a motor control center (MCC) switch that is upstream of the disconnect panel and wearing appropriate PPE to prevent a worker from becoming part of the energized electrical circuit. The preferred injury-prevention method with this electrical hazard would be to isolate the disconnect panel at the upstream MCC with a LOTO procedure. LOTO would deenergize the electrical circuit and eliminate the hazard associated with the fuse replacement. Planning and execution of LOTO will be discussed later in this chapter. This example shows the importance of planning in the process of creating a safe workplace. If workers were not aware that changing a fuse in an energized circuit could injure them, they would only find out after feeling the shock or were lucky enough to wake up in the hospital.

You may be thinking, who would do such a thing? This happens more often than you might think. Thousands of workers are killed each year from electrical injuries, so proper training and prevention are not the norm. Knowing the hazards and determining corrective action requires training and guidance. It cannot be stressed enough that planning is important to safety. Use whatever resources are available such as regulations, standards, and industry best work practices to develop a plan for a safe workplace. Some managers may say this takes too much time and will cost too much to implement. A few qualified workers developing and implementing a safety plan will be far less expensive than millions of lost revenue from a plant closure, injured worker compensation payments, increases in insurance, or explaining to a spouse or family why dad or mom are not coming home tonight.

Implementation

Determining workplace hazards, documenting safe work procedures, purchasing safety equipment, developing safety guards, deciding on appropriate training requirements, and deciding on personnel needs are all great activities, but the workplace is not any safer unless action is taken to implement the plan. The safety plan should include a procedure to introduce the new safety policy and program to all employees so they understand the reason for changes. People tend to adapt to bad situations, so they become complacent and figure what they have been doing is normal whether or not it is safe. Correcting the situation can be accomplished through employee training on why the change is necessary and by fostering a cultural change to safe work practices.

People do not like change! There have been many studies and books that document resistant behavior. People can change, however, if it is made personal. How is this accomplished? People need a motivating factor to change. Motivating factors include money, improved self-worth, job enrichment, and special job perks. Money is not always the best motivator because shortly after people receive a bonus or a raise they tend to look for another incentive. Improved self-worth is a good motivator. Let someone who is doing a good job know about it and that it is helping the company meet its goals. This can be done on a one-on-one basis or by presenting a plaque or certificate to the employee at a meeting or special ceremony. Job enrichment is another good motivator provided it does not stretch the employee's time to thin. Job enrichment is giving a motivated worker more work responsibility. This may be done through giving him or her responsibility to lead a team or participate in a special task force to implement or follow up on a portion of the safety program. This can lead to the employee looking for a monetary incentive if he or she feels extra effort is being taken for granted. Special job perks may or may not act as a motivator, depending on the type of perk. Some companies set up fitness plans at local gyms for employees to participate in exercise programs, and others have set up certificate programs for different stores or restaurants. This may work for some employees but not all. For example, an employee may not be interested in participating in an exercise program so this perk would be of no value. Choose a perk that can work for the interests of most employees, but be flexible and make adjustments as needed.

Once the implementation process starts, do not go back. False starts and giving the impression that the program is going to be delayed give a bad impression. Employees quickly pick up on this attitude and view the change as a passing fad that can be dismissed. If employees are trained to do an activity safely, make sure they have the correct tools and safety gear to follow the new established procedure. Should they choose not to do the activity safely, then be consistent with warnings and reprimands. Policy should be implemented across the entire organization and at every level. Supervisors and managers need to be role models and not exceptions to the rules.

Safety Meetings

Weekly safety meetings and daily tailgate meetings are valuable tools to reinforce the importance of working safely. Weekly safety meetings should be used to go over company safety issues or industry issues. Present lessons learned from near-miss incidents. Just because an accident was avoided does not mean it will end so well the next time. Near-miss incident reports can highlight what went wrong and why. Was the proper equipment used? Was there adequate training for the employees? Was the safety procedure or work instruction incorrect? What were the environmental conditions at the time of the incident? One or more of these questions may yield a clue to what caused the near-miss incident. The accident pyramid can be viewed as several near-miss incidents leading to one minor accident. Several minor accidents lead to a serious accident or even a fatality. If a cause can be determined for the near-miss incident, then update the safety procedure or work instruction and publish the improvement. Reduce the chances for more near-miss incidents and the possible injuries that may occur as the odds stack against the organization.

Daily tailgate meetings were mentioned earlier with climbing activities, but they are worth mentioning again. Use the daily meetings for constructive information about how the job is going and what safety issues may have been encountered during the previous day's activities. Make notes on what those safety issues were and how they can be avoided during today's activities. This is the time to present which job activities are scheduled and go over safety procedures, work instructions, training requirements, required safety equipment, and tools that may be needed. If anyone has a question on these or other related items, stop and clarify what is needed. Save time and cover the issue before leaving the operations building. It may be a long trip back later if something was missed during the meeting.

Monitoring

The first run through a program will be a great start, but there will be hazards missed or procedures that may need adjustments. There are also changes in industry equipment and safety standards as an industry matures. Mature industries such as mining, agriculture, and steel construction probably do not see many changes, right? These industries have seen many new breakthroughs in technology and their associated safety issues over the past 20 years.

The wind-power industry is no exception to these rapid changes in technology. If a technician mentioned in 1990 that she would be working on a 4-MW wind turbine in 2010, her peers would have thought she fell on her head too many times. Articles are being published today that highlight 10-MW wind turbines being developed for deployment off the northern coast of Europe over the next few years. There is a philosophy of continuous improvement that is always driving organizations to develop larger and more-efficient equipment. A safety program should be treated the same way. There is always room for improvement. An organization should not sit back and say its safety record is good enough. This may be the case if it has not had any injuries for several years, but most organizations do not fall into this category. Government regulations and national standards organizations recommend review of a safety program on at least an annual basis. Take time during the year to audit portions of the safety program so that a full review is not needed at the end of the year. Use the audits to highlight what is working and what needs improvement. Use feedback from weekly safety meetings and tailgate meetings to research and correct issues that are highlighted as possible issues. There may be new safety equipment developments or recommended procedure changes presented at trade shows or in national trade journals that may help with identifying improvements. As mentioned earlier, use whatever resources are available to help with the safety program. Learn from others' mistakes. We do not have enough time in life to make every mistake.

Work Condition Analysis

Approaching a work activity that will be performed on an electrical system requires proper training and planning to ensure safety. The planning process should have been completed during the development of the organization's safety plan hazard-and-risk analysis. The hazards associated with routine and nonroutine activities should have a **job-safety analysis (JSA)** that lists the conditions of both an energized circuit and a deenergized circuit. The analysis should also include the approved method to bring the circuit to the deenergized state so work can be performed safely. If the circuit must be worked on while energized, then special precautions must be documented in a work permit to ensure the activity can be performed as safely as in the deenergized state. If the hazard-and-risk analysis has not been completed, it must be completed before starting the electrical activity. Include this information along with work procedures, required tools, and equipment in the tailgate meeting. Keep the following items in mind when analyzing an electrical work activity: hazards, PPE, personnel qualifications, and the deenergized or energized condition.

The level of hazard associated with electrical activities varies with the energy available to the conductors that will be exposed during the procedure. If an accidental contact were to occur with an exposed conductor, the amount of energy available to the circuit will determine the shock, electrocution, or arc-flash hazard present. An **arc flash** is the discharge of energy through an air gap between electrical phases or between an energized conductor and ground. Think of an arc flash as the discharge of energy that is observed when a person performs electric welding. Electric welding uses an energized electrode and a grounded metal workpiece. The electrical potential between the electrode and workpiece causes a discharge of energy that can be seen as light as well as a transfer of metal between the electrode and workpiece. This same discharge can occur when a tool or a person comes into contact or close proximity with an energized and exposed conductor. The amount of energy available to create the arch flash can cause severe burns, dismemberment, and even death if the worker is not properly protected.

Training, planning, and appropriate PPE are the ways to prevent these injuries. Training requirements for a **qualified person** to work around conductors of 50 volts or higher include:

- techniques that are necessary to distinguish between different conductors during an activity,
- an understanding of methods to determine nominal voltage and the required approach boundaries for voltage levels that may be encountered during the activity, and

- the necessary skills and understanding to assess the hazards, plan preventive measures, and perform the activity safely.

Approach Boundaries

Approach boundaries are the distances that should be observed when working around live electrical conductors. These conductors may be in power-distribution panels, control, consoles, or mounted overhead on utility poles. Any of these conductor locations may apply to a wind farm or other industrial setting. **Figure 3-1** shows a few examples of nominal voltage for conductors and the associated approach boundary that should be observed. NFPA 70E article 130 lists three categories of approach boundaries: limited, restricted, and prohibited. Boundary definitions are as follows:

- Limited approach—general work area of the conductor but not close enough to contact accidentally.
- Restricted approach—close enough to the conductor that accidental contact is possible.
- Prohibited approach—considered to be direct contact with conductor.

These boundaries are set up as a radial distance from a live conductor to a theoretical point in space around the conductor. Qualified electrical workers are the only personnel who should enter these boundaries to ensure that the work activity can be preformed safely. The only exception to this would be an unqualified worker entering the limited approach area under direct supervision of a qualified worker. To ensure these boundaries are followed, an employer needs to set up guidelines and procedures to prevent inadvertent entry of unqualified workers during work activities. Prevention may be accomplished through the use of barricades and signage along with appropriate training for unqualified workers to recognize the use of these prevention measures.

The employer should also include in the safety plan for live electrical work the necessary steps that qualified workers must follow to ensure their safety during these activities. The procedure should include:

- training requirements for the specific task to be accomplished,
- documented justification why this activity should be accomplished while the circuit or system is energized instead of isolating the circuit,
- risk analysis of the voltages to be encountered during the activity,
- means of approval for justification and hazard analysis before the activity is performed, and
- use of approved PPE for the levels of voltage and energy that will be encountered during the activity.

Documented justification for energized electrical work should include a work permit that should be followed to ensure the necessary steps to perform the activity safely have been followed. This process ensures that the qualified worker and management are aware of the work activity and that they have determined the live work is a necessary. Appendix C shows a sample **electrical work permit** that may be used to document the planning process.

Hazard Risk Category

Each work activity has associated risks that have been summarized in federal regulations and national standards documentation such as NFPA 70E. **Figure 3-2** lists a summary of work tasks that may be encountered during an electrical activity and the associated **hazard risk category (HRC)**. The use of a HRC for these tasks enables a quick reference of the necessary PPE that should be used to protect workers from potential injuries during the activity. **Figure 3-3** summarizes some of the PPE recommendations for each HRC.

Nominal Voltage to Ground	Suggested Minimum Approach Distance					
	Condition A		Condition B		Condition C	
	m	ft	m	ft	m	ft
0–150	0.9	3.0	0.9	3.0	0.9	3.0
151–600	0.90	3.0	1.0	3.5	1.2	4.0
601–2,500 V	0.9	3.0	1.2	4.0	1.5	5.0
2,501–9,000 V	1.2	4.0	1.5	5.0	1.8	6.0
9,001 V–25 kV	1.5	5.0	1.8	6.0	2.8	9.0
>25 kV–75 kV	1.8	6.0	2.5	8.0	3.0	10.0
>75 kV	2.5	8.0	3.0	10.0	3.7	12.0

© Cengage Learning 2013

FIGURE 3-1 *Electrical working clearances*

Information listed here is for reference only. For complete details, consult OSHA 29CFR 1910.303 Tables S-1 and S-2 or NFPA 70E Article 130

Condition A: Exposed live parts on one side and no live or grounded parts on the other side of the working space or exposed live parts on both sides effectively guarded by suitable insulating material. Insulated wire or insulated bus bars operating at 300 V or less are not considered live parts.

Condition B: Exposed live parts on one side and grounded parts on the other side.

Condition C: Exposed live parts on both sides of the work space without guarding and the operator located between parts.

Task on Energized Equipment	Suggested HRC
Equipment Rated Below 240 V	
Circuit-breaker operation with covered panel	0
Work with energized conductor, including voltage testing	1
Removal or installation of circuit breaker	1
Panels and Switchboards Rated Greater Than 240 V but Less Than 600 V	
Circuit-breaker operation with covered panel	0
Circuit-breaker operation with covers removed	1
Work with energized conductor, including voltage testing	2*
600-V Class Switchgear	
Noncontact inspections outside NFPA restricted-approach distance	2
Circuit-breaker operation with covered panel	0
Work activity with exposed control circuits energized above 120 V	2*
National Electrical Manufacturers Association (NEMA) E2 Motor 2.3 kV Through 7.2 kV	
Noncontact inspections outside NFPA restricted-approach distance	3
Circuit-breaker operation with covered panel	0
Work activity with exposed control circuits energized above 120 V	3
Metal Clad Switchgear, 1 kV Through 38 kV	
Noncontact inspections outside NFPA restricted-approach distance	3
Circuit breaker operation with covered panel	2
Work activity with exposed control circuits energized above 120 Volts	4

© Cengage Learning 2013

FIGURE 3-2 *Examples of electrical activities and associated hazard risk category (HRC) classifications*

Information listed for reference only. For an expanded listing of electrical activities and associated HRC, consult the latest National Fire Protection Association publication, NFPA 70E Article 130.

Personal Protective Equipment

PPE requirements for electrical work may include the items listed in the HRC examples, but they also include insulating equipment such as:

- blankets,
- line hoses,
- gloves,
- sleeves, and
- covers.

These devices along with the PPE listed in national standards ensure that a protective barrier is between the worker and any exposed electrical conductors that may be in the work area. As with all PPE, these devices should be inspected prior to each use. Considerations for appropriate PPE should be listed in the safety procedure for each electrical activity. Always choose the appropriate insulating equipment as recommended by the company safety procedure or national standards. Choose an insulating barrier such as a glove, blanket, or cover that meets or exceeds the voltage level that will be encountered. **Figure 3-4** shows examples of voltage ratings for insulating gloves. Protective gloves are categorized into six classifications based on approved voltage levels of protection. Glove classification can be determined quickly based on the glove color-coding system and tags.

Hazard Risk Category (HRC)	Suggested Clothing and PPE
HRC 0	
Flame-resistant (FR) clothing	Untreated natural fiber long-sleeve shirt and long pants
FR PPE	Safety glasses or goggles
	Hearing protection
	Leather work gloves
	Safety shoes
HRC 1	
FR clothing	AR long-sleeve shirt
Minimum arc-rated (AR) 4 cal/cm^2	AR long pants
	AR coveralls
	AR face shield or flash suit hood
	AR jacket, parka, or rain gear
FR PPE	Hard hat
	Safety glasses or goggles
	Hearing protection
	Leather work gloves
	Leather safety shoes

© Cengage Learning 2013

FIGURE 3-3 *Examples of hazard risk category and personal protective equipment (Continued)*

Information listed for reference only. For a complete listing of HRC and PPE for electrical activities, consult latest National Fire Protection Association publication NFPA 70E Article 130.

HRC 2	
FR clothing	AR long-sleeve shirt
Minimum AR 8 cal/cm²	AR long pants
	AR coveralls
	AR full face shield or flash suit hood
	AR jacket, parka, or rain gear
FR PPE	Hard hat
	Safety glasses or goggles
	Hearing protection
	Leather work gloves
	Leather safety shoes

HRC 2*	
FR clothing	AR long-sleeve shirt
Minimum AR 8 cal/cm²	AR long pants
	AR coveralls
	AR flash suit hood
	AR balaclava hood
	AR jacket, parka, or rain gear
FR PPE	Hard hat
	Safety glasses or goggles
	Hearing protection
	Leather work gloves
	Leather safety shoes

HRC 3	
FR clothing	AR long-sleeve shirt
Minimum AR 25 cal/cm²	AR long pants
	AR coveralls
	AR flash suit hood
	AR flash suit jacket
	AR flash suit pants
	AR jacket, parka, or rain gear
FR PPE	Hard hat
	FR or AR hard hat liner
	Safety glasses or goggles
	Hearing protection
	AR-rated gloves
	Leather safety shoes

HRC 4	
FR clothing	AR long-sleeve shirt
Minimum AR 40 cal/cm²	AR long pants
	AR coveralls
	AR face shield or flash suit hood
	AR jacket, parka, or rain gear
FR PPE	Hard hat
	FR or AR hard-hat liner
	Safety glasses or goggles
	Hearing protection
	AR-rated gloves
	Leather safety shoes

FIGURE 3-3 (Continued)

© Cengage Learning 2013

Tag Color	Class	Proof Test Voltage		Maximum Usage Voltage	
		AC	DC	AC	DC
Beige	00	2,500	10,000	500	750
Red	0	5,000	20,000	1,000	1,500
White	1	10,000	40,000	7,500	11,250
Yellow	2	20,000	50,000	17,000	25,500
Green	3	30,000	60,000	26,500	39,750
Orange	4	40,000	70,000	36,000	54,000

© Cengage Learning 2013

FIGURE 3-4 *Voltage classifications for rubber insulating gloves*
Refer to the latest revision of OSHA 29 CFR 1910 for more details.

When purchasing electrical insulation devices, ensure they meet or exceed the ratings and specifications required for the activity. The manufacturer should be able to answer any questions on national standards ratings for their devices. **Figure 3-5** shows an example of markings on insulating gloves. The equipment must have the appropriate voltage rating, national standard specification number, batch number, and manufacturer contact information. Verify that this information is available and legible during each inspection process. Another item to verify is the testing lab's inspection date listed on the equipment. If any of this information is missing, notify the company safety coordinator for proper disposition of the item.

These activities show the necessary steps involved with working with and around energized electrical systems. In most cases, electrical activities should not be performed while the system is energized. Working on an energized system should only be considered when deenergizing the system can create a greater hazard to personnel or others who depend on the system. The company safety policy should indicate the necessary procedures and training required for workers to isolate, deenergize, disable, and create an electrically safe work condition. Lockout–tagout is the method used to create and secure a zero-energy condition.

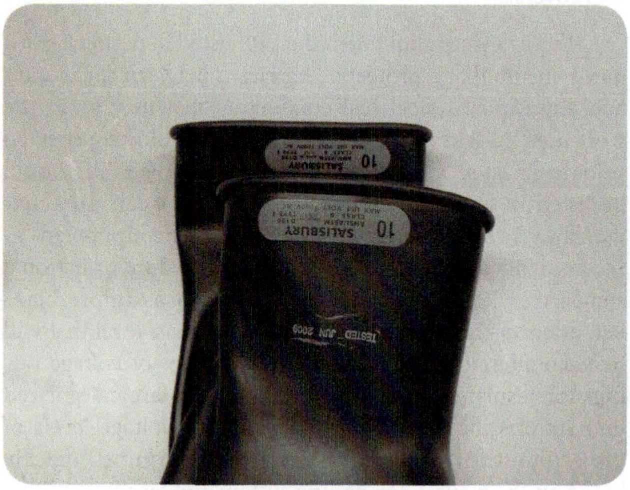

FIGURE 3-5 *Markings on insulating gloves*

© Cengage Learning 2013 and a W. Kilcollins photograph

Lockout–Tagout

LOTO refers to the application of a lock or tag to a system that has been deenergized and brought to a zero-energy condition. **Zero-energy condition** means that all energy sources have been isolated, deenergized, and disabled to prevent reenergizing. This method can be used on any system that can be brought to a state of zero energy. LOTO examples listed in this section will highlight electrical systems, but later sections in this chapter will refer to mechanical and chemical systems. The company safety policy should list the necessary training and skill level required for an employee qualified to perform LOTO procedures. The training for a LOTO-qualified person may include the following:

- employer responsibility;
- employee responsibility;
- LOTO device recognition;
- method of LOTO device installation;
- LOTO procedures, including execution and release;
- policy-enforcement practices;
- recognition of energy sources;
- control of energy sources;
- understanding of equipment operation;
- approved methods to discharge stored energy;
- grounding requirements to prevent reenergizing;
- methods to determine voltage levels;
- procedures required for working with affected employees;
- procedures required for working with contractors on site; and
- procedures for simple and complex LOTO applications.

The safety policy should also document the steps required to execute a LOTO procedure and the steps to release the equipment or system back to normal service. This procedure should include the requirements necessary to notify management and any affected employee. An **affected employee** is anyone who works directly with the equipment or system or may be in an area adjacent to the controlled work area. Steps to implement a LOTO activity may include:

- perform an informational meeting with personnel performing the LOTO activity, affected employees, managers, support personnel, and contractors (if required);
- plan for the shutdown (account for all employees who enter the control area);
- determine all sources of energy feeding the equipment or system (written procedures and a checklist will improve the accuracy of this activity);
- isolate all energy sources and bring them to a zero-energy condition;
- apply locks or tags to the energy sources as approved by the LOTO policy (all employees performing work within the LOTO control area must apply a lock or tag);
- try to start the equipment or system after all energy sources are deenergized (equipment should *not* start); and
- perform maintenance or service activity.

Steps to release the LOTO and return the equipment or system back to service are as follow.

- plan for the release of LOTO (account for all employees working in the controlled area),
- ensure all work is completed and equipment is ready for return to service (inspect for misplaced tools, missing guards, etc.),
- make sure each employee removes his or her own lock or tag,
- remove LOTO devices and reenergize sources,
- run equipment or system to ensure proper functioning, and
- return equipment to normal operation.

These are many of the steps required to execute a LOTO and the subsequent steps to return the equipment to normal operation. Following these steps and company-specific safety procedures for each piece of equipment or system that requires LOTO will ensure an electrically safe work condition (refer to NFPA 70E, OSHA 29 CFR 1926.147, and OSHA 29 CFR 1910.147 for more details).

Case Study

Background

The scheduled semiannual maintenance activities for a utility-scale wind farm were to be completed with a supplemental team of outside contract personnel. Company policy for contractor personnel was to obtain all relevant training-matrix information for contractor personnel who were to be assigned to the activity. Qualification materials were to include working at height, wind-turbine systems, maintenance fundamentals, LOTO, hazard communication, electrical, fluid power, and any other relevant experience to ensure that personnel could perform their intended assignments efficiently and safely.

The contractor team assigned to support the wind-farm operations staff had qualifications for working at height, tower rescue, mechanical maintenance, and limited electrical experience. It was decided this level of training would be adequate provided the contractors had electrical support from the wind-farm operations staff. Daily job briefings would be used to assign activities, cover safety procedures, and establish maintenance teams which included an operations staff member. This activity would ensure the daily maintenance activities would be performed safely.

Incident. Contract employees were assigned the task of changing an oil lubrication system filter while the operations staff member was performing electrical activities in another area of the wind turbine. A typical maintenance procedure to change a similar lubrication oil filter might be as follows:

- Obtain an oil sample from the reservoir for lab analysis.
  - Open bleed valve and manually activate the pump motor contactor to pump gear oil into a clean sample container.

- After obtaining the sample, close the bleed valve, seal and mark the sample container as required.
- Apply LOTO device(s) to the pump motor contactor.
- Open drain-gate valve on canister assembly and drain oil into a clean container.
- Remove cover from the canister.
- Remove oil-filter element after the oil canister is drained.
- Replace with the new oil-filter element.
- Close drain-gate valve and replace canister cover.
- Inspect and return drained oil to lubrication oil reservoir.
- Remove LOTO device(s) from the pump motor contactor.
- Clean the area of any spilled oil and oil-soaked rags.
- Discard waste materials in accordance with company policy.

The contractor team leader decided to speed up the process of draining the oil-filter canister by using the pump to drain the oil. After collecting the oil sample, the contractor turned off the pump contactor and proceeded to reverse two phases of the three-phase 400-V_{AC} conductors feeding the pump motor. The contractor did not realize that turning off the contactor deenergized the terminals on the bottom of the assembly while the top terminals would remain energized. The contractor placed a standard screwdriver (noninsulated shank) on one of the top terminals to loosen a phase conductor feeding the contactor. While performing this activity, the screwdriver slipped and created a short between two of the phases. A subsequent arc flash and melting of the phase conductor occurred before the upstream circuit breaker tripped and deenergized the circuit. The operations staff member was alerted to the incident when power was disconnected from the control cabinet he was servicing. A contractor employee received a minor flash burn to his hand and temporary blindness from the flash intensity. Vision returned to normal within a few minutes after the incident.

Conclusion. The contract technician thought the process could be speeded up by reversing the pump motor direction. Swapping two of the three phase conductors will reverse the direction of a three-phase motor, but this was not an approved procedure for draining the oil canister. If this individual had been a qualified electrical technician, he would have known:

- which conductors were still energized after opening the contactor,
- which upstream circuit breaker would have deenergized the contactor feed circuit, and
- that a simple voltage test would have confirmed that the terminals were still energized.

The contract technician was fortunate to not have been electrocuted or severely burned by the arc flash created during the shorting of the two phases. This incident shows how important it is to follow established safety procedures and not attempt activities beyond a person's training and skill level.

MECHANICAL SYSTEMS

A wind turbine is made up of several mechanical systems with different functions. These systems include:

- drive train,
- blade pitch,
- yaw drive,
- braking,
- fluid power,
- cooling, and
- heating.

The main purpose of a wind turbine is to convert kinetic energy from moving air into rotational energy that can be coupled to an electrical generator. Creating rotational energy from the wind is accomplished through the use of aerodynamic blades attached to a rotor assembly. **Figure 3-6** shows examples of fixed-pitch and variable-pitch rotor assemblies. Systems designed with variable-pitch blades use large bearing assemblies and drive systems to position the blades to control torque and the rotational speed (in revolutions per minute, or RPM) supplied to the generator. Mechanical drive for pitch position control can be accomplished through a couple of methods: the use of electric motors to drive large gear assemblies or the use of a hydraulic drive system connected to a mechanical linkage to pitch the blades. Utility-sized wind turbines use a yaw drive system to position the rotor assembly into the wind for optimum electrical output. The yaw drive system uses a large bearing mounted between the bedplate and the tower along with one or more geared motor drives meshed with a large gear attached to the tower. Other mechanical systems in a wind turbine are used as a drive-train parking-brake system, manipulate air dampers for cooling, or circulate fluids for cooling, lubrication, and power. The same care must be used when working around mechanical systems as with electrical systems. Training to understand the hazards associated with mechanical systems and planning activities to eliminate these hazards will prevent accidents. Government regulations establish requirements for employers to develop a safety plan and implement safeguards to prevent injuries from mechanical systems just as they do for working at heights and electrical systems (refer to OSHA 29 CFR 1926.307 and 29 CFR 1910.219 for more details).

Types of Hazards

Understanding the hazards associated with mechanical systems will enable a technician to develop and implement safe work practices around these systems. Some of the hazards and safeguards associated with mechanical systems are summarized in **Table 3-1**. Refer to your company's safety policy, work procedures, and job-safety analysis for the recommended steps to prevent injuries around mechanical hazards. The planning step in working around any hazard cannot be mentioned enough. Before any work activity is

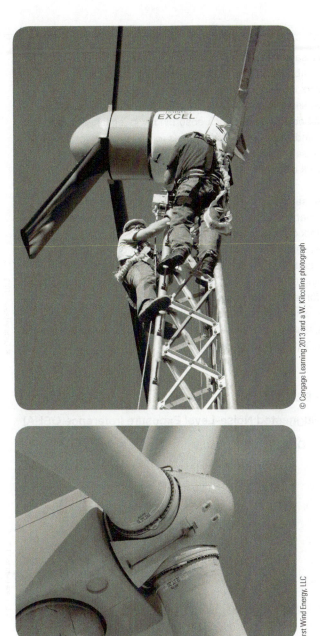

FIGURE 3-6 *Fixed-pitch and variable-pitch rotor assemblies*

performed, technicians, safety coordinator, and supervisor should meet to ensure that everyone understands the assignment and is aware of the necessary precautions to use during the activity.

Table 3-1 lists a summary of hazards associated with several of the mechanical systems in a wind turbine. It does not include all of the hazards that may be encountered within every system. Consult the company's safety policy and the documentation provided by the wind-turbine manufacturer for specific systems that are installed at your site. Use this information to prepare for activities on or around these systems.

TOOL SAFETY

Hand tools and power tools are necessary to perform many activities in a wind turbine. Knowing the proper use of a tool is as important as the selection process of tools. Tool selection is often listed in work procedures developed by company engineering departments, or it is left to the judgment of qualified employees. If the information is not available and you are not sure of the proper tool for an activity, then ask your supervisor. The proper tool will enable an activity to be completed efficiently and safely. Tools should be inspected for proper function and required safety features before each use. The company policy, national standards associations, manufacturer's literature, or federal regulations should be consulted if there is a question on tool safety. Never use a tool that is broken or not intended for a particular job—for example, do not use a screwdriver as a pry bar. This author has seen several expensive pieces of equipment damaged by improper tool use that could have been avoided if the technician had taken an extra few minutes to get the proper tool. Some of the damaged equipment required thousands of dollars and days to repair when only an extra 10 minutes would have been required to get the proper tool. Equipment damage is not the only concern with improper tool use. Injuries and some deaths could have been avoided if the proper tools were used. Take the time to get the right tool for the job (refer to OSHA 29 CFR 1926 subpart I and OSHA 29 CFR 1910 subpart P for more details).

TABLE 3-1

Summary of Mechanical System Hazards		
System	**Hazard**	**Safeguard**
Drive train	Moving parts, entanglement, crush, pinch	Covers, lock assemblies, no loose clothing or safety gear
Gear box	Moving parts, entanglement, crush, pinch	Covers, lock assemblies, no loose clothing or safety gear
Gear drive	Moving parts, crush, pinch	Covers, lock assemblies
Brake assembly	Moving parts, entanglement, pressurized fluid, spring energy, crush, pinch	Covers, bleed valves, lock assemblies
Cooling	Moving parts, pressurized fluid, crush, pinch, burns	Covers, bleed valves, interlocks
Fluid power lubrication	Moving parts, pressurized fluid, injection, crush, pinch, burns	Covers, bleed valves, interlocks

© Cengage Learning 2013

Activity	Personal Protective Equipment
Torque inspection of tower flange bolts with manual bolting tools	Safety glasses, safety shoes, cut-resistant gloves, hard hat, full body harness, shock absorbing twin-leg lanyard, ladder safety device, and appropriate work clothes for temperature conditions
Torque inspection of tower flange bolts with power bolting tools	Safety glasses, safety shoes, cut-resistant gloves, hard hat, hearing protection, full body harness, shock-absorbing twin-leg lanyard, ladder safety device, and appropriate work clothes for temperature conditions
Visual and infrared (IR) inspection of power cables and electrical cabinets without entering cabinets in the DTA	Safety glasses, safety shoes, hard hat, hearing protection, and appropriate work clothes for temperature conditions
Visual inspection of decking, ladder, and cable supports	Safety glasses, safety shoes, hard hat, cut-resistant gloves, full-body harness, shock-absorbing twin-leg lanyard, ladder safety device, and appropriate work clothes for temperature conditions
Lubrication of bearing assemblies	Safety glasses, safety shoes, hard hat, cut-resistant gloves (climbing), chemical-resistant gloves (lubrication activities), full-body harness, shock-absorbing twin-leg lanyard, ladder safety device, and appropriate work clothes for temperature conditions
UPS battery inspection in the DTA	Safety glasses, safety shoes, hard hat, appropriately rated insulated rubber gloves, leather outer glove, and appropriate FR-rated work clothes for cabinet HRC and temperature conditions

© Cengage Learning 2013

FIGURE 3-7 *Examples of work activities and suggested PPE*

General PPE

The choice of PPE for hand tools and power tools depends on the particular tool and activity, but in most cases it is recommended to use ANSI-approved safety glasses for eye protection. Some activities such as grinding or gas welding also require the full facial protection of a shield. Hearing protection is also recommended for activities with equipment that generates excessive sound levels such as the power units for tower bolting. Consult your company's hearing conservation policy, OSHA 29 CFR 1910.95, or 29 CFR 1926.52 for further details on hearing protection. **Table 3-2** shows exposure duration limits and sound levels that should be considered when choosing hearing protection. **Figure 3-7** shows examples of work activities and suggested PPE. Refer to the latest revision of OSHA 29 CFR 1910 regulations for further details on PPE requirements.

Hand Tools

A hand tool is any tool that requires manual force to perform its function. This includes screwdrivers, wrenches, hammers, pry bars, and chain hoists, among others. **Figure 3-8** shows examples of hand tools that may be used during maintenance activities for a wind turbine.

Inspection Items

Inspect tools for damage before each use. Inspect for defects such as cracks and missing or deformed components. If any of these problems are encountered, remove the tool from

TABLE 3-2

Suggested Noise-Level Exposure (reference OSHA)	
Daily Time Exposure in Hours	Sound Level (Decibels)
8	90
6	92
4	95
3	97
2	100
1½	102
1	105
½	110
¼ or less	115

© Cengage Learning 2013

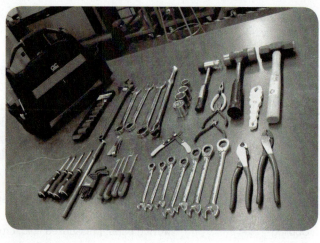

© Cengage Learning 2013 and a W. Klicollins photograph

FIGURE 3-8 *Examples of hand tools used for wind-turbine maintenance*

service and request a replacement. When in doubt, check with your supervisor or safety coordinator (refer to OSHA 29 CFR 1926.301).

Power Tools

Power tools include any tool that uses an external power source such as—pneumatics, hydraulics, or electricity—to perform its function. These include drills, grinders, bolting tools, pneumatic wrenches, power hoists, and others. **Figure 3-9** shows examples of power tools that may be used during maintenance activities for a wind turbine.

Inspection Items

Inspect power tools before each use to ensure that they function properly and that all safety features are in place as required by manufacturers' literature and government requirements. If a power tool has a damaged power cord, hose, safety guard, or tool attachment, remove it from service and request a replacement from the supervisor or safety

coordinator. **Figure 3-10** shows examples of defects that may be encountered with power tools. Do not use a power tool with a missing guard or other safety device (refer to OSHA 29 CFR 1926.302 and NFPA 70E Article 110.9). As with hand tools, do not use a power tool or attachment for a purpose other than its intended design. For example, never use die-cast (typically, chrome-plated) sockets with power-bolting

FIGURE 3-9 *Examples of power tools that may be used for wind-turbine maintenance*

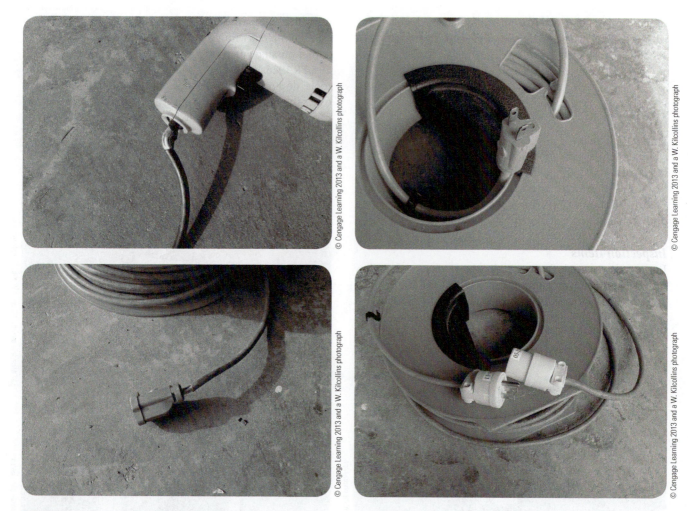

FIGURE 3-10 *Examples of defects with power tools*

equipment. Use manufacturer-recommended impact sockets that have a black oxide finish for power-bolting tools. The die-cast sockets will fracture under the high-stress loads developed by power tools. Die-cast sockets should only be used with hand tools.

RIGGING

Web straps, chain slings, cable slings, and other rigging equipment are not just used with large cranes for lifts. These devices may be used to lift components that weigh 100 pounds (45.4 kg) or less out of tight locations in the nacelle or to raise tools and components up the tower with a hoist. Safely lifting components requires an understanding of proper lifting techniques and the use of acceptable rigging equipment. **Table 3-3** lists some guidelines for lifting tools, equipment, and components with installed tower hoists.

A variety of hoist assemblies may be used for lifting items up the tower. Some wind turbines have a hoist mounted at the yaw deck that is used to bring items up the inside of the tower, whereas others have a hoist mounted in the nacelle on a rail assembly or a swing arm that may be used to bring items up the outside of the tower. A variety of manual hoists may be used to lift items within the hub or the nacelle during some replacement activities. **Figure 3-11** shows examples of hoist equipment that may be used at a wind farm. Inspecting rigging components is as important as properly selecting these components.

Inspections

Make a habit of inspecting rigging components along with other safety equipment at the beginning of the work activity. This habit will ensure the equipment taken to the job site will be acceptable for use and save a return trip to the operations and maintenance shop for a replacement (refer to OSHA 29 CFR 1926.251 and OSHA 29 CFR 1910.184 for more details). Inspection of rigging components should include a visual

TABLE 3-3

Lifting Guidelines	
Lift Activity	**Recommendations**
Tools, equipment, and supplies	Use a lift bag rated for the weight of the tools.
	Never exceed the bag rating.
	Never exceed the rating of the hoist.
	Never leave components hanging outside the bag to catch the deck opening.
	Cover sharp edges to prevent tearing the lift bag.
Oil pails and buckets	Never lift by the wire handle on the pail or bucket.
	Place the pail or bucket in a lift bag.
Long components	Use a choke hold with a sling for the lift.
	Never use the hoist cable or chain for the sling.
	Use features on the component to ensure the sling will not slide off.
	Have a technician climb up alongside the component to ensure it clears each deck opening.
Large components	Ensure the component will fit through the deck hatch openings.
	Have a technician climb up alongside the component to ensure it clears each deck opening.
Loaded tool box or pouch	Never lift the box or pouch by its handle.
	Place the box or pouch in a lift bag.
Component within the nacelle	Use an appropriate hoist rated for the component weight.
	Use an approved lift point within the nacelle.
	Use a small gantry assembly when available.
Component within the hub	Use an appropriate hoist rated for the component weight.
	Use an approved lift point within the hub.
	Use a hub crane assembly when available.

© Cengage Learning 2013

examination of all shackles, web straps, chains, cables, slings, and hoists for defects. **Figure 3-12** shows examples of rigging components that may be used to raise tools and service items up the tower. The inspection should include verification of a manufacturer's safety tags. These tags are required by federal regulation to be present on rigging components. Refer to the manufacturer's literature that is supplied with each component or government regulations for details.

Permanently Mounted Tower Hoists

Never lift components with internal tower hoists without an initial inspection of the hoist assembly, chain, or cable, especially if the cable or chain has been tied off down tower for storage. The cable or chain may have abraded against deck openings during operation of the wind turbine and created weak spots that may fail during use. The first person to climb the tower should inspect the cable or chain at each deck

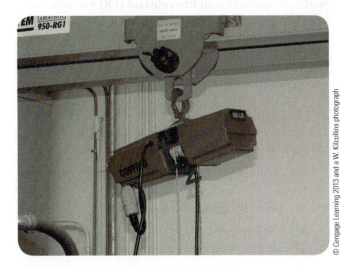

© Cengage Learning 2013 and a W. Kilcollins photograph

University of Maine at Presque Isle

FIGURE 3-11 *Examples of cable and chain hoist systems*

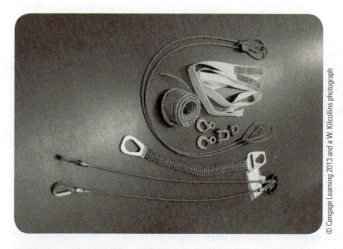

© Cengage Learning 2013 and a W. Kilcollins photograph

FIGURE 3-12 *Examples of rigging components that may be used to move equipment up the tower*

LIFTING SAFETY TIPS

- Lift with your legs not with your back.
- Maintain a straight back while lifting.
- Never twist your back when changing direction while carrying a load.
- Team lift for long or awkwardly shaped items.
- Use lifting equipment or team lifting for loads more than 45 pounds (20.4 kg).
- Never exceed the load capacity of the lifting components.
- Use lifting points capable of withstanding the anticipated load.
- Never stand under a load during a lift.
- Use tag lines to control an object being lifted.
- Use ANSI-approved head protection when there is an overhead hazard.
- Use ANSI-approved safety shoes.
- Use leather gloves to prevent cuts and abrasions when possible.

opening during the climb and then inspect the hoist assembly before the lift. Some wind-turbine towers have the hoist assembly mounted on a swing arm that is deployed through a door in the back of the nacelle. The hoist assembly and chain should be inspected before use. Follow the inspection items listed earlier for defects that may be encountered with each of these hoists.

HOT-WORK ACTIVITIES

Certain metal-inspection procedures require removing paint to view the underlying welded joint. These operations typically use small grinders or other abrasive tools that can create sparks. Any operation that can develop sparks is considered **hot work**. Other hot-work activities in a wind turbine include metal-cutting operations, welding, and heating. Hot-work activities always present a fire hazard to technicians whether in the shop or other areas around the wind farm. With proper precautions and following your company's hot-work permit process, this hazard can be minimized. Refer to OSHA 29 CFR 1910.39, 252; OSHA 29 CFR 1926.151, 352; and NFPA 51B for requirements of a fire-prevention and fire-protection plan. **Figure 3-13** is a sample hot-work permit that may help in developing your own permit procedure. Even if the best plan is not always enough to prevent a fire so there should be a planned egress to the outside of the facility. Hot-work activities within the upper levels of a wind turbine tower, especially the nacelle, may create a death trap if there has not been proper planning. There are only two ways out of a tower: a ladder descent or a climb up and out of the nacelle to the ground. Most utility-scale wind-turbine towers run between 200 and 300 feet (60 to 90 m) in height, so the trip to the ground can be problematic. If the lower ladder egress is blocked by smoke or flames, technicians need to have everything in place to exit the nacelle and descend to the ground safely. Planning for hot-work activities is only part of a fire-prevention and fire-protection plan.

Fire-Safety Plan

Fire protection and fire prevention may be important in industrial and residential settings, but it is even more important on elevated working platforms such as a wind turbine's. It does not take long for the nacelle to be consumed by fire, so you need to act fast to evacuate. A fire-prevention and fire-protection plan should include precautions to avoid fires such as the related safety procedures for each hot-work activity. These procedures should specify precautions such as removing combustible materials from the work area such as lubricants, dust, paper, and any other materials that could ignite easily. Covering adjacent painted surfaces and exposed plastic components with **fire-resistant (FR)** blankets, wearing FR clothing, and ensuring proper ventilation to prevent buildup of combustible gases. These are some of the precautions that should be included in your procedure. A few other items include having an appropriately rated fire extinguisher, a designated fire-watch person, and tower escape equipment.

Fire Extinguisher

Fire extinguishers are rated according to the type of fire they can safely extinguish. The types include the following:

A—combustible organic materials such as paper and wood,

B—flammable liquids and gases,

C—energized electrical devices, and

D—combustible metals such as magnesium.

These type ratings are good for the United States. Other countries may have different designations for the letters such

Company name:	
Contact person:	
Date of activity:	
Type of hot-work activity: (Check one)	Grinding Cutting Welding Riveting Other (describe)
Location of activity:	
Equipment number:	
Justification for hot-work activity (Repair, inspection, modification, etc.)	
Safety considerations for area where hot work will be carried out	Flammables: Liquids, gases, etc. Combustibles: Paper, wood, coatings, etc. Hazardous fumes due to coatings: Zinc, lead, cadmium, etc. List all that apply and method to prevent hazard.
Ventilation required?	Yes/No If yes, describe method to be used.
Confined space?	Yes/No If yes, attach confined space permit information.
List personal protective equipment to be used:	
Fire extinguisher(s)	Type: Size:
Tailgate meeting attendees	
Approvals	Approval date
Site manager: Lead technician: Safety coordinator:	

© Cengage Learning 2013

FIGURE 3-13 *Sample of hot-work permit*

as the United Kingdom, where type C is for combustible gases and E is for electrical fires. Verify the extinguisher is good for your application before it is needed. The typical fire extinguisher used around a wind turbine is the ABC type. This extinguisher uses a dry chemical powder made from highly refined baking powder. **Table 3-4** shows examples of fire-extinguisher types. Choosing the proper extinguisher type is only part of the equation. You will also need to determine the proper size extinguisher to ensure there is enough extinguishing material to put out any fire that may be produced during the hot work activity. Having an extinguisher that does not provide enough material to extinguish the fire

will only contain the fire for a short time without putting it out. A fire doubles in size every 30 seconds, so there is little time to think about what needs to be done. Your reaction has to be instinctive. When the fire is noticed, the designated fire-watch person needs to act quickly.

All wind technicians should receive fire-extinguisher training as part of their standard safety program. Remember PASS? If you have not had fire-extinguisher training, then take a course. In the meantime, remember that **PASS** stands for *pull* the pin, *aim* near the base of the fire, *squeeze* the trigger, and *sweep* side to side across the base of the fire. It is important not to hit the base of the fire when you first pull the trigger

TABLE 3-4

Types of Fire Extinguishers

	Class of Fire	Typical Fuel Involved	Type of Extinguisher
Class △ Fires A (green)	**For Ordinary Combustibles** Put out a Class A fire by lowering its temperature or by coating the burning combustibles.	Wood Paper Cloth Rubber Plastics Rubbish Upholstery	Water*[1] Foam* Multipurpose dry chemical[4]
Class ☐ Fires B (red)	**For Flammable Liquids** Put out a Class B fire by smothering it. Use an extinguisher that gives a blanketing, flame-interrupting effect; cover whole flaming liquid surface.	Gasoline Oil Grease Paint Lighter fluid	Foam* Carbon dioxide[5] Halogenated agent[6] Standard dry chemical[2] Purple K dry chemical[3] Multipurpose dry chemical[4]
Class ◯ Fires C (blue)	**For Electrical Equipment** Put out a Class C fire by shutting off power as quickly as possible and by always using a nonconducting extinguishing agent to prevent electric shock.	Motors Appliances Wiring Fuse boxes Switchboards	Carbon dioxide[5] Halogenated agent[6] Standard dry chemical[2] Purple K dry chemical[3] Multipurpose dry chemical[4]
Class ☆ Fires D (yellow)	**For Combustible Metals** Put out a Class D fire of metal chips, turnings, or shavings by smothering or coating with a specially designed extinguishing agent.	Aluminum Magnesium Potassium Sodium Titanium Zirconium	Dry powder extinguishers and agents only

© Cengage Learning 2013

*Cartridge-operated water, foam, and soda-acid types of extinguishers are no longer manufactured. These extinguishers should be removed from service when they become due for their next hydrostatic pressure test.

Notes:
(1) Freezes in low temperatures unless treated with antifreeze solution, usually weighs over 20 pounds (9 kg), and is heavier than any other extinguisher mentioned.
(2) Also called ordinary or regular dry chemical (sodium bicarbonate).
(3) Has the greatest initial fire-stopping power of the extinguishers mentioned for Class B fires. Be sure to clean residue immediately after using the extinguisher so sprayed surface will not be damaged (potassium bicarbonate).
(4) The only extinguishers that fight A, B, and C classes of fires. However, they should not be used on fires in liquefied fat or oil of appreciable depth. Be sure to clean residue immediately after using the extinguisher so sprayed surfaces will not be damaged (ammonium phosphates).
(5) Use with caution in unventilated, confined spaces.
(6) May cause injury to the operator if the extinguishing agent (a gas) or the gases produced when the agent is applied to a fire is inhaled.

because it may blow the combustible material around, creating a larger fire. This is especially true for flammable liquids. **Figure 3-14** shows images of PASS-method training steps. When fighting a fire, remember to always have your back to the exit. This ensures that you do not get trapped with the fire between you and the exit. If the hot-work activity is executed to plan, then there should be no need for the extinguisher, but a backup plan should always be available. Fire prevention and fire protection also includes an evacuation plan when the situation deteriorates beyond the control of an extinguisher.

Evacuation Plan

The site fire-prevention and fire-protection plan includes an evacuation plan for all facilities on the wind farm, including the wind turbines. As mentioned earlier, exiting a burning nacelle from a couple hundred feet in the air is problematic

at best. The evacuation plan for a wind turbine will include a rope or cable rescue system that will be transported up the tower each climb or be permanently mounted in the nacelle as part of the standard safety equipment. A word of caution: inspect your safety gear before each use at the beginning of the day to ensure it meets standard requirements. Rescue gear should also be inspected before each use to ensure it is functioning properly, it is not damaged, and has a tag that indicates the proper length. There have been instances where a mouse or squirrel has found a way into the storage area and made a nest of the rescue rope. Finding the nest when you are ready to use the rescue gear would be catastrophic. With regard to rope or cable length, using a system with a 250-foot rope on a 300-foot tower will not end well either. If the gear is permanently mounted in the tower, then take time to verify its function and condition. Plan ahead and be safe.

FIGURE 3-14 *PASS method training steps*

CONFINED SPACE

If you have ever worked around large industrial settings, then there probably were areas marked with signs that read "No Entry—Permit-Required Confined Space." What were these areas, and why were they marked to prevent entry? A **confined space** is defined as any area that is not intended for employee occupancy, is large enough for an employee to enter and perform work, and has a limited means of entry or egress. These areas may include electrical vaults, reaction vessels, storage silos, chemical tanks, and storage bins.

The wind-power industry has similar areas as well. They include concrete vaults for power connections, the hub area of the rotor assembly, and the interior of the blades. Each of these areas meet the confined-space definition. What is the distinction between a confined space and a *permit-required* confined space? A **permit-required confined space** has hazards that could cause a serious injury or death to anyone not trained or prepared for entry into these areas. These hazards may include a dangerous atmosphere, a sloped internal

configuration that could cause entrapment, or an engulfing hazard, among other health or safety hazards. A permit-required confined space may include one or more of these hazards, so an understanding of the possible hazards is essential before entering a confined space.

Figure 3-15 lists examples of permit-required confined space (refer to OSHA 29 CFR 1910.146 and NFPA 70E Article 130.6 B, C, and F). If these areas require entry for work activities, then federal regulations require an employer to develop and implement a plan for safe entry. The plan must include procedures for safe entry, training requirements for

Wind-turbine blades
Utility underground electrical vaults
Chemical reactor vessels
Water-treatment vessels
Transportation tanks (railway and road)
Cargo-ship or barge containment tanks

FIGURE 3-15 *Examples of permit-required confined space*

employees who must enter these areas, personnel requirements for entry, and a rescue plan for these areas should an emergency occur during an entry.

Confined-Space Entry Plan

The confined-space entry plan must address any hazards associated with the confined space such as dangerous atmosphere, internal configuration, engulfing, and other hazards. It must list each hazard and document the means to eliminate the hazard such as LOTO, ventilation, and PPE. The plan should also include the prevention methods for

unauthorized entry, the documented permit process that includes justification for the entry, required training for entrants, number of employees required for entry, their duties, and a documented rescue plan should an emergency occur during the entry. **Figure 3-16** shows an example of a confined-space entry permit for a wind-turbine hub entry. The sample permit lists action items that need to be addressed for a safe hub entry. The entry plan should also address work activities that will be performed in the space. Activities such as cleaning, painting, hot work, or other activities that may introduce an atmospheric or fire hazard should include a documented action plan to eliminate these hazards.

CONFINED SPACE ENTRY PERMIT

Date issued: _____ Time issued: _____ Anticipated duration of activity:_____ (hours)

Wind farm name: _____

Wind turbine number: _____

Lead technician or site manager _____

Proposed activity: _____

Technicians performing the activity (3 minimum):

Entrant: _____

Attendant: _____

Ground support: _____

Backup rescue requirements: _____

Technician training requirements current: _____

Other permits required? **Y/N** (i.e. Energized work permit or hot-work permit) _____

List all that apply: _____

Considerations: Weather conditions during the entry activity—wind speed, lightning alerts, icing conditions, etc.

Step	Activity	Y/N
1	Shut down of the wind turbine systems	
2	Lock out—Tag out of tower remote operation	
3	Lock out—Tag out of affected electrical or hydraulic systems	
4	Update on weather advisories	
5	All atmospheric testing requirements followed per company safety policy? Procedure number and revision: _____ Oxygen concentration within specified level: Toxic fume concentrations within specified level: Ventilation requirements: Natural ( ) or mechanical ( ) Attach data sheets or supplemental permits as necessary:	
6	Perform entry activity per company procedures	

Activity review and approval:

All required information and procedures have been reviewed and appropriate documents have been completed during the tail gate meeting.

Site manager: _____ Date: _____

Site lead technician: _____ Date: _____

Site safety coordinator: _____ Date: _____

FIGURE 3-16 *Sample of confined-space entry permit for hub or blade activities*

Hazardous Atmosphere

A hazardous atmosphere includes any of the following issues: flammable gas or mist, airborne combustible dust, deficient or enriched oxygen levels, and any other toxic or hazardous substance. Any of these hazards must be eliminated before a safe entry can be attempted in the confined space. Your employer will complete a study of the atmospheric hazards for all confined-space areas present at the facility during the development of the safety plan. Most companies do not have the expertise to do this type of study in house, so they may subcontract this type of activity. Understanding the activities at the facility and the natural reactions that may occur in a confined space will lead to the initial assessment of atmospheric hazards. For example, if the confined space is an underground vault with a moist atmosphere and exposed steel or iron components, the atmosphere is probably oxygen deficient. Why? The presence of moisture around iron tends to accelerate the formation of rust (iron oxide). With oxygen being depleted from the air in a closed space, the percentage of available oxygen to breathe will decrease, causing a condition known as **hypoxia**. An example of an airborne combustible mixture would be wood or grain dust in the air that could ignite easily if an open flame or spark was accidentally initiated in the confined space.

Employee training is important in understanding the possible hazards and performing activities to eliminate the hazards. The presence of atmospheric issues is determined by the use of a test meter that is designed for known contaminants that may be in the space. Test meters do not detect all possible hazards, so ensure that the correct test meter is used for the space to be entered. Your employer will have the possible atmospheric hazards listed on the entry permit along with the testing requirements for the space. **Figure 3-17** shows examples of atmospheric hazards that may be encountered with confined spaces. Knowing the atmospheric hazard is only part of the entry plan; the other part is to eliminate the hazard. Typical methods to eliminate an atmospheric hazard are to ventilate the area with a fan and duct system. This generally will eliminate the hazard. In some cases, the introduction of atmospheric oxygen could present an explosion issue, so the introduction of an inert gas such as nitrogen or argon to displace a flammable gas would be completed first and then ventilation would be performed. Atmospheric testing should be completed for each step of the ventilation process and during the entry to ensure that the atmosphere remains safe. If the atmospheric hazard cannot be reduced to an acceptable level, then entering the space with a **self-contained breathing apparatus (SCBA)** and other appropriate PPE may be necessary. SCBA gear is typically used by first responders or firefighters to safely enter areas with a known or possible hazardous atmosphere. Refer to the company safety plan for the confined-space entry procedure necessary for the work activity to be performed. This procedure should include the use of appropriate PPE and LOTO practices to ensure personnel safety (refer to OSHA 29 CFR 1910.146 Appendix B, atmospheric testing requirements).

Internal Configuration

The internal configuration of a confined space may include sloping walls to a small exit chute at the bottom of the area. This configuration is typical for feed hoppers used for plastics, ash, or wood by-product processing. Entry into this confined space should include a method to block the exit chute area to prevent an employee from becoming trapped or asphyxiated by falling into the small exit chute. This same type of condition would be present if a technician were to slide down into a blade assembly. The blade tapers to a smaller opening as you approach the tip. Becoming trapped or wedged into this space would create the same type of hazard.

Engulfing

Engulfing hazards at a wind farm are not typically a concern for wind technicians but are referenced here to show the industrial standard. You never know what activities you may be presented with during your career. Confined spaces that create an engulfing issue include storage silos or raw material bins filled or partially filled with a granular or thick liquid substance. Should an employee stand on or fall into the material the granular nature of the material would allow it to slide up around them. This material would then bury the employee and cause asphyxiation. You may recall the quicksand that has been shown in many movies over the years that has trapped unsuspecting victims who were not aware of their surroundings. If work is necessary in this type of confined space, then drain the engulfing material before entry. If draining is not possible, construct a platform and use a safety harness to avoid falling into the engulfing material. A possible engulfing hazard related to the wind

	Blade Assembly	Hub Assembly
Hazard	• Polyester resin out gassing from materials • Reduced oxygen available with extended work activities in the blade	• Hydrogen gas produced during discharge process of the backup battery pack
Corrective Action	• Ventilation fan • Limit work activities to a specified amount of time before requiring breaks	• Allow gases to escape after removing the entry panel for several minutes

© Cengage Learning 2013

FIGURE 3-17 *Examples of atmospheric hazards that may be encountered with a confined space such as a wind turbine hub or blade assembly*

industry would be working in excavations during construction activities. Not following the excavation requirements of sloping, benching, shoring, or shielding would expose a technician to an engulfing or crushing hazard if the excavation soil walls collapsed. Refer to the company's safety policy and OSHA 29CFR 1926 subpart P excavation regulations for further details.

Other Health or Safety Hazards

Other health or safety hazards may include moving components, exposed electrical conductors, biohazard, or other hazards that may injure or cause death to an individual who enters the space without proper safeguards. Entry into a wind-turbine hub without implementing approved LOTO procedures could end with the entrant spinning within the hub like a carnival ride. Aside from the carnival ride, the entrant would be exposed to moving mechanical components such as gears, tools, and loose consol covers that may have been removed for maintenance. This would not make for a good day. Refer to the hub entry procedure at your site to ensure proper LOTO requirements for a safe hub entry.

Training

Federal regulations require that employers train employees who participate in confined-space entry so that they can enter and perform their assigned activities safely. Employee training should include:

- identification of signage for confined-space areas;
- recognition of hazards associated with confined-space entry;
- procedures to eliminate the hazards;
- approved procedures for atmospheric-testing equipment;
- method and frequency of atmospheric testing;
- approved LOTO procedures;
- approved entry and egress techniques;
- duties of the entrant, the attendant, and the entry supervisor;
- requirements to prevent unauthorized entry;
- PPE requirements related to activities in the confined space;
- safeguard requirements related to activities in the confined space; and
- communication requirements for the confined-space entry.

Personnel Requirements

Every confined-space entry must have at least three employees participating in the activity. The person who is assigned the responsibility of entering the confined space is the **entrant**. His or her duty is to carry out the assigned activity such as cleaning, electrical work, mechanical work, adjustments, or other assigned duties. During the confined-space entry, an **attendant** is required to be at the entrance to prevent unauthorized entry and ensure that the entrant is safe during his or her activities. This may require atmospheric monitoring, fire watch for hot work, or monitoring of ventilation equipment. The attendant cannot enter the confined space while performing his or her responsibilities. If the attendant must enter the confined space, then another attendant must assume the first attendant's duties. This process must be followed even in an emergency situation. The attendant will notify the entry supervisor to call for the rescue team and then initiate the rescue if it can be performed safely. An **entry supervisor** is responsible for the confined-space entry activity. He or she is responsible for ensuring that the activity is completed safely and that all procedures are followed. Duties also include coordination of employee resources required for the entry including subcontractors, completion of required documentation, ensuring that rescue personnel are on standby, and debriefing employee resources after the entry is complete.

Rescue Plan

Federal regulations require that a confined-space entry also include a rescue plan should an emergency situation arise during the entry. Should the confined-space entrant become injured or have health issues, there must be a means to quickly extract him or her from the confined space. It takes less than a minute to incapacitate an entrant who is accidentally exposed to an oxygen deficiency or toxic gases. The rescue plan should include required first aid and CPR training for rescue team members, response time, extraction methods and equipment for each confined space, PPE requirements for the rescue team, and backup resources for rescue such as local fire and emergency medical technicians.

CHEMICAL SAFETY

Chemicals are a part of everyday life. Cabinets in our homes and workplaces hold a variety of cleaners, lubricants, and paints, among other items. Do you know the required safety measures for each chemical? What precautions should be used if one of these chemicals were to be splashed into someone's eyes? How about the maximum exposure limit for an 8-hour shift? Federal regulations require that employers train all employees who handle chemicals in the proper use and safety measures for these chemicals. This training is called **hazard communication (HAZCOM)**. This training is provided before an employee is assigned a task with chemicals and followed up on an annual basis for all affected employees. Requirements for the employer include developing a chemical safety plan, maintaining an inventory of each chemical in the workplace, documenting training program for employees, providing easy access for employees to the **material safety data sheets (MSDS)** for each chemical product and PPE, establishing emergency plans and equipment should an accident occur with chemicals, evaluating the effectiveness of the program, and meeting other requirements, depending on the types of chemicals used (refer to OSHA 29 CFR 1926.59 and OSHA 29 CFR 1910.1200).

Employee Training

A documented HAZCOM training program may list requirements such as who is responsible for training activities, the type and duration of training activities, schedules, requirements for retaining, and a list of job functions that require training. The training activity may include recognition of chemical classifications, how to use manufacturer-supplied documentation, reactions of different chemicals, a list of approved PPE for using each chemical, first-aid requirements for accidental exposure, emergency contact information, use of workplace-safety systems (alarm system, emergency showers, and eyewash stations), proper handling and storage, approved fire-suppression equipment, and emergency procedures should a spill or contamination occur. This information may be included in the company safety policy, JSA and work procedures. Another resource for safety information is supplied by the chemical distributor or manufacturer in the form of an MSDS.

Material Safety Data Sheets

Each chemical used at the wind farm or workplace should have an MSDS available for review during the daily safety briefing or tailgate meeting. If this is the first time using the material, take extra time to review and highlight some of the sections. Take special note of exposure limits, required PPE, first-aid measures, fire-suppression measures, and any other information related to the planned activity. If you have used the material before, do not be complacent: review the information and take a copy of the data sheet with you for reference. Ensure that you have the necessary items in the service truck as directed by the MSDS. If the MSDS requires a D-type fire extinguisher and your service truck is equipped with an ABC type, then do not take the material unless you have the correct extinguisher. This is the same for all other recommendations on the MSDS. If you suspect the MSDS is outdated, contact your supervisor, safety coordinator, chemical manufacturer, or do a search on the Internet for an updated copy of the MSDS. The MSDS is broken down into several sections to improve searches for specific information. These sections typically include:

- the material's common name or trade name;
- the manufacturer's contact information;
- a chemical description—liquid, gas, or solid, along with color and aroma;
- the chemical composition, whether directly stated or listed as trade secret (the manufacturer must provide

specific chemical information to medical provider for treatment in the case of emergency);
- properties—caustic, acidic, corrosive, reactive, explosive, high vapor pressure, flammability, flash point, and so on;
- first-aid measures and emergency contact information;
- routes of entry—inhalation, absorption, or ingestion;
- recommended PPE;
- fire-extinguishing method;
- threshold and maximum exposure limits listed as time-weighted values, concentrations, and so on;
- handling and storage;
- accidental release; and
- disposal information.

Appendix D shows an example of an MSDS for LGWM 2 grease. Being exposed to lubricant products such as grease during preventative-maintenance (PM) activities is standard practice for wind technicians. The sample MSDS shows the information necessary to work safely with this product.

Each section includes valuable information for developing safety procedures and planning activities to ensure that proper steps are taken to safeguard personnel and the environment when working with the materials. If an accidental release of a chemical occurs, follow your company's emergency action plan to ensure prompt containment, cleanup, and evacuation as necessary. Do not attempt containment or cleanup unless trained. Contact your supervisor and safety coordinator to activate the plan. Use available container signage and placard information such as United Nations (UN), Department of Transportation (DOT), or OSHA 29 CFR 1926.60, 1910.1201 reference materials for identification of material released and required safety precautions. **Figure 3-18** shows examples of placards and the use of reference materials to determine a material and associated safety measures. Your supervisor or safety coordinator will contact trained personnel and the local **hazardous material (HAZMAT) team** to coordinate containment and cleanup. Large releases will require notification of the federal **Environmental Protection Agency (EPA)** and your state's **department of environmental protection (DEP)** for assistance and compliance information.

Class	Type	U.S. *Code of Federal Regulations* (49 CFR)
1	Explosives	173.50
2	Compressed gases	173.115
3	Flammable liquids	173.120
4	Flammable solids	173.124
5	Oxidizers	173.127
6	Poisons	173.132
7	Radioactive materials	Subpart 1
8	Corrosive liquids	173.136
9	Miscellaneous	173.140

FIGURE 3-18 *Examples of hazardous materials placard information*

SAMPLE PROBLEM

A maintenance team is planning an activity to replenish lubricant in the main bearing assembly. Determine the recommended PPE, fire-extinguishing measures, cleanup of accidental environmental release, approved disposal method, and emergency contact information. Use MSDS for LGWM 2 grease listed earlier for this activity.

SOLUTION

Section 1: Emergency contact information

Phone: (+44) 08 45 46 47 (The Netherlands)

Section 5: Firefighting measures

Extinguisher: Extinguish with powder, foam, carbon dioxide, or water mist. Do not use jet of water.

Section 6: Accidental-release measures

Use the same personal protective equipment as stated in section 8. Sweep up or collect spills for possible reuse or transfer the material to suitable waste containers. See section 13 for instructions on disposal. Wipe up minor spills with a cloth. Prevent spillage from entering drains or surface water.

Section 8: Exposure controls and personal protection

Respiratory protection is not required unless in areas with insufficient ventilation.
Wear appropriate PPE: gloves and goggles.
Wash hands before breaks and restroom breaks.

Section 13: Disposal considerations

Disposal should be in accordance with applicable regional, national, and local laws and regulations.

ENVIRONMENTAL WORK HAZARDS

Working outside on a sunny summer day gives you an appreciation for those occupations where time is spent outside on a daily basis. These occupations include forest rangers, commercial fishers, ranchers, construction workers, and wind technicians. There are many days when it is great to work outside during the year. This is a great selling point for any outdoor occupation and one that this author shares frequently with students and acquaintances. There are also a small percentage of days that are not so enjoyable and the work must be completed under cold, wet, and extremely windy days. If you are working around wind turbines, chances are you will enjoy the wind most of the year. A considerable amount of time and money has been spent finding the windiest locations around the globe to install wind turbines. These sites give the greatest

return on investment for utility-scale wind turbines. The wind is a great renewable resource for energy, but it is not very forgiving on the human body. The environmental temperature extremes of working outdoor are only part of the considerations to develop safety plans and implement safe outdoor work practices. Other considerations should include hazardous weather, insects, wildlife, and transportation to remote locations in the mountains and at sea. Yes, at sea. Many wind-farm developers are looking to the consistent and higher-velocity winds that are present several miles off the northern coast of Europe and the northern coasts of the United States. These offshore locations are being developed with much larger turbines that are set on pilings or floating platforms, depending on the depth of the water. Transportation to these turbines is typically by surface vessel but may require helicopter travel for periods of rough seas. Whether the work is in the high plains of Texas, farmlands of the central United States, or off the coast of New England, it can be rewarding and safe as long as preventative measures are in place to reduce or eliminate the hazards. The next sections of this chapter will present several environmental factors, wildlife, and transportation considerations for working on wind turbines.

Weather Systems

Listen to the weather channel or the local news each day to hear about weather systems that are moving around the country. These systems bring high- and low-pressure regions along with warm and cold fronts. The transitions between the regions and fronts give us wind and rain as the atmosphere mixes around these disturbances. The differences in temperature and pressure within these regions determine the available energy. This can be seen with weather patterns such as nor'easters, tropical depressions, typhoons, and hurricanes. The available energy for these disturbances is an indication of wind speeds, amount of precipitation, and lightning. Any of these weather events can destroy equipment and property and take lives. **Table 3-5** lists some of the hazards associated with severe weather phenomena.

Any of these hazards could happen as a severe weather pattern tracks through a region of the country. This author has been in devastated areas after severe weather passed through regions of the United States and Caribbean countries. These areas were left without many of the modern conveniences to which our society has become accustomed: no running water, no food, no electricity, no transportation, and no access to medical facilities. People caught outside during these events are often killed. Preparation for these events and taking early warning measures are the way to reduce or eliminate these hazards. Many wind farms subscribe to weather services that provide early warnings of severe weather, especially lightning. Wind turbines make great lightning rods. Some of these weather services provide a warning when lightning is detected within 60 miles (96.6 km) of a farm and again when lightning is within 30 miles (48.3 km) of the farm. This enables technicians

TABLE 3-5

Severe Weather Hazards

Phenomena	Hazard
High winds Micro burst Tornado Blowing snow Blowing sand	Falling trees Damaged power lines Reduced visibility Capsized vessels Damaged turbines, vehicle, or buildings Damage to exposed skin and eyes Respiratory problems with inhalation of sand
Lightning	Equipment damage Electrocution Fires
Heavy rain	Flash floods Damaged roads Reduced visibility Drowning Damage to tower foundation Flooded tower, vehicle, or building
Heavy snow	Damaged tower or building Reduced visibility Avalanche Blocked roads and park egress
Hail	Damaged tower, equipment, vehicles, or buildings Blocked roads and park egress
Freezing rain	Ice accumulation Damaged power lines Turbine blades Blocked roads and park egress Damaged surface vessel Slippery surfaces

© Cengage Learning 2013

to evacuate a wind-turbine tower and the farm before they are subjected to the risk of a lightning strike. Many blades and towers have been damaged by direct lightning strikes. Any technician in a tower during a strike may be killed.

Freezing rain also presents danger to wind turbines, utility towers, power lines, and trees because of the added weight. This danger does not vanish as soon as the temperature rises above freezing. As the temperature transitions through the freezing point—32 °F (0 °C)—it can present a lethal hazard to anyone in the area. Falling ice has been known to damage trees, vehicles, and buildings in the vicinity of tall structures. Most wind farms include ice hazards in their safety programs to ensure that technicians are not near the towers during the time that they are shedding ice. In most cases, the wind turbines are shut down automatically or manually to prevent the falling ice from being thrown from the moving blades or falling from one blade onto another as they move. Some ice blocks weigh several hundred pounds, so they create considerable damage when they fall on something.

Wind-farm safety programs should include contingency plans for many of these weather issues. Other natural events such as volcanic eruptions, ash clouds, and earthquakes also may need to be considered, depending on the farm location. After all, some wind technicians work on wind farms in Hawaii. Other environmental exposure issues that should be considered include cold and hot temperature extremes.

Cold-Weather Exposure

Working outside in cold climates is not new. Many societies have adapted to living and working in colder climates for months at a time. Workers have adjusted the number of layers they wear and their types of clothing to enable them to maintain body temperature. Old weather clothes have been constructed from animal skins, animal coats (wool, down), multiple layers of natural fibers (cotton), and synthetic fibers (polyester, polypropylene, etc). Each material can be effective in maintaining body temperature, but the choice has to be practical for the activity. An animal skin coat might be acceptable for trekking across the snow in Antarctica but impractical to climb a 265-foot (80-m) tower.

The cold-weather gear required for working in a wind turbine has to be lightweight, durable, and fire resistant. Several cold-weather gear suppliers in North America and Europe market gear specifically for work in cold temperatures. **Figure 3-19** shows an example of cold-weather gear. The choice of gear and accessories depends on the temperature extremes that will be encountered during the outside activity. The selection will be different for work at 32 °F (0 °C) than it would be at 0 °F (−17.8 °C) or colder. The selection is very important to maintain comfort over a range of activities, whether meticulous troubleshooting activities to strenuous activities such as climbing the tower ladder.

Consider this: air is a great thermal insulator, and water is a poor one. Water is a good conduction medium to

FIGURE 3-19 *An example of cold-weather gear*

First Wind Energy, LLC

TABLE 3-6

Cold-Weather Layering Suggestions		
Fabric Location	**Material**	**Purpose**
Against the skin	Fleece or cotton[1]	Wicks moisture from skin
Middle layer I	Cotton[1] or thick layer of synthetic fiber[2]	Wicks moisture away from body and maintains layer of air for insulation
Middle layer II	Wool, cotton, or synthetic shell filled with down or synthetic fibers	Maintains layer of air
Shell outer layer	Water-resistant canvas[3] or water-resistant synthetic fibers	Prevents wind-driven air and water from entering inner layers

Head, hands, and feet should use a similar layering strategy, with water-repellent leather as an optional outer shell.

[1]When performing electrical work requiring arc-flash cal suits, use natural fibers for all layers in the cold-weather gear. The energy released from an arc flash or arc blast will cause the synthetic fibers to melt and adhere to the body. This would create a secondary burn injury from the accident that could be worse than the initial arc-flash exposure.

[2]There are several suitable synthetic fabrics with a variety of trade names on the market. Any of these may be used for different layers with most outdoor activities.

[3]Electrical work not requiring cal-suit protection synthetic fibers may be considered for inner layers of the cold-weather gear. A heavy cotton fiber or canvas fabric with an FR-rated chemical treatment is recommended for the outer shell for fire protection.

© Cengage Learning 2013

transfer heat from the body, so the ability of clothing to vent perspiration during periods of strenuous activity is critical. If moisture builds up in the fabric, it will become wet and reduce the thermal insulation value. Very few fabrics can maintain their insulation value after they become wet. Wool and synthetic materials such as polyester fleece are a couple of exceptions. Using layers of material to trap air and provide an effective insulation barrier between the cold and the skin will maintain body temperature over a longer period of time than using a single layer. **Table 3-6** lists suggestions for layers that may be effective for cold-weather activities. The number of layers should be increased or decreased, depending on the outside temperature and amount of physical activity anticipated during the job assignment.

Proper planning and preparation will protect the body during exposure to cold-weather conditions, but insufficient preparation can create a hazard. What happens when the body's core temperature drops? What are some of the signs of this drop in temperature? A drop in the body core temperature is known as **hypothermia**. Hypothermia can lead to the brain shutting down the body's extremities to protect itself. If actions are not started to reverse the core temperature drop, the body will shut down completely. Some of the symptoms that indicate onset of hypothermia include:

- shivering,
- weakness in the limbs,
- numbness in extremities (hands and feet),
- disorientation and
- sleepiness.

Final stages of hypothermia include:

- unconsciousness, and
- cardiac arrest.

The initial stage of hypothermia manifests as shivering. The body involuntarily begins contracting muscles to generate heat and raise the body temperature. This is the point at which action should be taken to prevent further medical issues. Outside activities should be limited to brief periods of exposure time during cold temperatures followed by time in a warm area to prevent hypothermia or frostbite. **Frostbite** is caused by freezing water in body tissues. It typically occurs in extremities such as hands and toes. It will also occur in exposed skin around the eyes, nose, cheeks, and ears because these areas may not be covered adequately by cold-weather gear. Signs of frostbite are tingling of extremities, loss of sensation, discoloration, and blackening of the area as the tissue dies. Warm the area slowly and do not massage because it will damage the frozen tissue. Seek medical attention for advanced frostbite with areas of discoloration and blackening to prevent infection or gangrene.

One other concern with cold-weather exposure is **dehydration**. Most people do not recognize dehydration as a cold-weather issue because they do not perspire and show obvious water loss as they would on a hot day. Moisture loss is not just through the skin. Breathing cold, dry, air decreases the moisture in the lungs each time we exhale. This loss of moisture causes dehydration—just like sweating on a hot day. Because there is no sensation of moisture loss, people typically do not realize they are dehydrated until severe symptoms occur such as headaches, dizziness, and fainting (syncope). It is very important that technicians be trained in first aid, CPR, and rescue techniques as part of a company's safety program. As can be seen with cold-weather exposure, there are times when knowing how to save a life becomes extremely important. Exposure to hot temperatures also requires life-saving experience to prevent severe injuries and death.

Hot-Weather Exposure

Working in warm climates such as Texas or California is a pleasant change after spending time in northern regions for maintenance activities. Carrying around 20 or so pounds of cold-weather gear is exhausting when walking around on the ground. Climbing towers and working in cramped areas makes the effort even more strenuous. Warm weather does not require the extra 20 pounds of clothing to prevent hypothermia, but it does present other health hazards. Hazards with warm climates include hyperthermia and overexposure to the sun.

> **HYPERTHERMIA:** effects include heat stroke, heat cramps, heat exhaustion, and dehydration.

Hyperthermia is an elevation of the body's core temperature because of excessive physical exertion in a hot environment. The human body cools itself as perspiration evaporates on the skin's surface. For evaporation to occur, water must absorb energy from its surroundings. The energy required for evaporation is known as the *latent heat of evaporation*. Excessive physical exertion on a hot, dry, day allows the body temperature to increase through accelerated metabolic processes (burning of food calories) and dehydration that will reduce the body's ability to produce perspiration. Hyperthermia can also occur on a hot and humid day because of metabolism and excessive moisture in the air preventing the perspiration on the skin from evaporating. Either situation can have lethal consequences if not recognized and prevented. Symptoms of hyperthermia include heat rash, muscle spasms, dizziness, increased pulse rate, and syncope. If these symptoms continue without intervention, then a person will go into cardiac arrest. Move a victim to a cool shaded area, administer fluids if the person is conscious, and seek medical attention. To prevent the hazards of hyperthermia, take frequent breaks in cool shaded areas and drink plenty of fluids. Water works well, but in large quantities it can create an imbalance in the body's electrolyte levels. To counter this, drink plenty of fluids such as sports drinks that have added electrolytes to maintain the body's electrolyte balance.

> **OVEREXPOSURE TO THE SUN:** sunburn and skin cancer.

Overexposure to the sun can cause minor skin irritation or blistering if protective measures are not taken. To reduce the effects, apply a generous coating of sunscreen with a minimum sun protection factor (SPF) of 30 to exposed skin, cover up with light-colored clothing, use polarized sunglasses, and minimize time in direct sunlight. These precautions will protect the skin from short-term sun damage from the two main forms of ultraviolet light (UVA and UVB) and the long-term possibility of skin cancer from gamma radiation.

The company's safety plan for hot climates should include training to recognize heat-related injuries, related first-aid measures, work procedures that include work breaks, and PPE to reduce exposure.

Wildlife Hazards

Working around wind turbines may also present hazards because of wildlife encounters. Towers and other human-made structures are a great place for rodents, insects, and reptiles to protect themselves from predators and foul weather. Being aware that these creatures can be in or around wind turbines can reduce the hazard of an encounter. Lizards and snakes will congregate under steps, in drainage culverts, under pad-mount transformers, in brush piles, and on rocks in the early morning. Sunning themselves on rocks or concrete allows them to raise their body temperature after the cool night. Hornets, wasps, and bees have also been known to build hives in electrical cabinets, in nacelles, and in tower sections. Rodents will enter the wind turbine during construction or when the access door is open during maintenance activities.

Insects

Hazards with hornets, wasps, spiders, and scorpions is through their stings and bites or the surprise of an encounter. Hornets, wasps, and bee stings not only can cause severe pain through stings but also may cause allergic reactions. An allergic reaction can be so severe with some individuals that it may cause swelling of the tongue and throat along with anaphylactic shock. This reaction swells the tissues around the throat and potentially closes the airway, causing suffocation and death to the victim if an injection of epinephrine (commonly through an EpiPen device) is not administered during the initial signs of the reaction. If a person knows of this possible reaction, ensure that plans include an EpiPen in the first-aid case of the service truck as well as on the technician.

Spiders and scorpions can create a hazard to an unsuspecting technician through bites and stings. These creatures can inject a **neurotoxin** into the victim that will interfere with normal functioning of the nervous system. This reaction may shut down normal pulmonary functions or paralyze limbs. To prevent life-threatening complications, the victim should be transported to the nearest medical facility to be treated with an antitoxin. Take note of the specific creature that inflicted the bite or sting so the medical provider can administer the proper antitoxin.

Reptiles

Few lizards present genuine poisoning dangers, although they can seem menacing if surprised and unable to flee. Snakes, however, are another matter. Rattlesnakes can be found in many areas of the United States. They are found not only in hot arid regions but also in forested areas as far north as Massachusetts. If you encounter a snake, keep your distance. Do not expose yourself to an unnecessary bite. A coiled snake can strike rapidly and as far as two-thirds of its length. A snake will attempt to flee if it is not cornered or threatened. Snake bites from the subfamily of venomous snakes known as pit vipers can cause life-threatening health issues; these include the common rattlesnake, Southern Pacific rattlesnake, Mojave rattlesnake, copperhead, and cottonmouth or water moccasin. Depending on the species, the bite can introduce one of several toxins into the body that can break down tissues and organs, cause hemorrhaging, or damage the nervous system. To reduce the chance of a snake bite during an encounter, wear heavy long pants and heavy boots. If you or a colleague is bitten, seek immediate medical attention. Take note of the type of snake to assist the medical provider in determining the appropriate antivenom. When administered quickly enough, antivenom can reduce the chance of death significantly. Venomous snakes produce different toxins, depending on their species. These toxins along with the body's reaction are:

> hemotoxic breakdown of tissues, degeneration of organs, and disruption of normal blood clotting (rattlesnakes);
>
> hemolytic breakdown of red blood cells, which releases hemoglobin into surrounding tissues (copperheads and cottonmouths); and
>
> neurotoxic shutdown the nervous system (tropical venomous snake species).

Offshore Project Safety

Working on offshore wind-tower projects requires the same level and detail of safety planning and training as for onshore projects. The major difference with these projects is the commute to work. Transit to and from offshore wind towers or utility structures is provided by surface vessels or helicopters. Each transportation method can be performed safely in a variety of weather conditions. When rough seas are created by a weather disturbance, the helicopter is the preferred method of transportation. Emergency planning for each transportation method considers cold-water survival as a vital factor in safety. If something goes wrong, you are going to get wet. How well you prepare your team for the water encounter will make the difference in surviving the emergency and being rescued.

Surface-Vessel Transport

Planning, construction, and service of offshore projects are performed with the use of surface vessels. This method of transportation is the most cost-effective long-term investment. There will be days when the waves are rough and not very comfortable because of cold winds and freezing saltwater spray splashing over the deck, but this is a safe and reliable transportation method. Safety planning and training for surface-vessel transportation should include the benefits of a positive attitude as well as training for cold-water survival, exposure gear, **personal flotation device** (**PFD**), and the use of a survival raft.

Should an emergency arise that requires abandoning the surface vessel because of capsizing or fire, it is extremely important to have the proper training, attitude, and gear available to increase the chances of survival in cold water. **Figure 3-20** shows examples of survival gear that may be used during an evacuation at sea. The proper attitude in any emergency is critical in surviving the situation. It also helps others in the emergency to stay focused on the tasks necessary to assist injured colleagues and set activities in motion to contact rescue teams. Maintaining composure also enables better problem-solving activities when situations arise

FIGURE 3-20 *Examples of survival gear that may be used during evacuation at sea*

© Cengage Learning 2013 and a W. Kilcollins photograph

that may not have been planned or practiced before. Off-shore wind farms are typically located close enough to shore that rescue after an emergency takes place within hours by another company vessel or the coast guard. The task of the surface vessel or flight crew is to stay afloat, warm, and to perform first aid as needed before help arrives.

Cold water will drain the body of warmth faster than cold air because water is a great conductor of thermal energy. This attribute of water will cause hypothermia to occur rapidly if precautions are not followed. Planning for a water evacuation from a surface vessel should include having extra clothing available that can be donned to add layers of thermal insulation along with a survival suit that will keep water away from the body. **Table 3-7** shows examples of other survival gear that should be available for a surface evacuation. Other techniques to slow down the onset of hypothermia include entering the water slowly if possible, keeping legs together, holding a PFD with arms across the chest and against the body, and group huddling with other colleagues. Jumping into the water adds the shock factor to the body along with the possibility of forcing water into the survival suit and limiting its effectiveness. The PFD is used to increase buoyancy and reduce the amount of effort required to stay on the surface. The group huddle improves the ability to maintain body temperature by reducing everyone's exposure to cold water. The group huddle also increases buoyancy and increases the visibility of the survivors to surface vessels and search aircraft. If a life raft is available, enter the raft from the surface vessel without entering the water or get into the life raft as quickly as possible to reduce exposure time to the cold water. Once in the raft, assist others entering the raft and attend to injured personnel. As quickly as practical, remove wet clothing and dry off to improve your ability to stay warm. Other activities include taking a head count of survivors and taking an inventory of available supplies, sets of communication equipment, and rescue aids.

There are many more details that need to be considered when executing survival activities at sea during an emergency. The suggestions listed here briefly cover several items of evacuation, survival, and equipment needs. More details would be presented in a surviving at sea training program. The **United States Coast Guard (USCG)** requires anyone working offshore on floating platforms to be trained in ocean survival techniques. Other countries have similar requirements, so review these requirements and get the required training to meet local regulations and improve the chances of survival.

Helicopter Transport

Transportation to an offshore wind project with a helicopter can be faster and typically smoother than a surface vessel. The view is also improved over that of the surface trip. Working around aircraft can be exciting and rewarding, but some precautions must be kept in mind to prevent injuries. Approaching a helicopter with a moving rotor requires full attention and a bent posture to prevent contact with the rotor. Being struck by a moving rotor will cause severe head trauma and potentially decapitation. Keep arms down to prevent contact with the rotor or rear stabilizer assembly to eliminate the risk of a limb amputation. Approach a helicopter only from the side and take a direct path to the access doors after receiving clearance from the pilot or copilot. During the approach, be aware of flying debris that may have been picked up by the rotor wash. Wearing safety glasses or goggles is recommended to eliminate eye injuries and improve visibility. After entering the helicopter, sit down, fasten the safety harness, and don the radio headset. The headset reduces noise exposure from the elevated sound levels of the aircraft and enables communication with the flight crew.

During takeoff, passenger conversation in the cabin should be limited to tasks at hand with no distractions to the flight crew other than for flight-related safety items. These may include other air traffic that may not be seen by the flight crew or an issue such as smoke or fire in the back of the cabin. The flight crew will provide a full safety briefing before takeoff that may include other items such as the location of fire extinguishers, egress procedures, and other safety equipment onboard. This safety briefing is a Federal Aviation Administration (FAA) requirement before each flight.

Cruising flight is the time when work assignments for the day can be discussed, equipment procedures can be reviewed, and any issues related to the drop procedure to the wind-tower or work platform can be detailed.

On station, conversations should be related only to the task of placing the technicians on the platform. Follow the flight crew's instructions and be prompt so that time over station can be minimized. The flight crew is working to maintain station with possibly wind gusts and other weather

TABLE 3-7

Examples of Survival Gear

Gear	Application
Personal flotation device (PFD)	Buoyancy device
Personal locator beacon	Radio transponder to assist rescue team in locating survivors
Emergency radar reflector	Assist surface vessels in locating survivors
Distress beacon	Visual assistance in locating survivors
Exposure suit	Enable extended exposure to cold water
Whistle	Aid in communicating with other survivors
Life raft	Increase survival time by limiting exposure to cold water
Water proof flashlight	Aids in communication, visibility, and night activities
Survival supplies	Supplies of water, food, first aid, VHF radio, and other vital supplies

© Cengage Learning 2013

FIGURE 3-21 *Technician ready for retrieval from wind-turbine deck*

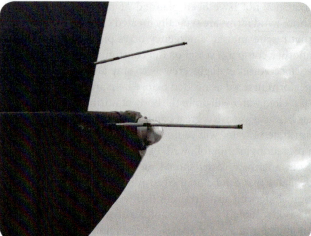

FIGURE 3-22 *Static discharge wick on an aircraft*

issues in the area. Make the pilot's job easier by preparing in advance what needs to be lowered or raised and in the required order. **Figure 3-21** shows a technician ready for retrieval from a wind-turbine deck. One thing to keep in mind during the transfer process is buildup of static electricity on the helicopter. The first technician down may get a shock because of the buildup of a static charge during flight. Aircraft are equipped with static discharge wicks to dissipate most of the charge, but there may still be residual electrical charge that has not dissipated. **Figure 3-22** shows an example of a static discharge wick located on an aircraft.

In the event of an emergency during flight, the crew will brief passengers on the activities that will be performed during **ditching**, or the controlled landing of an aircraft into water. Time before ditching should be used by trained passengers or available crew members to make sure the doors will open and to locate PFDs, life rafts, survival equipment, and any other necessary items to ensure survival at sea until rescue. These items should be ready and available for deployment as soon as the helicopter is in or near the water. If the flight crew can maintain altitude long enough for deployment of the survival equipment and evacuation of most of the crew, the activity can be carried out without the hazard

of the moving rotor. If this cannot be accomplished, then extreme caution must be used not to come in contact with the moving rotor or tail stabilizer.

If controlled ditching is not possible, be prepared to exit the helicopter quickly when it comes to rest in the water and the rotor has stopped. The orientation of the helicopter on contact with the water will determine how long the aircraft will be able to stay afloat. Egress from the helicopter after impact will be aided by the doors being ajar. Should the airframe deform during impact, the door latches may not open if there is binding with the frame. If the helicopter rolls on its side or completely upside down after landing on the water, egress will take place underwater. Now here is the interesting part. Do not panic: wait for the cabin to fill with water before preparing to exit. As long as the water is being forced into the cabin, you will not be able to control your position. Having your seat belts secure will keep you and the other occupants from being forced around and becoming injured. This is the same advice given to passengers in a car that has accidentally been submerged in water. Once the cabin has filled, remove your seat belt and float out of the helicopter to the surface. If possible, assist others to safety and push survival gear out of the cabin and to the surface.

A few notes of caution are in order. Fuel may be floating on the surface and present the possibility of a fire. Take care to swim out away from the debris and any fuel slick and fire to prevent injury. It is possible to become disoriented as to which way is up after leaving the cabin in poor visibility or at night, so watch for escaping air bubbles. Air bubbles will always move upward to the surface, so use them as your guide. Once you and the others are in the water, collect the survival gear and inflate the raft. Everything from this point on is going to be similar to the surface-vessel evacuation. These items include but are not limited to entering the life raft, assisting others, performing required first aid, and making an emergency distress call.

TYPES OF LUBRICANTS

Lubricants are classified in several different categories according to their origin. These include animal, vegetable, mineral, petroleum-based, and synthetic. Animal and vegetable oils have been used for thousands of years. Animal oils have been harvested directly from carcasses in such forms as whale oil or through the rendering of animal fat to produce oils or grease. Vegetable oils have been processed for centuries as well. These oils are from plant varieties such as olives, sunflowers, castor beans, and soy beans, along with other grains or nuts. These oils were useful in the development of many ancient technologies because they were readily available and easy to process. Several early civilizations, such as the Egyptians and Romans, used grease formulated from animal and vegetable oils to improve the efficiency of mechanical components. The Romans processed olive oil with ground limestone to produce grease to lubricate wheel sleeves and other mechanisms. Further discussions on modern grease formulation will be presented later in the chapter.

Animal- and vegetable-based lubricants were used extensively up through the Industrial Revolution. Even today, some applications use vegetable-based lubricants. These applications include food processing because of the nontoxic nature of vegetable oils. These lubricants are not typically used in nonfood industrial applications because they tend to break down to form organic acids that are easily displaced by water and are susceptible to bacterial growth. Each of these issues creates the need for constant cleaning and replenishing activities to avoid equipment damage and product contamination.

Petroleum-Based Lubricants

Other categories of lubricants are mineral- and petroleum-based products. These two groups are refined from crude oil. Crude-oil–based lubricants have been used for many applications over the years as well as the vegetable- and animal-based lubricants. Early sources of crude oil were from surface pools. These surface pools include the famous La Brea tar pits in Southern California. Commercial drilling for crude oil began in the mid-1800s and since then petroleum-based lubricants have been widely used in different forms for their stability. Petroleum-based types can be categorized into four types, depending on the geographic origin of the stock. These stocks include paraffinic, naphthenic, aromatic or mixed base, and asphaltic. **Paraffinic crude oil stock** is found in the Appalachian basin of the United States in states such as Pennsylvania and Ohio. This type of crude oil stock is thermally stable and has a relatively high viscosity index. The **viscosity index** is the measure of an oil's stability with changes in temperature. A high viscosity index means that its viscosity does not change much as temperature increases or decreases. It is a very good base stock for lubricants such as motor oils. The presence of wax in this base stock makes it less suitable for low-temperature applications. At lower temperatures, the wax will precipitate and reduce its lubrication properties.

Naphthenic crude oil stock is found around the Gulf of Mexico. This type of crude oil stock is less thermally stable and has a lower viscosity index. It contains virtually no wax and flows well at lower temperatures. This oil also makes a good base for low-temperature lubricant applications, and the low wax content makes the oil less likely to form carbon deposits in elevated-temperature applications.

Aromatic crude oil stock or **mixed base stock** can be found in many areas such as the Middle East, Africa, and Asia. This crude-oil base has varying properties, depending on the amount of paraffinic or naphthenic in the stock. This base stock may also contain varying quantities of asphaltenes, resins, and aromatics. The amount of refining necessary to get the required lubricating properties will depend on its composition of base stock.

Asphaltic crude oil stock comes from oil sands such as the ones located in Canada and Venezuela. These are very thick tarlike deposits referred to as *crude bitumen* that typically must be heated to flow. These asphalt-based oils have been used to produce heavy and inexpensive lubricants for low-speed open-gear systems found in large equipment. This type of lubricant may be brushed on and left for extended periods with minimal runoff. This stock is typically used to produce roofing materials, road-paving materials, and fuels for industrial heating and power generation.

Lubricants may be formulated from a combination of crude base stocks to develop the properties that are required for a particular application. Examples include the addition of naphthenic oil base to paraffinic oil base to decrease the viscosity index and make the formulation suitable for a lower temperature than the original lubricant. A formulation of one of these base stocks with an asphaltic base would produce a high-viscosity lubricant that could be brushed onto the open gears of a turbine yaw system or the drive assemblies of a large crane. A lubricant that would stay in place for an extended period of time without runoff would be very important. **Figure 4-4** shows an example of the open gears for a wind-turbine yaw-drive system.

FIGURE 4-4 *Open gears in a wind-turbine yaw-drive system*

REpower—Ralf Grömminger Photograph

Synthetic Lubricants

Formation of synthetic compounds is another method of developing lubricants to meet the demands of extreme environments. **Synthetic compounds** are created from multiple chemical reactions of discrete atoms or molecules to form a molecule with desired characteristics. The manufacturing process of synthetic oils provides a lubricant that is more stable through a wide range of temperatures and pressures than natural petroleum-based lubricants. This improves the efficiency of modern industrial equipment that is required to work under extreme temperatures, heavy loading, and high speeds. Synthetic-based lubricants were originally developed to meet the demands of high-performance engines during World War II. Another added benefit of synthetic lubricants is that its uniform molecule size further reduces friction more than natural petroleum-based products. Other benefits include protection from oil breakdown, deterioration, and fire resistance, which extends operational life and improves safety over natural petroleum lubricants.

Viscosity Classifications

Viscosity is a measure of the force to shear molecular attraction within a fluid and allow it to flow. If the attraction requires excessive force to move molecules relative to each other, the fluid would be considered highly viscous. One example of a common highly viscous fluid is heavyweight gear oil. If the attraction requires little force to move molecules relative to each other, the fluid would be considered to have a very low viscosity. Gasoline is an example of a very-low-viscosity liquid.

The use of viscosity classifications for lubricants has been defined by several organizations in the United States and other countries. These organizations include the **Society of Automotive Engineers (SAE)**, the **American Society for Testing and Materials (ASTM)**, the **American Petroleum Institute (API)**, the **American Gear Manufacturers Association (AGMA)**, and the **International Organization for Standardization (ISO)**. Each of these organizations uses a different scale to quantify a lubricant's viscosity, but each compares the time a set amount of lubricant will flow through a fixed orifice at a given temperature. An **orifice** is an opening or hole of a set diameter in a fitting. A simplified apparatus that may be used to perform this test is shown in **Figure 4-5**. The test apparatus includes a controlled temperature bath, lubricant reservoir, fixed orifice, and receiver container of a set volume. Typical units used for viscosity include **viscosity grades (VG), Saybolt universal seconds (SUS), Saybolt seconds universal (SSU)**, and **centistokes (cSt)**. **Figure 4-6** shows a comparison of the different viscosity values.

Lubricant Additives

A variety of chemical additives can be combined with lubricants to enhance their overall performance. These **lubricant additives** include corrosion inhibitors, oxidation inhibitors,

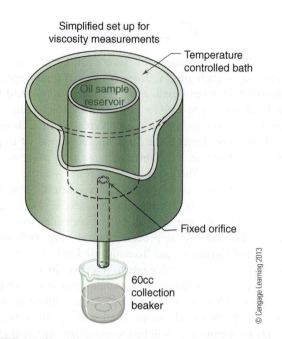

Simplified set up for viscosity measurements — Temperature controlled bath — Oil sample reservoir — Fixed orifice — 60cc collection beaker

© Cengage Learning 2013

FIGURE 4-5 *Simplified apparatus for testing lubricant viscosity*

demulsifiers, detergents, dispersants, antifoam, viscosity index improvers, and extreme pressure additives.

Corrosion Inhibitors

Corrosion inhibitors are added to lubricants to improve their ability to stick to metals. The ability of a lubricant to provide a continuous protective film will prevent air and moisture from contacting and oxidizing the metal's surface. This additive will extend the life of the metal components by preventing the continuous buildup and breakdown of oxides as the components come into contact during motion. The oxidation of iron or steel components is known as rust. This can be seen as a reddish-brown rough surface that forms on gears or other iron parts that are exposed to the atmosphere. Semiannual lubricant testing to determine gearbox oil condition would show excessive amounts of iron in the sample if a breakdown of gear-tooth surfaces was occurring. Lubricant analysis as a preventative-maintenance activity will be discussed later in the chapter.

Oxidation Inhibitors

Oxidation inhibitors are not the same as corrosion inhibitors even though the term *oxidation* refers to the chemical process of materials combining with oxygen to form an oxide. Oxidation inhibitors are added to lubricants to prevent the breakdown of the hydrocarbon molecule that makes up the lubricant. When a lubricant is exposed to elevated temperatures in the presence of air, the hydrocarbon molecules will break down to form organic acids. The presence of these organic acids can be seen in the thickening of the lubricant and the formation of corrosion on metal surfaces exposed to the lubricant. The addition of oxidation inhibitors retards the normal oxidation process that takes place in lubricants.

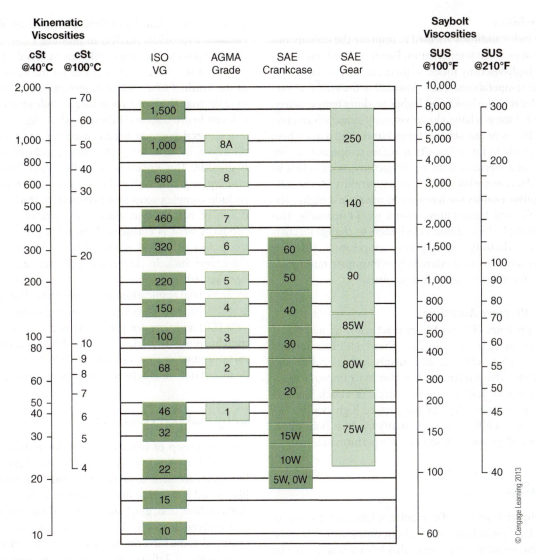

FIGURE 4-6 *Comparison of different viscosity values*

Demulsifiers

Contamination of lubricants with small amounts of water prevents the lubricant from forming a protective film. You have probably seen a drop of oil on water. Remember the oil spreads out over the water to form a very thin film. This same dispersion prevents the lubricant from separating the moving components. The addition of demulsifiers enables the water to separate from the oil and allows the lubricant to protect the moving parts.

Detergents

Detergents are added to lubricants to break up deposits that form on metal components because of exposure to elevated temperatures. These deposits are typically associated with internal combustion engines, but they may occur with other mechanical systems if elevated temperatures form sludge from the combination of contaminants and oxidized lubricant.

Dispersants

Dispersants are added to lubricants to hold sludge and microscopic particles in suspension after the detergent additive breaks them down. Holding the sludge and particles in suspension enables the filtration system to capture them as they pass through during circulation of the lubricant. If the particles are too small for the filter to capture them, they will remain in suspension until the next scheduled lubrication replacement. Lubricant filtration systems will be discussed later in the chapter.

Antifoam

Pumping lubricant over gears and around bearing elements may allow air to become entrapped. This entrapped air can create foam in the lubricant that inhibits its ability to protect moving parts. An antifoam additive enables the air bubbles to break up quickly to prevent damage from the disruption of the needed lubrication film.

Viscosity Index

Viscosity index additives are used to improve the stability of a lubricant at elevated temperatures. Paraffin-based oils have a relatively high-viscosity index, so their viscosity is stable over a range of temperatures. Their downfall is the wax that creates problems at the lower and higher working temperatures. Naphthenic base oils have a lower viscosity index, which gives them a better pour rate at lower temperatures and makes them less susceptible to form deposits at higher temperatures. The drawback of a lubricant with a low-viscosity index is that its viscosity decreases with increase in temperature. A viscosity index additive enables the lubricant to maintain a higher viscosity at elevated temperatures than it would normally. This is done through the addition of polymers to change its characteristics of viscosity with respect to temperature. This type of additive has been used extensively to formulate multigrade lubricants for internal combustion engines.

Extreme Pressure Additives

Extreme pressure (EP) additives react with the metal surfaces to create a protective film. These additives include chemicals such as chlorine and metals such as sulfur and phosphorous. The purpose of the film is to improve load-carrying capacity and reduce the effects of shock loading, which may create welding of the surface high points. Extreme pressure additives are commonly used for both lubricating oils and greases to improve their performance.

Grease

Some industrial applications require a lubricant to stay in place for long periods of time without dropping or running out of a mechanical component. This lubricant may also be needed to prevent contaminants from entering the mechanical component or to maintain a supply of lubricant for a system in a remote location. This type of industrial application requires a lubricant that has a high viscosity to stay in place, but it will lubricate the moving components like lower-viscosity oil. Grease has the high-viscosity property of staying in place yet lubricates mechanical components at approximately the viscosity of the base oil when stressed. Grease is considered a **plastic fluid** because of the change in viscosity that occurs when it is stressed. The viscosity of grease will also decrease with time, which makes it **thixotropic**. Lubricating grease is formed by the addition of a **thickener** material to a mineral or vegetable oil. One thickener for grease is metallic soap. Soap is formed through **saponification** of oil with an alkali. An **alkali** is a substance with a relative high pH value. A pH test is used in chemistry to determine the extent an aqueous (water) solution is acidic or alkaline. The **pH scale** runs from 0 to 14, with 0 being extremely acidic and 14 being extremely alkaline. Household bleach has a pH around 13, and stomach acid has a pH level around 1. Distilled water is used as a neutral pH reference because it has a pH level of 7. Early bathing soaps were produced by heating animal fat to a liquid and then adding an alkali such as lye (sodium hydroxide, NaOH) to produce a chemical reaction that forms soap. The type of alkaline material added to hot oil to form a soap thickener depends on the desired properties of the finished lubricant. Alkaline substances typically used to make soap thickener for grease include sodium hydroxide, calcium hydroxide, and lithium hydroxide.

General-purpose grease used for most industrial applications includes soap thickener produced using lithium hydroxide. This grease is typically called *lithium grease* or *all-purpose grease*. Other thickeners used to formulate grease include complex soap and nonmetallic thickeners. The addition of barium and aluminum salts to soap thickeners produces what is considered **complex grease**. Nonmetallic thickeners are added to the oil base to form a paste. These thickeners include clay, mica, polyurea, and tar. Clay has been added to thicken oil and produce inexpensive grease for industrial applications for many years. This type of grease is still used in low-performance applications. Extreme pressure (EP) additives for grease include graphite, copper, and molybdenum disulfide. These additives prevent metal-to-metal contact during a shock load when oil would normally be forced out from between contacting surfaces.

Dry Lubricants

Another group of lubricants used for reducing friction in industrial applications is known as **dry lubricants**. These lubricants are used to separate the moving components just like the oil and grease discussed previously. These dry lubricants include lead, graphite, polytetrafluoroethylene (PTFE), and molybdenum disulfide. Polytetrafluoroethylene is the chemical name for a fluorocarbon known under the DuPont trade name of Teflon. This material has been used as a dry lubricant, grease additive, and nonstick coating for many applications, including cookware.

Compatibility

The type of lubricant used for a particular application in a wind turbine will be recommended by the turbine manufacturer. For instance, the grease used in the main shaft bearing will be different from that used for a generator bearing. The main shaft bearing runs at relatively low speed, has high loads and lower temperatures, and is exposed to the environment, whereas the generator bearings run at higher speeds higher temperatures, and lower loads. **Figure 4-7** shows a chart of suggested grease products for different bearing applications. Caution must be taken to ensure that lubricants are never mixed. Should a mix occur with incompatible lubricants, the bearing assembly could be damaged because of inadequate lubrication. If a lubricant is mixed accidentally, care should be taken to thoroughly clean all lubricant from the components, rinse with a degreaser, and replenish the system with the recommended lubricant.

SKF bearing grease selection chart

Bearing working conditions	Temp	Speed	Load	Vertical shaft	Fast outer ring rotation	Oscillating movements	Severe vibrations	Shock load or frequent start-up	Low noise	Low friction	Rust inhibiting properties	Description	Temperature range (*1)		Thickener / base oil	Base oil viscosity (*2)
													LTL	HTPL		
LGMT 2	M	M	L to M	O	−	−	+	−	−	O	+	General purpose industrial and automotive	−30 °C −22 °F	120 °C 250 °F	Lithium soap / mineral oil	110
LGMT 3	M	M	L to M	+	O	−	+	−	−	O	O	General purpose industrial and automotive	−30 °C −22 °F	120 °C 250 °F	Lithium soap / mineral oil	120
LGEP 2	M	L to M	H	O	−	O	+	+	−	−	+	Extreme pressure	−20 °C −4 °F	110 °C 230 °F	Lithium soap / mineral oil	200
LGFP 2	M	M	L to M	O	−	−	−	−	−	O	+	Food compatible	−20 °C −4 °F	110 °C 230 °F	Aluminium complex / medical white oil	130
LGEM 2	M	VL	H to VH	O	−	+	+	+	−	−	+	High viscosity plus solid lubricants	−20 °C −4 °F	120 °C 250 °F	Lithium soap / mineral oil	500
LGEV 2	M	VL	H to VH	O	−	+	+	+	−	−	+	Extremely high viscosity with solid lubricants	−10 °C 14 °F	120 °C 250 °F	Lithium-calcium soap / mineral oil	1 020
LGLT 2	L to M	M to EH	L	O	−	−	−	O	+	+	O	Low temperature, extremely high speed	−50 °C −58 °F	110 °C 230 °F	Lithium soap / PAO oil	18
LGGB 2	L to M	L to M	M to H	O	−	+	+	+	−	O	O	Green biodegradable, low toxicity	−40 °C −40 °F	90 °C (*3) 194 °F	Lithium-calcium soap / synthetic ester oil	110
LGWM 1	L to M	L to M	H	−	−	+	+	+	−	−	+	Extreme pressure, low temperature	−30 °C −22 °F	110 °C 230 °F	Lithium soap / mineral oil	200
LGWM 2	L to M	L to M	M to H	O	O	+	+	+	−	−	+	High load, wide temperature	−40 °C −40 °F	110 °C 230 °F	Complex calcium sulphonate / synthetic (PAO)/mineral oil	80
LGWA 2	M to H	L to M	L to H	O	O	O	O	+	−	−	+	Wide temperature (*4), extreme pressure	−30 °C −22 °F	140 °C 284 °F	Lithium complex soap / mineral oil	185
LGHB 2	M to H	VL to M	H to VH	O	+	+	+	+	−	−	+	EP high viscosity, high temperature (*5)	−20 °C −4 °F	150 °C 302 °F	Complex calcium sulphonate / mineral oil	400
LGHP 2	M to H	M to H	L to M	+	−	−	O	O	+	O	+	High performance polyurea grease	−40 °C −40 °F	150 °C 302 °F	Di-urea / mineral oil	96
LGET 2	VH	L to M	H to VH	O	+	+	O	O	−	−	O	Extreme temperature	−40 °C −40 °F	260 °C 500 °F	PTFE / synthetic (fluorinated polyether)	400

(*1) LTL = Low-temperature limit
HTPL = High-temperature performance limit
(*2) mm²/s at 40 °C / 104 °F = cSt.
(*3) LGGB 2 can withstand peak temperatures of 120 °C / 250 °F
(*4) LGWA 2 can withstand peak temperatures of 220 °C / 428 °F
(*5) LGHB 2 can withstand peak temperatures of 200 °C / 392 °F

+ = Recommended   **O** = Suitable   **−** = Not suitable

Basic bearing grease selection

Generally use if: Speed = M, Temperature = M and Load = M → **LGMT 2** General purpose

Unless:

Expected bearing temperature continuously > 100 °C / 212 °F	LGHP 2	High temperature
Expected bearing temperature continuously > 150 °C / 302 °F, demands for radiation resistance	LGET 2	Extremely high temperature
Low ambient −50 °C / −58 °F, expected bearing temperature < 50 °C / 122 °F	LGLT 2	Low temperature
Shock loads, heavy loads, frequent start-up / shut-down	LGEP 2	High load
Food processing industry	LGFP 2	Food processing
"Green" biodegradable, demands for low toxicity	LGGB 2	"Green" biodegradable

Note: – For areas with relatively high ambient temperatures, use LGMT 3 instead of LGMT 2.
– For special operating conditions, refer to the SKF bearing grease selection chart.

Bearing operating parameters

Temperature

L =	Low	<50 °C / 122 °F
M =	Medium	50 to 100 °C / 122 to 230 °F
H =	High	>100 °C / 212 °F
EH =	Extremely high	> 150 °C / 302 °F

Speed for ball bearings

EH =	Extremely High	n.dm over 700 000
VH =	Very High	n.dm up to 700 000
H =	High	n.dm up to 500 000
M =	Medium	n.dm up to 300 000
L =	Low	n.dm below 100 000

Speed for roller bearings

	SRB/TRB/CARB	CRB
H = High	n.dm over 210 000	n.dm over 270 000
M = Medium	n.dm up to 210 000	n.dm up to 270 000
L = Low	n.dm up to 75 000	n.dm up to 75 000
VL = Very Low	n.dm below 30 000	n.dm below 30 000

Load

VH =	Very high	C/P < 2
H =	High	C/P ~ 4
M =	Medium	C/P ~ 8
L =	Low	C/P 15

Available pack sizes

	LGMT 2	LGMT 3	LGEP 2	LGFP 2	LGEM 2	LGEV 2	LGLT 2	LGGB 2	LGWM 1	LGWM 2	LGWA 2	LGHB 2	LGHP 2	LGET 2
SKF SYSTEM 24			•	•		•								
35 g tube			•			•								
200 g tube					•		•							
420 ml cartridge	•	•	•	•	•	•	•	•	•	•	•	•	•	
1 kg can	•		•	•						•	•	•	•	
5 kg can	•		•	•							•	•	•	
18 kg can	•		•											
25 kg can	•		•											
50 kg drum	•	•	•	•		•				•	•			
180 kg drum	•	•	•	•		•								
50 g (pt m) syringe														•

SKF bearing greases

Highest quality grease for bearing lubrication

Guarantee of consistent quality as each product is manufactured at one location to the same formulation

A complete product programme for general and specific bearing lubrication requirements

International standardisation of the SKF grease testing methods and equipment

Worldwide product availability through the SKF dealer network

www.mapro.skf.com
skf.com/lubrication

SKF Maintenance Products

Publication MP3401E · 2009/08   © SKF 2009
® SKF, CARB, SYSTEM 24 are registered trademarks of the SKF Group

The contents of this publication are the copyright of the publisher and may not be reproduced (even extracts) unless prior written permission is granted. Every care has been taken to ensure the accuracy of the information contained in this publication but no liability can be accepted for any loss or damage whether direct, indirect or consequential arising out of the use of the information contained herein.

SKF Maintenance Products

FIGURE 4-7 *Suggested grease products for bearing applications*

PM AND LUBRICANT LIFE

Preventative and **predictive maintenance** play important roles in wind-turbine management. Knowing how each system functions and analyzing changes can reduce operation expenses and eliminate costly downtime. Predicting lubricant life is a vital component of this overall strategy. Lubricants change with time and service conditions, so monitoring these changes can flag wear issues with system components before a failure occurs. Taking samples of lubricants at each scheduled maintenance cycle will allow trend analysis of changes. Lubricant analysis has been used successfully for decades in the railway industry to monitor oil performance in expensive locomotive engines. Since World War II, the military has monitored engine oil and hydraulic fluids to maintain the reliability of fleets of vehicles, aircraft, and naval vessels. Commercial trucking and civilian aircraft fleets have also taken advantage of this valuable tool. The relative low cost of oil analysis has eliminated the need for costly equipment repairs and unscheduled downtime. The wind-energy industry is no different. Wind-turbine manufacturers have invested billions of dollars in manufacturing fleets of turbines, so preventative and predictive maintenance is a must to ensure uninterrupted service for wind-farm developers and owners who rely on the equipment.

Oil Analysis

Oil analysis has been primarily used to monitor the condition of motor oils, but this valuable tool also can be used to monitor the condition of gearbox oil and hydraulic fluids. The changes in motor oil are similar to other lubricant oils and fluids. These changes can be tracked in the following areas: base oil condition along with additive levels, contaminent levels, and wear levels of various metals. Each indicator will give valuable information as to how the system is functioning. Another valuable resource is the testing laboratory that is performing the analysis. Choose a lab that has a proven track record in the industry and technicians with experience. Any lab with the proper equipment can analyze the sample provided, but having technicians who understand your equipment and who have years of data from similar equipment can add insight into interpreting the results. We cannot all be experts, so choose an experienced lab that can be a team player in making your operation the best.

Oil Condition

Lubricating oil is formulated with viscosity and additive levels to enhance its performance in a specific application such as a wind-turbine gearbox or motor gear drive. Additives such as corrosion inhibitors, viscosity index improvers, and conditions such as extreme pressure were previously mentioned. In some cases, the oil has been formulated from a synthetic base. The use of synthetic oil improves its lubricating performance and life, but it also increases the initial cost. Oil analysis can be used to monitor the lubricants' viscosity and additive levels just like the manufacturer did at the refinery. The viscosity of lubricating oil increases throughout its life because of the evaporation of volatile components and the degradation products that may form longer polymer chains. Reduced additive levels can also make the oil less effective in lubricating the moving parts. Oil analysis can determine if the oil viscosity is still within acceptable limits and which additives may be replenished to maintain its desired properties. Follow the recommendations from the testing lab or equipment manufacturer for additive quantities to replenish the oil back to required levels. **Figure 4-8** lists some of the typical additives that may be present in an oil sample.

Appendix E shows an example of a report listing elements and contamination quantities present in the oil sample. The appendix also shows a bar graph that may be used to compare changes to past reports. Several reports may be necessary to determine trends in quantities. You cannot use one report to make a judgment call on what is happening in a system unless you have other systems to compare or maximum values that show something is obviously wrong. If the levels are that far out, then there will probably be some other indicator that has alerted technicians to investigate. These

Iron	Silver	Calcium
Aluminum	Antimony	Magnesium
Chromium	Silicon	Barium
Copper	Sodium	Molybdenum
Lead	Boron	Potassium
Tin	Zinc	
Nickel	Phosphorus	

FIGURE 4-8 *Typical elements that may be listed on an oil analysis report*

may be an over temperature alarm, float valve indicating low lubricant level, or an indication of a plugged filter. Barring these types of issues, a periodic oil analysis will aid in predicting the typical life cycle of oil. It will also enable tracking of contaminants and machine wear products that may enter the system. Monitoring contaminants and wear products will yield other vital clues as to issues that may be occurring in the system.

Fluid Contaminants

Contaminants can enter a lubricating system through vents as well as be applied as coatings to the interior of the system during manufacturing. Contaminants may include water, dust, dirt, paints, corrosion-inhibiting coatings, and sealants, as well as organic acids that form during lubricant breakdown. Systems with large volumes of lubricant are vented to prevent pressurization when the system heats up during normal operation. When the system cools down, the vent allows air to reenter the system. If the vent does not have an air filter or anhydrous filter cartridge, it will draw in airborne contaminants as well as water vapor. An **anhydrous** cartridge (anhydrous – without water) is filled with an absorbent material to trap water vapor as it passes through the assembly.

Some medium-sized turbine manufacturers have installed air-filtration devices or a combination filter that includes an anhydrous cartridge on the drive-train gearbox to prevent the introduction of these contaminants. **Figure 4-9** shows examples of air-vent filters that may be used on drive-train gearboxes. Dust, dirt, and organic contaminants can be listed on the oil-analysis report as silica or other elements that do not match the additive lists. You may be wondering how dust and organic contaminants can get 300 feet up in the air to enter a nacelle or cooling vents. Well, this author has cleaned several heat exchangers and replaced air filters that were plugged with pollen, seeds, and dust. It is amazing how far the wind and the convection draft in the tower can carry particulates. It happens, and an air-filter defense can prevent contamination of lubricants and cooling systems. These air-filter assemblies should be inspected and replaced if needed during each preventative-maintenance cycle. Anhydrous filter elements will change color as they absorb water vapor. Some elements change from amber to dark green, whereas others change from pink to blue as they become saturated. Consult the manufacturer's information for the element used in your system.

First Wind Energy, LLC

© Cengage Learning 2013

First Wind Energy, LLC

FIGURE 4-9 *Examples of air-vent filters used on drive-train gearboxes*

Wear Indicators

Other elements that may show up in varying amounts may be wear products from gears, sleeves, bearings, seals, and other moving components. Several oil-analysis snapshots over time will give a good indication as to normal wear rates for components such as gears, shafts, and bearings. When these rates change dramatically or spike from one test to another, a thorough investigation should be scheduled to determine the suspect component. **Wear product** elements show up in the oil-analysis report as iron, chromium, silver, lead, tin, and copper. Seal damage shows up as organic material particles on the analysis report. If a seal is wearing, then particles may show up on the filter cartridge or manifest as a leak around the pump motor or filter canister. When significant levels of wear products are discovered, the machine should be inspected per the manufacturer's recommendation and another oil sample should be drawn to ensure that there was not an accidental contamination in the previous sample. If nothing is discovered during the inspection, the machine should be monitored closely until the next oil sample has been analyzed and confirms the previous results or verifies a possible sample contamination.

FLUID-FILTRATION SYSTEMS

Lubrication system filtration systems are used on large gearboxes, lubricant cooling systems, and hydraulic power units. The **filtration systems** are used to prevent contamination of the lubricant, pumps, manifolds, actuators, and other components of the systems. These filtration systems include filter elements made of pleated paper or fabric, screens, and magnet dipsticks to remove iron wear products and other contaminants that may enter the circulation system. **Figures 4-10 and 4-11** show typical filtration elements along with cartridge and canister assemblies that may be seen in a wind-turbine lubricant circulation system.

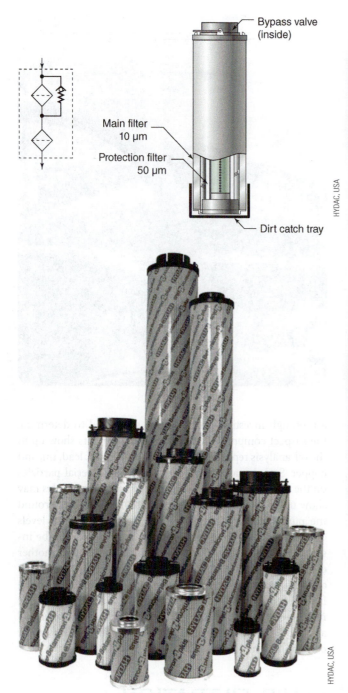

Bypass valve
(inside)

Main filter
10 μm

Protection filter
50 μm

Dirt catch tray

HYDAC, USA

FIGURE 4-10 *Typical filtration elements*

First Wind Energy, LLC

University of Maine at Presque Isle

FIGURE 4-11 *Cartridge and canister assemblies in a wind-turbine lubricant circulation system*

TYPICAL COMPONENTS TO LUBRICATE

Wind turbines have several large bearing assemblies to enable the nacelle, drive train, and blades to move independently during normal operation. Without lubrication of these assemblies, the turbine would not function properly. Components that rely on grease for lubrication include yaw bearings, yaw-drive gears, main shaft bearings, generator bearings, pitch bearings, and pitch drive gears. **Figures 4-12 through 4-16** show examples of these assemblies. Each assembly will present a different operating environment for the lubricating grease, so follow the component manufacturer's recommendation for each assembly. The open-gear systems of the yaw and pitch drives will require grease that must stay in place for long periods of time without the dropping or sling-off created by movement of the assembly. The turbine manufacturer will have a recommended grease to fit this application. Some wind-turbine manufacturers use a single large **hydraulic actuator** or three smaller actuators to move the blades in the rotor assembly so the drive-linkage components will need to be lubricated during each preventative-maintenance cycle.

Figure 4-17 shows a hydraulic drive linkage used to pitch the blades. Greasing each bearing or drive assembly is a portion of the lubricant requirements for a preventative-maintenance cycle. Lubricant oil reservoirs also need to be inspected for adequate reserve levels.

FIGURE 4-12 *Example of a yaw-drive bearing assembly that requires lubrication*

FIGURE 4-13 *Example of a yaw-drive gear assembly that requires lubrication*

Yaw drive system pinion and ring gears

FIGURE 4-14 *Example of a main bearing assembly that requires lubrication*

FIGURE 4-15 *Example of a blade-pitch bearing assembly that requires lubrication*

FIGURE 4-16 *Example of a blade-pitch drive assembly that requires lubrication*

Lubricant and Fluid Inspections

Lubricant preventative-maintenance activities include a visual inspection of the drive-train gearbox, if equipped, and the gear-drive motor assemblies used with the yaw-drive and pitch-drive assemblies. Some new wind-turbine models have been designed with a direct-drive system to eliminate the need for main gearbox maintenance. The design philosophy with this major change is to eliminate a major system wear component and eliminate the high-frequency sound produced by the generator running around 1,400 to 1,800 RPM. The trade-off of the maintenance and sound with this design is having a very large generator assembly. **Figures 4-18 through 4-20** show examples of the drive component assemblies and their fluid-level sight glass.

FIGURE 4-17 *Example of a hydraulic drive linkage system used to pitch wind-turbine blades*

FIGURE 4-18 *Example of drive train gearbox showing the fluid-level sight glass*

FIGURE 4-19 *Example of yaw motor gear drive showing the fluid-level sight glass*

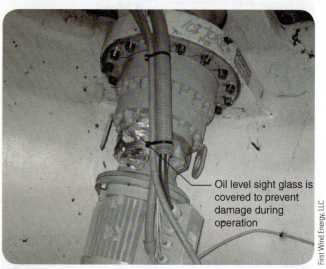

Oil level sight glass is covered to prevent damage during operation

FIGURE 4-20 *Example of pitch gear motor showing the fluid-level sight glass*

A fluid level **sight glass** is a glass tube located on the side of a reservoir that shows the level of fluid in the system. The sight glass is also a useful inspection tool to determine the color of the fluid. Fluid that has darkened from the original color may be contaminated or degraded because of other factors. Blade-pitch systems that incorporate hydraulic actuators are controlled by a hydraulic power unit. These systems also require maintenance and a visual inspection of fluid levels. Further details on fluid power systems will be covered in Chapter 5 on hydraulic control systems.

TYPICAL LUBRICANT SYSTEMS

Lubricant-supply systems in wind-turbine applications include manifold blocks and feed lines for the yaw-drive and blade-pitch bearing assemblies. Examples are shown in **Figures 4-21 and 4-22.** Other lubricant-feed systems may include a grease-pump assembly and open-cell foam follower gears meshed with the main yaw-drive gear. An example of this type of pump system is shown in **Figure 4-23.** Most bearing assemblies encountered in a wind-turbine system will have a grease nipple or **Zerk fitting** for use with a manual grease gun. These bearings will also have a grease trap to capture waste grease as it is forced out of the bearing assembly during the addition of grease and during normal operation. These trap assemblies should be cleaned to prevent overflow during each maintenance cycle. **Figure 4-24** shows different waste-grease traps.

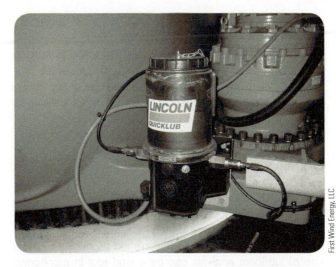

FIGURE 4-23 *Example of an automatic pump system for grease*

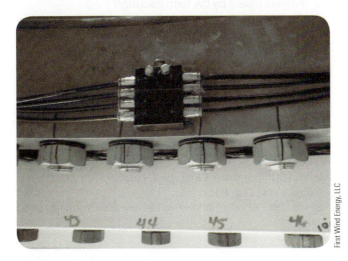

FIGURE 4-21 *Example of manifold block and feed line for yaw-drive and blade-pitch bearing assemblies*

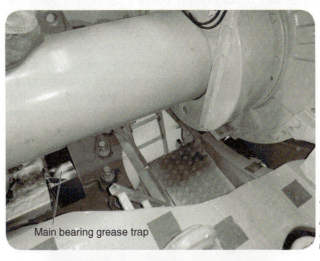

Main bearing grease trap

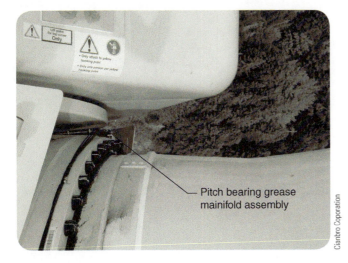

Pitch bearing grease mainifold assembly

FIGURE 4-22 *Example of manifold block for blade-pitch bearing assembly*

Bearing grease trap

FIGURE 4-24 *Different waste-grease traps*

SUMMARY

This chapter introduced the concept of friction in moving components found in mechanical systems. Unwanted friction has been an issue for mechanical designers dating back to ancient civilizations. Each civilization has developed unique ways to reduce or use friction to benefit its needs. Lubrication systems can be used to reduce or eliminate friction and its associated issues. With the use of lubrication comes the need to understand appropriate lubricants and methods to apply the lubrication. The field of tribology endeavors to do this.

The use of lubricants requires preventative and predictive measures to ensure that the lubricants chosen for machinery are maintained and replenished as necessary. The use of lubricant analysis can be a vital tool in monitoring equipment condition and making scheduled repairs or replacement of components before a component failure creates unwanted downtime.

Some wind-turbine components and systems that require preventative-maintenance activities were highlighted in this chapter to assist the reader in gaining an appreciation of lubrication. Properly lubricated components will ensure a long and profitable service life for any large mechanical system such as a wind turbine. The goal in preventative maintenance is to keep the equipment running and to minimize downtime and keep customers happy. After all, they have invested several millions of dollars in their equipment.

REVIEW QUESTIONS AND EXERCISES

1. List some applications that use friction to enable equipment to function properly.
2. What is tribology?
3. Of the three lubrication film conditions, which is the most effective in reducing friction and wear in a bearing assembly?
4. Explain why a lubricant with a high viscosity index would be beneficial for an application with a wide operating-temperature range.
5. What is the purpose of a dispersant additive in lubricating oil?
6. List examples of dry lubricants.
7. What is the AGMA?
8. List reasons why oil analysis is important in determining the condition of a machine.
9. What is meant by the term *viscosity*?
10. The rise in what wear element would indicate gear damage?
11. List some of the elements shown on a typical oil-analysis report.
12. List some components found in a wind turbine that require periodic grease replenishment.

REFERENCES

Mark's (11th ed): *Mark's Standard Handbook for Mechanical Engineers*, 11th edition.

Teamrip.com (2011): http://www.teamrip.com/viscosity_chart.html

CHAPTER 5

Fluid Power

KEY TERMS

accumulator
active brake
actuator
ball valves
bulk modulus
check valves
compressible
compressor
cooling system
directional control valve
filter element
flow control
fluid transfer
fluid power
heating system
horsepower (HP)
hydraulic
hydraulic fluid

hydraulic head
hydraulic motor
hydrodynamic system
hydrostatic system
incompressible
linear actuator
passive brake
pneumatic
positive displacement
power unit
pressure
pump
reservoir
sensor
solenoid
specific gravity
specific weight

OBJECTIVES

After reading this chapter and completing the review questions, you should be able to:

- Describe the difference between hydraulic and pneumatic systems.
- Describe what is meant by *compressibility* of fluids.
- Describe the difference between fluid power and fluid transfer.
- Describe properties of fluids used in fluid-power systems.
- Describe some of the properties of fluid-power systems.

- Describe the function of components used in fluid-power systems.
- Describe the use of fluid-power systems in wind turbines.
- Describe the use of fluid power in tools used for wind-turbine maintenance.
- Describe preventative-maintenance requirements for fluid-power systems.

INTRODUCTION

Fluid systems come in two types: fluid transfer and fluid power. **Fluid transfer** is used to move fluids around a system for lubrication, as a medium for heat exchange, or for processing. **Fluid power** is used to provide power for motion control in a variety of industrial, transportation, and agricultural applications. Fluid-power components may come in compact units for use in industrial robotic applications or in very large units like those used in construction and mining equipment. A basic fluid-power system comprises a fluid, power source, control system, and actuator. This basic system may be used to lift heavy components such as the boom on an excavator, to apply clamping force to a disc brake assembly, or to move the control surfaces of an aircraft wing. The brute force or finesse of the system depends on the size and control of the system components chosen. The purpose of discussions in this chapter is to further develop a technician's understanding of fluid power. Understanding how fluid-power systems function will improve the proficiency and safety of work activities.

Figure 5-1 shows examples of fluid-power applications for construction and wind-power applications. Wind turbines have used fluid-power systems to control such systems as the drive train parking brake, yaw drive, yaw brake, or pitch control of the blades. Blade pitch with fluid power may be accomplished by a single large linear hydraulic actuator connected to a common blade-control linkage or through the use of individual actuators, one connected to each blade. An **actuator** may create linear motion through the use of a cylinder and rod assembly or rotary motion through the use of a hydraulic motor. **Figure 5-2** shows examples of fluid-power actuators. Before discussing the components and maintenance requirements of a fluid-power system, we will review some of the principles of fluid power.

Fluid-Power Principles

As the term implies, fluid power uses a pressurized or moving fluid to transfer power within the system. The term *fluid*

covers liquids and gases. Typical fluid-power systems use a **hydraulic fluid** that comprises oil or a mixture of water and glycol along with additives that were mentioned in Chapter 4 on lubricants. The purpose of the hydraulic fluid is multifold. It is used to transfer power through a combination of flow and pressure, lubricate system components, and transfer heat generated by friction in the system. Fluid-power systems that use gases such as air are considered **pneumatic**. The major difference between the systems is the nature of the fluid. **Hydraulic** fluids such as oil, water, glycol, and other liquid types are virtually incompressible. **Incompressible** means you can pressurize the fluid and there will be virtually no change in the volume. Gases used in pneumatic systems are considered **compressible**. Compressing a volume of gas will

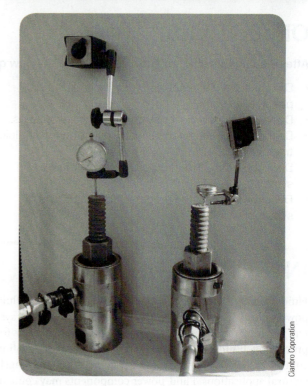

FIGURE 5-1 *Examples of fluid-power applications for construction and wind power*

FIGURE 5-2 *Examples of fluid-power actuators*

increase its pressure and decrease its physical size. You would be able to compress a large volume of gas into a small fixed container with a pump, but it would be impossible to compress the same volume of liquid into that container. A **pump** is a mechanical device used to energize a fluid through an increase in pressure or fluid flow or a combination of these. The term **compressor** is typically used to describe a pneumatic pump.

The physical relationship for a gas system can be expressed as

$$PV = nRT$$

where P is pressure, V is volume, and T is temperature, and the values of n and R are constants for the gas that relate the other variables. **Pressure** is defined as force over an area:

$$P = \frac{F}{A}$$

Earth's atmosphere is made of a mixture of gases such as nitrogen and oxygen, among others. The pressure exerted by the atmosphere at sea level is approximately 15 pounds per square inch (psi). That means that the atmosphere exerts

15 pounds of force on every square inch (in.²) of our body or surroundings. Look at the volume of air that can be compressed into a basketball with a volume of approximately 1 cubic foot:

$$1 \text{ft}^3 = 1{,}728 \text{ in.}^3$$

For this calculation, we will not consider the change in temperature (T) of the air as it is compressed and the value of R is the same from the large volume of air to the compressed volume in the basketball. As air is compressed, it will increase in temperature from the addition of energy during the compression process. The opposite occurs when a gas is allowed to increase in volume. Energy is absorbed from the surroundings to enable the gas to expand. This can be seen when dew or frost forms on the outside of a gas cylinder after some of the gas is released. The equation can then be written as

$$P_I V_I = P_F V_F$$

where P_I is atmospheric air pressure (15 psi), V_I is the initial air volume, P_F is the final air pressure (45 psi), and V_F is the final air volume (1,728 in.³). The air volume that can be compressed into the basketball is then

$$15 \text{ psi } (V_f) = 45 \text{ psi } (1{,}728 \text{ in.}^3)$$

$$V_I = \frac{[45 \text{ psi } (1{,}728 \text{ in.}^3)]}{15 \text{ psi}} = 5{,}184 \text{ in.}^3 \quad \text{or} \quad 3 \text{ ft}^3$$

The air volume in this example was compressed to one third its original volume.

Hydraulic fluids such as oil cannot be compressed like the air in this example. The equation used to determine the compressibility of an oil looks like this:

$$\frac{\Delta V}{V} = -\frac{\Delta p}{B}$$

where *B* is the **bulk modulus** of the fluid or the measure of its incompressibility (psi or Pa), *V* is the fluid's initial volume (in.³ or m³), Δp is the change in pressure (psi or Pa), and ΔV is the change in volume (in.³ or m³). The minus sign in the equation indicates the volume will decrease because of an increase in pressure. **Figure 5-3** shows the physical compression of a theoretical cube of fluid.

Consider the change in oil volume in a hydraulic power-bolting tool. Working pressure for the tool is 8,000 psi, so $\Delta p = 8{,}000 - 15$ psi = 7,985 psi. With this large change in pressure, the difference may be considered 8,000 psi with minimal error in the final answer for volume change. Using the bulk modulus for oil, 250,000 psi, the change in volume can be expressed as

$$\frac{\Delta V}{V} = \frac{-8{,}000 \text{ psi}}{250{,}000 \text{ psi}} = -0.032$$

or a 3.2% decrease in volume. Compare this oil-volume change at 8,000 psi to the one calculated for air with only a 30-psi change. The volume change of a pneumatic fluid—its compressibility—should not be considered a problem but an asset. One benefit of a pneumatic system over a hydraulic system is faster response time created by the expansion of the compressed gas into useful work. It would take longer for a hydraulic pump to move hydraulic fluid into the same size of actuator for the same travel distance. The other side of this argument would be the comparison of two equal-sized cylinders: one pneumatic and the other hydraulic. The pneumatic cylinder could not provide the same output force as its hydraulic

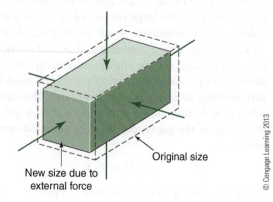

FIGURE 5-3 *Physical compression of a theoretical cube of fluid*

New size due to external force

Original size

© Cengage Learning 2013

SAMPLE PROBLEM

Determine the change in pressure (Δp) required to create a 5% decrease in oil volume. Use International System (SI) units of pressure (pascal) for the calculation.

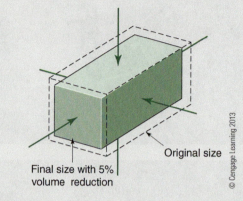

Final size with 5% volume reduction

Original size

© Cengage Learning 2013

SOLUTION

Change 5% to a decimal form:

$$\frac{5\%}{100\%} = 0.05 = \frac{\Delta V}{V}$$

Using SI units for the calculation requires converting the bulk modulus units from pounds per square inch (psi) to pascal (Pa):

$$1 \text{ psi} = 6{,}895 \text{ Pa}$$
$$= 6.895 \text{ kPa}$$

so

$$250{,}000 \text{ psi} \left(\frac{6.895 \text{ kPa}}{1 \text{ psi}} \right) = 1{,}723{,}750 \text{ kPa}$$

$$\frac{\Delta V}{V} = -\frac{\Delta p \text{ (kPa)}}{1{,}723{,}750 \text{ kPa}}$$

$$= -0.05 \text{ (decrease)}$$

$$-0.05(-1{,}723{,}750 \text{ kPa}) = \Delta p$$

$$= 861{,}875 \text{ kPa}$$

A 5% decrease in oil volume requires an increase in pressure of 861,875 kPa.

cousin. The slower actuation time would also improve motion control if that was a requirement for the application. Another property of fluid-power systems is *force multiplication*.

Force Multiplier

Force multiplication is a factor in using hydraulics systems with earthmoving equipment, automobiles, and aircraft control systems. Increasing the size of equipment, such as with excavators or forestry equipment, requires increasing mechanical advantage without a large increase in physical size. Solely using longer lever arms and cables becomes impractical. Applying force to hydraulic fluids with different actuator piston areas can increase mechanical advantage

SAMPLE PROBLEM

The pressure change (Δp) in the previous problem was 861,875 kPa. Convert this value to another common SI pressure unit known as a *bar*.

SOLUTION

Conversion for kPa to bar:

$$1 \text{ kPa} = 0.01 \text{ bar}$$

$$1 \text{ bar} = 14.696 \text{ psi}$$

Converting kPa to bar is then

$$861{,}875 \text{ kPa} \left(\frac{0.01 \text{ bar}}{1 \text{ kPa}} \right) = 8{,}618.75 \text{ bar}$$

while reducing physical size. **Figure 5-4** shows an example of mechanical advantage or force multiplication between two piston assemblies. Piston 1 has the application of 50 pounds force over a 2-in.² piston that produces a 500-pound output force at piston 2 with a 20 in.² The expression for mechanical advantage that relates the two piston areas is

$$F_{out} = \left(\frac{A_{out}}{A_{in}} \right) F_{in}$$

If the area of piston 2 were equal to piston 1, then the output force would not increase. This increase in mechanical advantage comes at a price. Like lever systems, input travel is farther than the corresponding output travel. **Figure 5-5** shows the relationship of mechanical advantage for a lever system. What travel distance is required for the 50-pound input force to get 500 pounds force at the output side to move 1 in.? This distance can be determined using the expression

$$d_{in} = \left(\frac{A_{out}}{A_{in}} \right) d_{out}$$

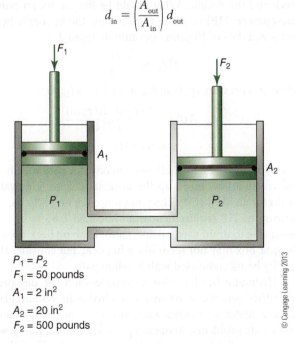

$$P_1 = P_2$$
$$F_1 = 50 \text{ pounds}$$
$$A_1 = 2 \text{ in}^2$$
$$A_2 = 20 \text{ in}^2$$
$$F_2 = 500 \text{ pounds}$$

FIGURE 5-4 *Example of mechanical advantage or force multiplication between two piston assemblies*

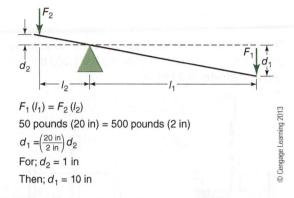

$$F_1 \, (l_1) = F_2 \, (l_2)$$

50 pounds (20 in) = 500 pounds (2 in)

$$d_1 = \left(\frac{20 \text{ in}}{2 \text{ in}} \right) d_2$$

For; $d_2 = 1$ in

Then; $d_1 = 10$ in

FIGURE 5-5 *Relationship of mechanical advantage for a lever system*

where d is the travel distance of the piston:

$$d_{in} = \left(\frac{20 \text{ in.}^2}{2 \text{ in.}^2} \right) 1 \text{ in.}$$
$$= 10 \text{ in.}$$

It takes 10 inches of movement on the input side to get 1 inch of travel on the output side. The larger input travel distance may not be an issue, but it must be considered. Another principle to consider for hydraulic systems is hydraulic head.

Hydraulic Head

Hydraulic head is the pressure (psi) created by the weight of a column of fluid. Remember the earlier discussion of atmospheric pressure? Approximately 15 psi exerted on our body or Earth's surface is due to the weight of the column air above us. It does not appear to be an important factor, does it? Not for air, but how about water, which weighs 0.0361 pounds per cubic inch? A 100-foot (1,200-in.) column of water is 43.32 pounds or 43.32 psi. There is more than 80,000 feet of atmosphere above us, producing around 15 psi.

How about the hydraulic head for oil? This can be determined using the weight per unit volume (**specific weight**) of oil compared to an equal volume of freshwater. This ratio is known as the **specific gravity** of oil [$sg_{(oil)}$] and can be expressed as

$$sg_{(oil)} = \frac{\gamma_{(oil)}}{\gamma_{(water)}}$$

where $\gamma_{(oil)}$ is the specific weight of an oil (56 lbs/ft³) and $\gamma_{(water)}$ is the specific weight of freshwater (62.4 lbs/ft³), so

$$sg_{(oil)} = \frac{(56 \text{ lbs/ft}^3)}{(62.4 \text{ lbs/ft}^3)}$$
$$= 0.897 \text{ or } 0.90$$

The specific gravity of a fluid can be found in an oil manufacturer's documentation or engineering reference manuals. Multiplying the $sg_{(oil)}$ by the specific weight of freshwater from above (0.0361 pounds/in.³) gives a specific weight of oil to be 0.0325 pounds/in.³

How is this information useful? Consider the pressure created in a hose because of oil draining from the wind-turbine gearbox. What pressure will be created by the 300-foot

SAMPLE PROBLEM

Determine the pressure supplied to the system by piston 1 if the output force of piston 2 is 1,000 pounds and its diameter is 3 inches.

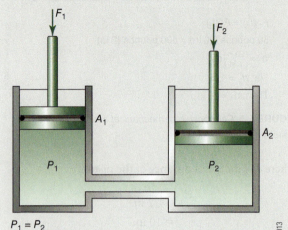

$P_1 = P_2$
$F_1 = ?$ pounds
$A_1 = 2$ in^2
$D_2 = 3$ in
$F_2 = 1000$ pounds

SOLUTION

First determine the area of piston 2. Diameter is 3 inches, so the area is

$$A = \frac{\pi D^2}{4}$$

$$= \frac{3.142 \,(3 \text{ in.})^2}{4}$$

$$= 7.069 \text{ in.}^2$$

Next, determine the pressure requirement to produce 1,000 pounds force on the piston output rod. Pressure is a function of the piston area and the applied force. This can be expressed as

$$P = \frac{F}{A}$$

$$= \frac{(1,000 \text{ pounds})}{7.069 \text{ in.}^2}$$

$$= 141.47 \text{ psi}$$

Piston 1 will need to generate 141.5 psi to produce 1,000 pounds of force from piston 2. This pressure also can be supplied by a pump that powers the system. Pumps are discussed later in the chapter.

SAMPLE PROBLEM

Determine the output force in newtons (N) if the input force is 100 N and the diameters of piston 1 and 2 are 125 mm and 1250 mm, respectively.

SOLUTION

Area of piston 1 is

$$A = \frac{\pi D^2}{4}$$

$$\frac{\pi (0.125 \text{ m})^2}{4} = 0.01227 \text{ m}^2$$

Pressure produced by 100 N force on the area of piston 1 is

$$P = \frac{F}{A}$$

$$\frac{100 \text{ N}}{0.01227 \text{ m}^2} = 8,150 \text{ N/m}^2 = 8,150 \text{ Pa}$$

Area of piston 2 is

$$A = \frac{\pi (1.25)^2}{4} = 1.227 \text{ m}^2$$

Output force of piston 2 is

$$F = P(A)$$

$$= 8,150 \text{ N/m}^2 \,(1.227 \text{ m}^2) = 6,642 \text{ N}$$

Output force produced at piston 2 is 6,642 newtons, with 100 N supplied to piston 1.

---

to overcome this hydraulic head to refill the gearbox with fresh oil? The pump design for this system has to include the 117 psi to support the column of oil between the service truck and the nacelle. What would be the minimum pump **horsepower (HP)** required to overcome the hydraulic head and give a flow of 10 gallon per minute (gpm)?

$$\text{HP}_H = \frac{(pQ)}{1,714}$$

where p is pressure (psi) and Q is fluid flow (gpm):

$$\text{HP}_H = \frac{117 \text{ psi } (10 \text{ gpm})}{1,714}$$

$$= 0.68 \text{ HP}$$

It would take around 0.7 HP to overcome the hydraulic head and create a flow of oil up the hose at 10 gpm. This value works for an ideal 100% efficient pump. For an 85% efficient system, the HP requirement would be 0.68/0.85 = 0.8 HP. Being able to drain used oil and pump fresh oil back up into the gear box may not seem like a big deal, but consider this activity being completed with 5-gallon pails.

Hydraulic head is also a consideration for designers when they determine the layout of a hydraulic system. Placing an actuator at too great a distance above or below a pump may create problems. Actuators placed above the pump will decrease the fluid pressure in the actuator, so it will not have adequate force for the application. Actuators placed below

---

elevation between the nacelle (gearbox) and the service truck storage tank? Water weighs 0.0361 pounds/in.^{3}, and oil is 90% of the weight of water. Pressure in the hose is equal to 0.0361(0.9) $\times$ 300 ft $\times$ 12 in./ft = 117 psi. Does not seem like much? Consider that the pressure in a domestic water system is around 30 psi. How about the pump size required

the pump will cause an increase in pressure that may cause an excessive output force or cause the actuator to leak because of excessive pressure on the seals.

Fluids

Hydraulic fluids are typically hydrocarbon-based materials with additives like those used for lubrication oils. It may be either a natural or a synthetic base, depending on the working pressures and temperatures of the application. Other considerations for hydraulic fluid include environmental impact and fire resistance. Hydraulic systems that are used in agricultural, forestry, and marine applications typically are mineral oil or vegetable oil blends to reduce their impact on the environment should a spill occur. The mining industry requires oils that are fire resistant because of applications involving tight spaces with possible ignition sources. If a hydraulic hose breaks and sprays fluid onto the hot cutting tools of a mine excavator, the resulting fire or explosion would be catastrophic. For this reason, hydraulic fluids for the mining industry are typically formulated from a solution of water and glycol to eliminate a potential fire or explosion. Land-based wind-turbine projects use hydraulic fluids that are found in other industrial applications. Offshore wind-turbine projects will require hydraulic fluids that match existing marine-industry environmental standards. The challenge for wind-turbine manufacturers will be to match their equipment needs of lubrication, temperature, and pressure with those fluids that produce a minimal impact on the environment should there be an accidental release.

FLUID-POWER COMPONENTS

A fluid-power system consists of four basic device categories as defined by general function: energize the fluid, control the fluid, convert the fluid power to work output, and support the system. Each category includes several devices that allow the designer to tailor the hydraulic circuit to best meet the needs of the application. Understanding these categories will enable a technician to appreciate each function.

Energize Fluid

Fluid systems are used for two basic functions as previously mentioned in the chapter: transport and power. **Hydrodynamic systems** use the energy associated with fluid movement to transport fluids in a system. Pumps used for this type of process do not develop a significant amount of pressure for fluid transfer. A typical pump used for this type of system would be a centrifugal pump. **Figure 5-6** shows an example of a centrifugal pump. This type of pump may be used with a power-conversion cooling system. The fluid flows through a series of electronic cooling blocks to remove heat energy

FIGURE 5-6 *A centrifugal pump*

generated during equipment operation. Heat generated by the equipment is transferred to the cooling fluid causing it to increase in temperature. The warmed fluid is then run through a liquid–air heat exchanger (radiator) to dump the heat energy to the surrounding air.

Figure 5-7 is a conceptual diagram of an electronic liquid-cooling system. This style of pump demonstrates *nonpositive displacement*. Blocking the outlet port of this style pump would not significantly increase system pressure. A blocked outlet port would cause the fluid to leak around the impeller assembly and thus prevent a significant buildup of pressure in the system. **Hydrostatic systems** use pressurized fluid interacting with a surface such as a piston or vane to create a force and motion. Pump designs used with a hydrostatic system include piston, gear, and vane. These pumps are considered *positive displacement*. **Figure 5-8** shows an example of a positive-displacement pump and the ANSI symbol. **Positive displacement** means that as long as the pump assembly is moving, the fluid is being displaced. For example, a piston-type pump is a positive-displacement pump. Each time the piston retracts in the cylinder, it pulls in a fixed volume of fluid to fill the cavity formed by the motion, and when the piston extends it forces the fluid out of the cavity through the outlet port. The assembly is made with very tight tolerances, so there is virtually no space between the piston and cylinder for the fluid to leak. This type of pump design would continue to build pressure with a blocked outlet port until a rupture in a seal or line would allow the pressure to decrease. This type of failure would be extremely dangerous and costly. A person in the way of this rapid release of high-pressure fluid could be burned, injected with fluid, or injured by flying debris.

System Control

Fluid-power systems use a variety of system control valves: fluid direction, velocity, pressure, and maximum system

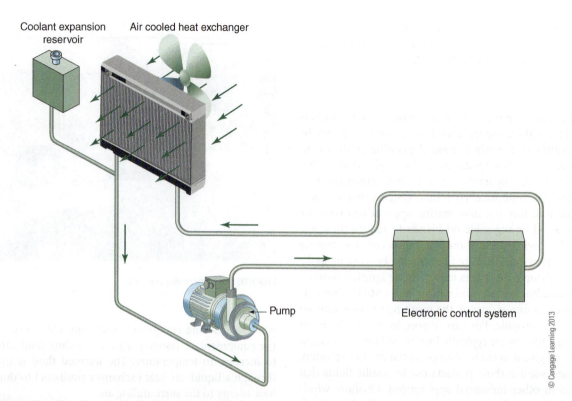

Coolant expansion reservoir

Air cooled heat exchanger

Pump

Electronic control system

© Cengage Learning 2013

FIGURE 5-7 *Simplified diagram of a liquid-cooling system for power electronics*

pressure. Fluid direction can be controlled with a variety of valve designs such as spool valves, ball valves, and check valves. **Figure 5-9** shows examples of fluid-control valves.

Fluid Direction

Directional control valves are designed with a sliding machined rod (spool) that directs fluid within an enclosed block assembly. The valve may be controlled manually, electrically, or mechanically. Manual directional control valves have a handle or foot pedal that can be moved by an operator to direct flow to system components. Manual directional control valves are not used in wind turbines, but a service

© Cengage Learning 2013 and a W. Kilcollins photograph

© Cengage Learning 2013 and a W. Kilcollins photograph

© Cengage Learning 2013 and a W. Kilcollins photograph

FIGURE 5-8 *Example of a gear type positive-displacement pump*

FIGURE 5-9 *Examples of fluid-control valves*

technician may see them in other industrial or equipment applications.

Electrically operated directional control valves use a **solenoid** to position the spool within the valve body. As you may recall, the solenoid operates on the principle of electromagnetism. The solenoid is made with a coil of magnetic wire positioned around a spring-loaded iron or steel rod. The spring holds the rod and spool in a rest position; when the wire coil is energized, it pulls the rod toward the center of the coil and thus shifts the spool to redirect fluid flow. Service technicians will encounter solenoid-controlled directional valves in wind turbines that are controlled by the programmable logic controller (PLC). Caution must be used around any automatic device such as a solenoid-controlled valve. If the PLC is functioning, it can change the state of the valve or pump whenever program conditions are satisfied. *Always disable, relieve pressure, and lockout–tagout (LOTO)* the fluid-power system as required by company policy or the equipment manufacturer's recommendation before disassembling hydraulic circuits. The third spool-valve control method is mechanical: the spool assembly will be attached to a cam, lever, or roller that will shift the spool when it is acted on by an external force. **Figure 5-10** shows examples directional control valves. Service technicians typically may not see mechanically operated directional control valves in wind turbines, but they may encounter them in other industrial settings or on fluid-power tools.

As you can see, directional control valves come in several actuation styles. These valves also come in several control design styles from two-way up to five-way, depending on the control sophistication required for the application. Examples of ANSI symbols for two position directional control valves are shown in **Figure 5-11**. The two-way and three-way solenoid-control versions are styles that service technicians will typically encounter with wind-turbine fluid-power systems.

Ball valves are used to restrict or block the flow of fluid in a system. Typical wind-turbine uses may include preventing the loss of system fluid during service work or in a drain assembly for ease of service. If a ball valve is used to prevent fluid flow during service work, then ensure that precautions are in place to prevent inadvertent opening of the valve by other technicians. If there is a question of control, ensure that LOTO procedures are implemented. **Figure 5-12** shows an example of a ball valve and ANSI symbol.

Check valves are used in a fluid-power system to restrict flow to a predetermined direction. Check valves may be used to ensure that fluid does not flow back through a pump when it is idle. A check valve may also be used in conjunction with a flow-restriction device to enable the free flow of fluid around the device in one direction and controlled flow in the other direction. **Figure 5-13** shows an example of a check valve and respective ANSI symbol. **Figure 5-14** shows an example of check-valve placement within a hydraulic circuit.

Fluid Velocity

Fluid velocity is another control feature that is necessary for a hydraulic circuit. Controlling the activation speed

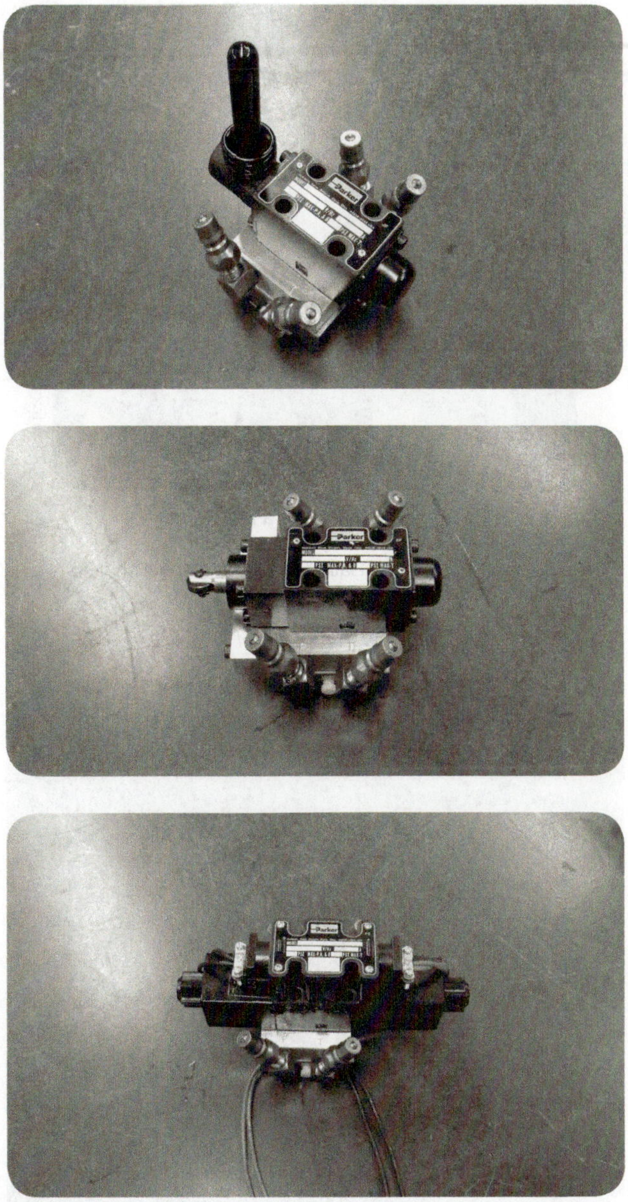

FIGURE 5-10 *Examples of manual, mechanical and solenoid operated directional control valves*

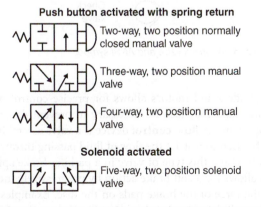

Push button activated with spring return

Two-way, two position normally closed manual valve

Three-way, two position manual valve

Four-way, two position manual valve

Solenoid activated

Five-way, two position solenoid valve

FIGURE 5-11 *Examples of ANSI symbols for directional control valves*

ANSI Symbol for a ball valve

Ball valve – type manual shutoff

FIGURE 5-12 *Ball valve and ANSI symbol*

ANSI Symbol for a check valve

Check valve

FIGURE 5-13 *Check valve and ANSI symbol*

of cylinders and motors allows for precise control of operations and may also be necessary to prevent damage to components. A **flow-control** device is a valve assembly that can be used to restrict the flow of fluid passing through the circuit. Using this type of valve in a parking-brake application allows for a soft brake action by progressively increasing the force of the brake pads on the disc. Examples such as this will be discussed in the application section later in

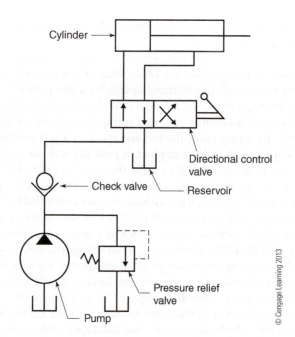

Cylinder

Directional control valve

Check valve

Reservoir

Pressure relief valve

Pump

FIGURE 5-14 *Check valve placement above hydraulic pump to prevent flow from circuit back into pump*

the chapter. **Figure 5-15** shows an example of a flow-control valve and the ANSI symbol used in a circuit diagram.

Fluid Pressure

Fluid pressure in a fluid-power system is typically controlled with an electronic pressure **sensor** to provide feedback from

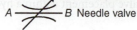
A   B Needle valve

FIGURE 5-15 *Flow-control valve and ANSI symbol*

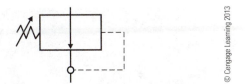

FIGURE 5-18 *ANSI symbol for variable pressure-reducing valve*

Some fluid-power applications may require a circuit branch to be at a different pressure level than others because of a specific operation. This can be accomplished by the use of a pressure reducing valve. **Figure 5-18** shows the ANSI symbol for a pressure-reducing valve. This valve may be field adjusted to the needs of an application. As with any field adjustments, follow the equipment manufacturer's recommendations to ensure system reliability.

Converting Output to Useful Work

Fluid-power systems are not useful without a way to convert the pressure and flow provided by the pump into useful work. As you recall, work is defined as force multiplied by the distance through which it is applied:

$$W = F \times d$$

where W is work (ft-lbs or N-m), F is the applied force (lbs or N), and d is the distance (ft or m). Earlier discussions focused on the force and pressure provided to a system through a piston and cylinder assembly. Our discussion now turns to output from a cylinder to produce useful work from a fluid that is under pressure. Fluid-power systems have a couple of different ways to provide useful work: linear motion and rotary motion. Linear motion is provided by a cylinder assembly like that discussed earlier in the chapter. Rotary motion is supplied by a device that has a design similar to a gear or vane pump, except that now the fluid causes the internal assembly to rotate because of system pressure and flow. The rotation of the internal assembly causes an output shaft to rotate to provide mechanical power. **Figure 5-19** shows an example of a **hydraulic motor** assembly and ANSI symbol. Rotary actuators are used in a variety of industrial and outdoor equipment applications. Examples include crawler drive systems for cranes and excavators along with a power takeoff assembly for farm equipment. Hydraulic motors can be unidirectional or bidirectional, depending on the requirements of the application.

Support Devices

Support devices, also known as *ancillary devices* may be used in fluid-power systems to provide fluid supply, conditioning, connections between devices, visual indication of system

FIGURE 5-16 *Electronic pressure sensor used in fluid-power circuits to provide feedback*

the circuit to the pump control system. **Figure 5-16** shows an example of an electronic sensor that may be used in fluid-power circuits to provide feedback. Pressure-relief valves are also used in hydraulic systems to prevent overcharging and subsequent damage. These devices can be designed as a fixed pressure value or with a variable adjustment that can be set in the field for the application. **Figure 5-17** shows an example of a pressure-relief valve along with its ANSI symbol.

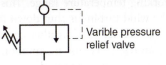

Varible pressure relief valve

FIGURE 5-17 *Pressure-relief valve and ANSI symbol*

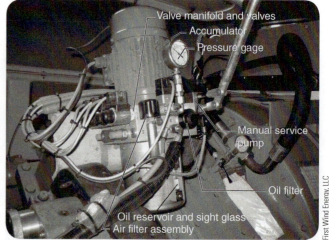

Fixed displacement
bi-directional

FIGURE 5-19 *Example of a hydraulic motor assembly with ANSI symbol*

condition, electronic feedback, backup supply of pressurized fluid, and sealing elements.

Fluid Supply

The pump assembly is connected to a **reservoir** or storage tank that is sized to hold an ample amount of fluid for system requirements. Many hydraulic systems are designed with an integral pump, reservoir, filters, and valves that collectively are known as a **power unit**. **Figure 5-20** shows an example of a power unit that may be used in a wind turbine. Note the components mentioned earlier along with an accumulator, the component that looks like a canister attached to the top of the unit. An **accumulator** is used to maintain a fixed volume of fluid under system pressure. Accumulator types and uses will be discussed later. Filters ensure that contaminants such as wear particles, dust, and moisture do not enter the pump and get pushed through the system. Particles of metal or dust may enter small ports or channels and create a blockage and subsequent damage. **Figure 5-21** shows examples of filter assemblies and generic ANSI symbol. The filter element may be made of a single layer of porous material such as paper, fabric, or metal. An element may also be assembled from multiple layers of pleated fabric to increase its effectiveness. The function of the **filter element** is to capture contaminants such as wear particles, dust, and organic material as they flow through the assembly to prevent them from reentering the system. Caution should be exercised when replacing filter elements to ensure that contaminants do not enter the pump inlet assembly. Consult the equipment manufacturer's instructions for the proper replacement element. Improper replacement parts may be incompatible with

FIGURE 5-20 *Power unit used in a wind turbine*

the fluid or not be sized properly to capture contaminants, which could damage the system.

Fluid Conditioning

Fluid conditioning may include heating and cooling systems. **Heating systems** may be used to maintain fluid temperature for equipment operating in colder climates. Systems operating in northern climates such upstate New York, Maine, and the Dakotas require heating systems to ensure that the hydraulic fluid is maintained within the manufacturer's recommended working temperature range. This is particularly important if the wind turbine is shut down for an extended length of time for service or if weather conditions produce subzero temperatures. It is not uncommon to see trucks and other outside equipment plugged in overnight or when not in use as the temperature drops below zero Fahrenheit (0 °F). Another conditioner used with fluid power is a cooling

FIGURE 5-22 *Cooling system heat exchanger used to reduce a fluid's temperature*

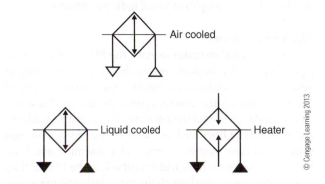

FIGURE 5-23 *ANSI Symbols for Hydraulic Heat Exchangers*

a

b

Filter/strainer

c

FIGURE 5-21 *(a, b) Examples of filter assemblies; (c) ANSI symbol for filter or strainer assembly*

system. The temperature of the fluid increases as it is compressed by the pump assembly and because of frictional losses as it flows through the system. The fluid is circulated through a **cooling system** to dissipate heat energy produced during operation. **Figure 5-22** shows an example of a cooling system heat exchanger used to reduce the fluid's temperature. **Figure 5-23** shows the ANSI symbols for heat exchangers.

Connections

No fluid system is complete without a method of directing the fluid flow between devices. The hoses and connections between the devices must be rated for system pressures, temperatures, and fluid compatibility. Connectors and hoses will have their rating information marked clearly on their components. If you need to change a component and there is no visible rating, do not assume it will work. The outcome may be fatal. Always use replacement parts that are specified for the

equipment. **Figure 5-24** shows examples of hoses, lines, and connectors that may be encountered during a fluid-power service activity. Important factors for hoses and connections include abrasion-resistant covers and mechanical support of fittings. If a hose assembly is required to flex during operation conditions, ensure it does not become abraded by contact with other components. If the condition may occur, place an abrasion sleeve over the hose or route it to prevent abrasion. Another consideration during service or maintenance activities is to ensure that system components are supported by appropriate means. Couplings such as unions or swivel elbows are not designed to carry the weight of most components. If a coupling is stressed because of vibration or because of a heavy component, it may fail. Downtime because of a broken connection is costly, not to mention the cleaning required when hydraulic fluid is sprayed over the inside of a nacelle.

Visual Indicators

Visual indicator devices that may be encountered during work activities with fluid-power systems include fluid-level indicators, flow gauges, pressure gauges, and temperature gauges. These are important when a visual status is required for work on the system. Each device can be used to visually confirm that the system is operating properly or has been reduced

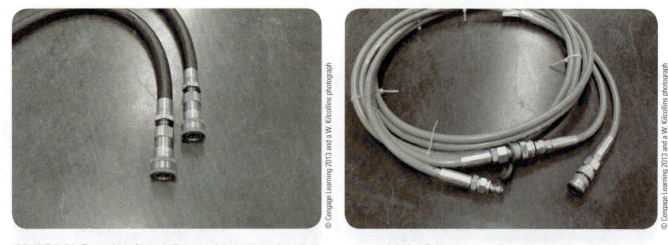

FIGURE 5-24 *Examples of hoses, lines, and connectors that may be encountered during fluid-power service activity*

to a zero-energy state for activities that require LOTO. **Figure 5-25** shows examples of visual indicator devices.

Electronic Sensors

Along with visual indicator devices, the PLC uses electronic equivalents for feedback to confirm the system is operating as designed. Some feedback devices or sensors were mentioned previously during the control system discussion. These devices include sensors for temperature, flow, pressure, position, and fluid level. Other feedback devices used in conjunction with an actuator may be a servo valve assembly and limit switches. A servo valve provides feedback to the PLC of fluid volume that is moving through the valve during piston extension or retraction. This information enables the PLC to determine the displacement of the cylinder during travel. This information along with feedback from limit switches would

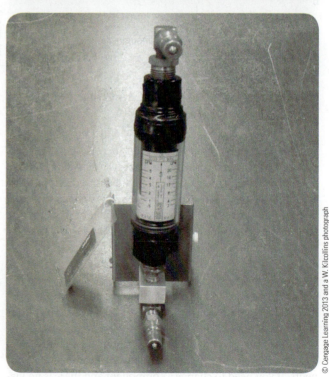

FIGURE 5-25 *Examples of visual indicator devices*

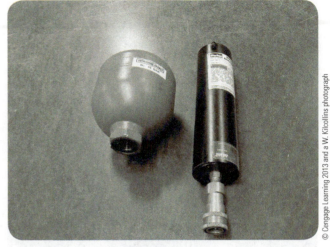

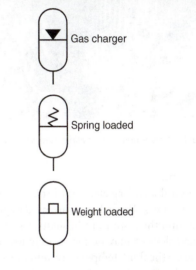

Gas charger

Spring loaded

Weight loaded

FIGURE 5-27 *Accumulator designs and their ANSI symbols*

FIGURE 5-26 *Feedback devices on a wind turbine*

enable the PLC to accurately position equipment components attached to a cylinder assembly. This feedback would be important for a hydraulic system that is used to pitch blades of the wind-turbine rotor assembly. **Figure 5-26** shows images of feedback devices that may be encountered during work on a wind turbine.

Backup Power

In the event of a grid outage or during an emergency shutdown of a wind turbine, the blades must be feathered out of the wind and the parking brake must be applied to stop the rotor from spinning. Some wind-turbine manufacturers use electrical equipment to position or pitch the blades, whereas other manufacturers rely on a hydraulic system for

this function. The standby power source for an electrical system is a battery pack.

What would the standby power system be for a hydraulic system? An accumulator assembly. A charged accumulator maintains a fixed volume of hydraulic fluid under system pressure to energize an actuator such as a cylinder when the pump fails or when there is a loss of grid power. **Figure 5-27** shows examples of accumulator designs along with their ANSI symbol. An accumulator may be designed with a spring-loaded piston, a gas charge above the piston, or bladder to maintain system pressure. These assemblies will force the fluid out of the accumulator when the pressure supplied by the pump decreases. An accumulator's size is selected by the volume of pressurized fluid that is necessary to operate required devices during loss of system pressure. Some power units have multiple accumulators integrated into the assembly when there are multiple actuators that need to be supplied with standby power. **Figure 5-28** shows a power unit with multiple accumulators. The use of accumulators in hydraulic circuits will be illustrated in the applications section of this chapter.

University of Maine at Presque Isle

FIGURE 5-28 *Power unit with multiple accumulators*

University of Maine at Presque Isle

FIGURE 5-29 *Example of blade-pitch system using single linear actuator for position control*

Seal Elements

Another useful ancillary element of the fluid-power system is a seal or packing. These devices are used to prevent leakage of fluid around the pistons, rods, shafts, valves, and connections. These devices may be metal, polymer, or composite, depending on the fluid, temperature, and pressure of the system. Examples of sealing elements include: O-rings, V-packs, loaded U-rings, and rod wipers.

FLUID-POWER APPLICATIONS

Many components can be used in hydraulic circuits to enable a wind turbine to perform as designed. We look now at the types of components and ANSI symbols from previous discussions to see how they function together. Typical wind-turbine applications that use fluid power include blade pitch, secondary brake, yaw drive, and yaw brake. An example of a service application that uses fluid power would be a power-bolting tool.

Blade Pitch

Blade pitch for a wind turbine needs to have sufficient control to position the blades for optimum wind-power collection while also allowing the blades to be positioned out of the wind when not in use. Our discussion will focus on a couple of

methods that may be encountered for blade control. The first method uses a main cylinder attached to a linkage that moves all three blades together. The second method uses three individual cylinders that move the blades in unison. Both designs require feedback components and software algorithms to enable the PLC to track and optimize blade position. Feedback and tracking systems are not the focus of this book, so discussion will be limited to the components that may be encountered during maintenance and service activities.

Single-Cylinder Pitch

The single-cylinder pitch system uses a central hydraulic cylinder mounted to the back of the gearbox in the nacelle with a piston rod that extends through the gearbox main shaft to a linkage assembly located in the rotor assembly. **Figure 5-29** shows an example of a blade-pitch system using a single **linear actuator** for position control. Control of the power unit and valve assembly maintains position of the cylinder during operation. When the blades are positioned, the spool valve closes to prevent the blades from moving. An accumulator in the circuit can be used to maintain pressure in the cylinder if leakage should occur around the piston or valve spool. The accumulator can also supply backup fluid under pressure to feather the blades in an emergency situation.

Multiple Cylinder Pitch

A pitch design that uses three individual cylinders to move the blades at the same time may use a single spool valve or

Technical specifications

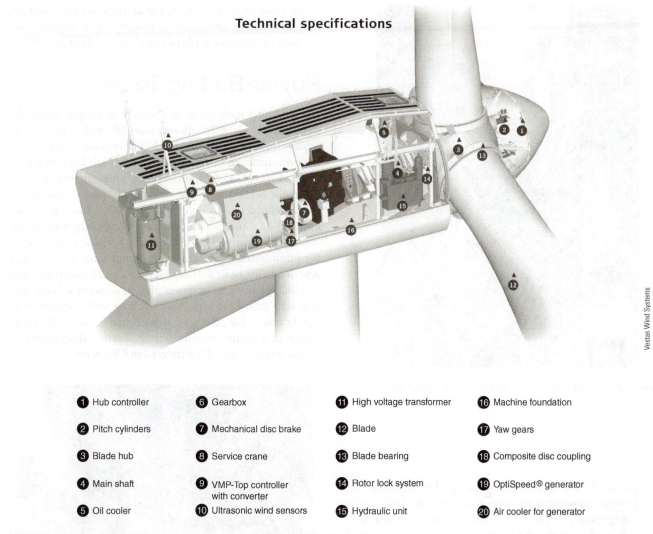

1. Hub controller
2. Pitch cylinders
3. Blade hub
4. Main shaft
5. Oil cooler
6. Gearbox
7. Mechanical disc brake
8. Service crane
9. VMP-Top controller with converter
10. Ultrasonic wind sensors
11. High voltage transformer
12. Blade
13. Blade bearing
14. Rotor lock system
15. Hydraulic unit
16. Machine foundation
17. Yaw gears
18. Composite disc coupling
19. OptiSpeed® generator
20. Air cooler for generator

Vestas Wind Systems

FIGURE 5-30 *Rotor assembly using three linear actuators for blade-pitch control*

three separate valves all controlled by the same electrical signal to activate the solenoid. **Figure 5-30** shows an illustration of a rotor assembly using three linear actuators for blade-pitch control. This type of system would also require an accumulator to supply pressurized backup fluid as necessary to maintain blade position in an emergency.

Parking Brake

The parking-brake system of a wind turbine consists of a large disc attached to the high-speed shaft of the gearbox and a caliper system that applies pressure to the disc with pads. This system is a large version of the disc-brake assembly that can be seen on an automobile. **Figure 5-31** shows an example of a wind-turbine parking-brake system used with a gearbox. Some wind-turbine systems such as direct-drive systems are designed without gearboxes. These systems would use a parking brake attached directly to the main shaft. The physical size of this brake would be much larger than the gearbox

design because the torque acting on the assembly would be significantly larger.

Yaw Brake

The function of the yaw brake for a wind turbine is to prevent the nacelle from shifting when the rotor assembly points directly into the wind. A couple of different methods can accomplish this function: one is a passive brake, and the other is considered an active brake. Some equipment manufacturers use a **passive brake** system that consists of a series of spring-loaded pistons to press down on a ring at the top of the tower to use static friction to maintain position. This assembly maintains the position of the nacelle when the yaw drive system is not operating. The pistons are made out of metal with a composite face that is positioned down against the tower ring. Another method of yaw braking used by other equipment manufacturers uses two, four, or more caliper assemblies to clamp onto a metal ring that is attached to the top of the

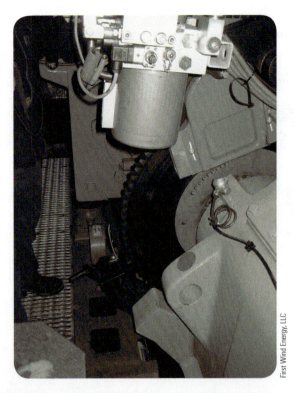

First Wind Energy, LLC

FIGURE 5-31 *Wind-turbine parking-brake system used with gearbox*

SAMPLE PROBLEM

Determine the output torque provide by the power-bolting tool with the following specifications: 6,000 psi fluid pressure, 1.5 in.² piston area, piston attached to a 3-in.-long pivot arm, and output drive.

SOLUTION

Determine the force created at the piston with the 6,000 psi:

$$F = P(A)$$

$$F = 6,000 \text{ psi } (1.5 \text{ in.}^2)$$

$$= 9,000 \text{ lbs.}$$

Calculate the torque generated with the 9,000-lb force applied to the 3-in. pivot arm:

$$T = F(d)$$

$$= 9,000 \text{ lbs } (3")$$

$$= 27,000 \text{ in.-lbs}$$

What is the torque value in foot-pounds (ft-lbs)?

$$1 \text{ ft-lbs} = 12 \text{ in.-lbs}$$

$$27,000 \text{ in.-lbs} \left(\frac{1 \text{ ft-lbs}}{12 \text{ in.-lbs}} \right) = 2,250 \text{ ft-lbs}$$

This torque value is not uncommon when tightening or testing large tower bolt assemblies or nut and stud assemblies for large bearings.

tower. This system is considered an **active brake** because it can open or close, depending on the braking action required. **Figure 5-32** shows an example of a yaw-brake caliper assembly.

Power-Bolting Tools

Power-bolting tools come in a variety of designs. Some designs use an electrical motor assembly and a gearbox to achieve high-output torque levels. Other power-bolting tools use pneumatics or hydraulics to achieve high levels of output torque. **Figure 5-33** shows examples of hydraulic-bolting tools. Hydraulic-bolting tools use a hydraulic power unit, hoses, and handheld tool to tighten large nuts or machine screws. **Figure 5-34** shows examples of hydraulic-bolting tools in use.

The bolting tool uses hydraulic pressure generated by the pump assembly to create an output force as previously described. The basic tool design uses a small piston attached to a linkage arm and output shaft that converts a linear motion into rotary motion. Power-bolting tool manufacturers supply charts for each tool model that relates tool output torque to a pump pressure setting. Further discussions on torque inspections will be covered in Chapter 6.

University of Maine at Presque Isle

FIGURE 5-32 *Example of yaw-brake caliper assembly*

Maschinenfabrik Wagner GmbH & Co. KG and PLARAD Bolting Technology, LLC

FIGURE 5-33 *Hydraulic-bolting tools*

FIGURE 5-34 *Hydraulic-bolting tools in use*

TYPICAL FLUID-POWER INSPECTIONS

Fluid-power assembly inspections are completed during each preventative-maintenance (PM) cycle or earlier if there is a suspected issue. These inspections are typically visual and will depend on the component or system being inspected. Some inspection areas common to wind turbines include fluid-reservoir levels, visual gauge function, valve function,

actuator motion, leaks, and filter-element replacement. These inspections should be listed on a checklist to ensure each is performed and documented for future reference. Documentation of PM activities will be discussed in later chapters to show how the data can be used for predictive maintenance. Some of these inspection activities will be summarized in the following sections.

Reservoir-Fluid Level

Reservoir-fluid levels are an important inspection during the PM cycle. Hydraulic systems that run low on fluid can overheat or fail to operate because of air bubbles in the system. Typically automated systems used with wind turbines have a variety of sensors to flag a pressure, temperature, and level change that is significant enough to indicate a problem. Do not wait for a tower to go off-line in the middle of the night when it gets one of these supervisory control and data acquisition (SCADA) flags. Be proactive and ensure that fluid levels are within specification each time there is a need to climb a tower for service or maintenance.

Gauge Inspection

Gauge inspections are a vital function during each service call or PM cycle. If a mechanical gauge or steam gauge, as some older technicians say, does not match the electronic sensor readouts there may be trouble ahead. Mechanical gauges are fairly robust and tend not to fail. Electronic sensors, on the other hand, tend to be a bit more sensitive to contamination and connection challenges. The majority of service calls that this author has been involved with resulted from poor electrical connections on sensors causing *false readings*. The PLC data that runs the system is only as good as the sensors providing the data. If a connection comes loose on a temperature probe or pressure probe, the PLC input sees an open-circuit signal. These sensors typically turn a linear change in resistance into a change in current that the PLC monitors. What is an open circuit with regard to a resistance reading? Does infinity sound familiar? If a PLC input is looking for a value of resistance around 100 ohms (Ω) and then it registers 999.99 Ω, it will shut down and wait for a technician to verify that the pressure or temperature of the system is corrected.

Valve Function

Checking the function of solenoid valves is another good practice with fluid-power systems. If a valve sticks, its spool assembly may be contaminated. Check for proper valve function and replace the valves as necessary. Sticking valves will create equipment issues and an eventual failure. This problem could be because of a plugged or damaged filter assembly.

Filter Inspection and Replacement

Inspecting the air-vent and fluid filters in a hydraulic system is an inexpensive means of ensuring consistent operation. Inspect the filter for contaminants and damage and replace during each PM cycle. Whenever a fluid filter is removed from a hydraulic line or manifold, it may introduce air in the system. Removal of the air can be accomplished by a process known as *bleeding*.

Bleeding Air from the System
Bleeding the system is accomplished by cracking (slightly loosening) a fitting connection upstream of the pump assembly or any location that would allow entrapped air to escape. **CAUTION: This process should be carried out by a qualified technician who is familiar with the system operating procedure. Damage to the fitting during this process can create a safety hazard and possibly damage equipment. Only loosen the fitting enough to allow a small amount of** oil to trickle out and vent any entrapped air that may have been introduced during the filter-element inspection or replacement. Once a smooth stream of oil is formed from the loosened fitting, it should be retightened and inspected for leaks. Clean the fitting, inspect, and tighten until there is no longer a fluid leak.

Leaks

All hydraulic components, lines, and fittings should be inspected for leaks. Slight leaks will show up as a soiled or dust-caked area around a connection or seal. Clean the area that is suspected of a leak and tighten the connection as necessary. Small leaks may not appear to be a problem, but they may later turn into larger leaks. Larger leaks could create an environmental issue and an expensive cleanup project. Leaks may also be locations where air can enter the system. Air bubbles compress with pressure and could cause issues when trapped in the small ports of valves and fittings.

Clean up any oil spills and properly dispose of the waste. Always consult the MSDS for the fluid that will be used. Follow the recommended safety precautions for hygiene, personal protective equipment, fire potential, and environmental impact.

Fluid Sampling and Analysis

The recommendation of fluid sampling and analysis covers hydraulic fluid as well as gearbox lubrication oil. The types of results that may be reported in an oil analysis were previously discussed in Chapter 4. The sample should be obtained from the system before the filter assembly and stored in a clean, well-marked bottle. Minimum recommended information for a fluid sample bottle includes wind-farm name, wind-turbine number, date sample was drawn, and technician's name or initials. If the hours of operation since the last oil sample are easily accessible, this would also be valuable information. Some oil-analysis laboratories can also examine the filter elements to determine the composition of particulate trapped in the pleats and fabric.

Comparison of Results
Use the sample analysis results and testing lab recommendations for the oil sample as a guide to possible corrective action. Never jump to conclusions if something appears to be out of tolerance with the sample. Go back and inspect the system and take another sample for comparison. Use other equipment on site and historical data that are available to base a decision on corrective action. Make a practice of following up on test results and watch for trends. These steps will be a big step toward improving the performance of the equipment and eliminating unscheduled downtime.

SUMMARY

This chapter reviewed several fluid-power topics such as fluid properties, fluid systems, system components, system circuit diagrams, and examples of wind-turbine hydraulic systems. These topics were intended to reinforce technicians' understanding of fluid-power systems so that they may work safely and efficiently. The chapter also included inspection recommendations that should be completed during each PM cycle. It was also recommended to visually inspect for some of these items during a service call to reduce downtime because of other equipment issues. The goal of PM is to improve equipment performance through analysis of inspection data and equipment performance.

REVIEW QUESTIONS AND EXERCISES

1. What is the difference between a hydraulic and a pneumatic system?
2. What is meant by *compressibility of a fluid*?
3. Calculate the reduction in a 100 ft³ volume of air if it is compressed from atmospheric pressure (15 psi) to 100 psi.
4. Calculate the reduction in volume of oil if it is compressed from 0.25 MPa to 1.5 MPa. Use the bulk modulus value derived earlier in the sample problem.
5. What is the difference between a fluid-power and a fluid-transfer system?
6. What is one of the required fluid properties for mining-operation hydraulic systems?
7. List a few fluid-power system properties.
8. What is the purpose of linear or rotary actuators in a fluid-power system?
9. List a few systems in wind turbines that use fluid power.
10. List several of the preventative-maintenance recommendations for hydraulic systems.
11. Sketch a simple hydraulic circuit using ANSI symbols that include the following: reservoir, pump, pressure-relief valve, four-way, two-position solenoid valve, double-acting cylinder, and spring-loaded accumulator. The four-way valve will be used to reverse the direction of travel for the double-acting cylinder, and the accumulator will ensure that the cylinder does not creep once in the extended position. The pressure-relief valve will ensure the system does not exceed application requirements. Refer to figure 5-14 for examples of these hydraulic components.

6

Bolting Practices

KEY TERMS

American National Standards Institute (ANSI)
American Society for Testing and Materials (ASTM)
American Society of Mechanical Engineers (ASME)
antiseize lubricants
bolt
compression
cross threading
elastic limit (E)
fastener
galvanizing
grade
head
high-pressure fluid
Hooke's law
hydraulic cylinder
International Organization for Standardization (ISO)
metal coatings
nut
pitch
power-bolting tools
proportional limit (P)
pump unit
reaction arm
reaction point

rock anchor rods
rupture strength
screw
shank
Society of Automotive Engineers (SAE)
star pattern
strain (ε)
stress (σ)
stud
tension
tension tools
thread
thread gauge
threads per inch (TPI)
torque multiplier
torque wrench
tower foundation rods
ultimate strength
Unified National Coarse (UNC)
Unified National Fine (UNF)
washer
yield strength
Young's modulus (modulus of elasticity)

OBJECTIVES

After reading this chapter and completing the review questions, you should be able to:

- Identify types and grades of bolts.
- Explain the mechanical strength differences between bolt grades.
- Explain the importance of using the proper bolt for an application.
- Explain the mechanics of bolted joints.
- Explain the use of coatings on bolts.
- Explain the use of lubricants in bolted joints.
- Identify the tools used for bolting applications.

- Describe the proper use of bolting tools.
- Explain the use of manufacturer tables in setting up tensioning and torque tools.
- Describe what happens to a bolt or stud when the torque or tension exceeds the maximum recommended level.
- Describe common practices in bolting maintenance.
- Recognize the safety hazards associated with using bolting tools.

INTRODUCTION

Fastening components together with bolts is not a new concept. Service technicians have been doing torque inspections on boilers, heavy machinery, and aircraft for many years. This chapter will develop a practical guide to bolting fundamentals related to the wind-power industry. The text will present fastener concepts, applications, and bolting equipment used on wind-turbine generators. The bolting-equipment section will introduce tooling for both torque inspection and tensioning. Safe operation and work practices with hand tools and power equipment is emphasized throughout the text.

FASTENER DESIGN

The term **fastener** refers to a general group of mechanical joining components that includes pins, rivets, screws, studs, and bolts. This chapter's discussion expands on threaded fasteners which include: screws, bolts, and studs. A **screw** or **bolt** is a threaded fastener with three sections in its design. Fastener configurations that technicians may encounter while working on or around wind turbines include hex bolts, machine screws, sheet metal screws (SMSs), self-drilling screws SMSs, carriage bolts, lag bolts, set screws, socket screws, and shoulder bolts. The bolt or screw **head** may be hex, square, spline, or a variety of cap configurations, depending on the application. The **shank** is the section below the head that is not threaded. Its length will vary, depending on the diameter and overall length of the bolt. In the case of a bolt 1 in. in diameter and 5 in. long, the threaded portion is typically slightly more than 2 inches, so the shank section would be a couple of inches long. Minimizing the threaded portion of the bolt will improve its strength and reduce the overall manufacturing cost. Consult your local fastener supplier for details on products they have available.

The threaded section of the bolt is used to engage a nut or internal thread in the mating component. The **thread** wraps around the bolt to form a helix with a cross section that resembles a symmetric V. **Figure 6-1** shows these bolt features. The thread helix may be left- or a right-hand lay, depending on the application. A left-hand lay has the designation "LH" on the end of the thread description. A right-hand lay is typically encountered with threaded fasteners

in the wind industry. A left-hand lay is used in applications where the rotation of the assembly may loosen and turn the fastener out of the joint. Thread count is specified by two methods: (1) number of **threads per inch (TPI)** or (2) the distance between thread peaks, or *pitch*. The thread categories are further broken down into two series: coarse [**Unified National Coarse (UNC)**] and fine [**Unified National Fine (UNF)**]. Coarse threads are used in most industrial applications where rapid assembly is necessary. Coarse threads also work well with fasteners that are made from soft metals and plastics. They are tolerant to adverse environmental conditions and can be easily assembled with minimal chance of cross threading. **Cross threading** is a condition in which the threads of the nut and bolt do not align properly; the threads jam and become damaged. Fine threads are used for applications that require greater strength and where there may be limited thread engagement. Fine threads are not typically exposed to harsh environmental conditions or contaminants.

Thread Standards

The **American National Standards Institute (ANSI)** and the **American Society of Mechanical Engineers (ASME)** standard B1.1 specifies thread count as threads per inch for American series fasteners. An example of TPI would be 1/2–13 UNC for a coarse thread and 1/2–20 UNF for a fine-thread series. The 1/2 refers to the nominal bolt diameter of 0.500 inches, and the 13 and 20 respectively refer to the number of threads per inch. The ANSI/ASME standard B1.13M and the **International Organization for Standardization (ISO)**

Unified National Coarse (UNC)	Unified National Fine (UNF)
6–32	6–40
8–32	8–36
10–24	10–32
1/4–20	1/4–28
5/16–18	5/16–24
3/8–16	3/8–24
7/16–14	7/16–20
1/2–13	1/2–20
9/16–12	9/16–18
5/8–11	5/8–18
3/4–10	3/4–16
7/8–9	7/8–14
1–8	1–14
1 1/8–7	1 1/8–12
1 1/4–7	1 1/4–12
1 3/8–6	1 3/8–12
1 1/2–6	1 1/2–12
1 3/4–5	
2–4.5	

© Cengage Learning 2013

FIGURE 6-2 *Examples of unified national thread series*

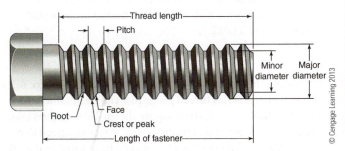

FIGURE 6-1 *Bolt features*

© Cengage Learning 2013

standard 898-1 set similar standards for the metric series of external threads. This series specifies the distance between thread peaks as **pitch**. An example of a metric thread specification would be M12–1.75 coarse and M12–1.25 fine. M12 is the nominal bolt diameter of 12 millimeters (mm), and 1.75 and 1.25 respectively refer to the distance between thread peaks in millimeters. **Figure 6-2** is a chart of the Unified National thread series, and **Figure 6-3** is a chart of the metric thread series. Comparison of U.S. customary units and International System (SI) metric units may be accomplished by a simple calculation knowing there are 25.4 mm in one inch. For example, 5 mm converted to inches would be: 5 mm × (1 in/25.4 mm) = 0.196 in.

The American thread and ISO series charts are valuable tools, but when presented with a fastener in the field you can measure the major diameter and use a **thread gauge** to

Nominal Size (mm)	Pitch (mm)
4	0.7
5	0.8
6	1
8	1
8	1.25
10	0.75
10	1.25
10	1.5
12	1
12	1.25
12	1.75
14	1.25
14	1.5
14	2
16	1.5
16	2
18	1.5
20	1
20	1.5
20	2.5
22	2.5
24	3
24	2
30	3.5
30	2
36	4
36	3
42	4.5
42	3
48	5
48	3
56	5.5
56	4
64	6
64	4

© Cengage Learning 2013

FIGURE 6-3 *Examples of the metric thread series*

SAMPLE PROBLEM

Determine the American thread series of a sample fastener from its physical characteristics.

A technician measures a bolt with a dial caliper and determines the outside thread diameter to be 0.494" and the number of threads in a 1/4" length to be around 3. **Figure 6-4** shows the dial-caliper measurement of the thread outside diameter. What is the thread series of this fastener?

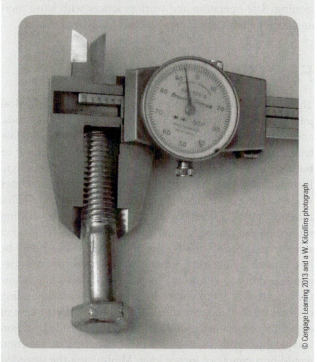

© Cengage Learning 2013 and a W. Kilcollins photograph

FIGURE 6-4 *Dial-caliper measurement of a thread's outside diameter*

SOLUTION

The nominal diameter of a 7/16" fastener is 0.438"; a 1/2" fastener is 0.500", and a 9/16" fastener is 0.563". From the measured outside thread diameter, the fastener is closest to a nominal 1/2" diameter. Now multiply the number of threads in the 1/4" length by 4 to determine the TPI for the sample fastener (3 × 4 = 12). The standard TPI series for a 1/2"-diameter bolt would be 13 UNC or 20 UNF. The closest TPI for our sample is 13, so the American thread series would be a 1/2–13 UNC.

determine the thread series you need to replace. **Figure 6-5** shows a metric series thread gauge that can be obtained from most hardware or industrial supply shops. These gauges take the guesswork out of determining what series you need to replace. Every technician's tool bag should have a dial or digital caliper and thread gauges.

FIGURE 6-5 *Metric series thread gauge*

Studs used to connect the rotor assembly to the hub adapter

FIGURE 6-7 *An application using studs to mount a rotor assembly to a hub adapter*

OTHER BOLTING COMPONENTS

A **stud** is a threaded fastener without a head. It may or may not have a shank section, depending on its length and thread-engagement requirement. **Figure 6-6** shows a couple of typical stud configurations, and **Figure 6-7** shows an application that uses studs to mount the rotor assembly to the hub adapter. Studs may also be used to mount blades to pitch bearings, generator subframes to bed plates, and gearbox pillow blocks to bed plates.

A **nut** is a mating component for a bolt or stud. It may come in several different configurations, but it typically has a hex profile. Another component typically used with a bolt or stud is a metal disc with a hole through it known as a **washer**. A flat washer is used to distribute the compressive force of the bolt and nut over the face of the joined components. Another variation of a washer that may be formed as a split ring is referred to as a *lock washer*. Lock washers come in a variety of forms with a main purpose of preventing the nut from loosening. **Figure 6-8** shows an example of a bolted joint including washers and a nut to complete the assembly. When the bolt and nut are tightened to the recommended torque level, the bolt acts like a spring in tension: its reaction force

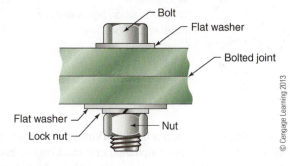

Bolt
Flat washer
Bolted joint
Flat washer
Lock nut
Nut

FIGURE 6-8 *A bolted joint*

on the joint clamps or compresses the components together. These forces are discussed later in the chapter.

FASTENER MATERIALS

Bolts and other fastener components are typically made of low- to medium-carbon steel. The strength of the bolt can be adjusted by the amount of carbon added to the iron during the steel-manufacturing process. Increasing the amount of carbon in the steel will increase its strength. Other methods of increasing steel strength are cold working, heat-treating, and adding other elements during the steel-manufacturing process. The classification given to a fastener's relative strength is considered its **grade**. The **Society of Automotive Engineers (SAE)** and the **American Society for Testing and Materials (ASTM)** are two U.S. organizations that set domestic standards for American thread-series fastener strength and the respective head marking to indicate the grade. **CAUTION: There are inferior strength fasteners available on the market today that appear to meet legitimate standards. If the price is too good to be true, then the fasteners probably do not meet the standards. Ensure the fasteners you use come from a reputable source. Personal safety and equipment longevity could be at risk.**

Examples of threaded stud configurations

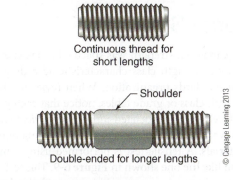

Continuous thread for short lengths

Shoulder

Double-ended for longer lengths

FIGURE 6-6 *Typical stud configurations*

TABLE 6-1

Comparison of SAE Grades and Strength Characteristics

Head Marking	Grade of Material	Nominal Size Range (in.)	Minimum Yield Strength (psi)	Minimum Tensile Strength (psi)
No markings	SAE grade 2 Low- or medium-carbon steel	1/4–3/4 More than 3/4 to 1½	57,000 36,000	74,000 60,000
radial lines	SAE grade 5 Medium-carbon steel quenched and tempered	1/4 to 1 More than 1 to 1½	92,000 81,000	120,000 105,000
6 radial lines	SAE grade 8 Medium-carbon boron steel or medium-carbon steel	1/4 to 1½	130,000	150,000

© Cengage Learning 2013

American thread-series head markings for fasteners include raised radial lines or alphanumeric markings or both. SAE standard J429 defines fastener-strength grades between 0 and 8 with raised radial lines. Grade 0 has a relative low strength, and grade 8 is the strongest. **Table 6-1** gives examples of SAE strength standards. The raised radial line scheme resembles spokes on a wheel. To read the marking scheme, count the number of raised radial lines and add 2 to the value. For example, a bolt head with three raised radial lines indicates the bolt meets the requirements of SAE grade 5. An ASTM example would be raised alphanumeric characters to indicate fastener strength such as the marking "A490" on the head of a bolt. The A490 indicates the bolt meets the ASTM specification of A490 type 1 medium-carbon alloy steel that has been heat-treated. If the alphanumeric symbol includes

a raised underline (i.e., "<u>A490</u>"), this marking indicates that the bolt is made from a stainless-steel alloy and meets the ASTM specification of A490 type 3 atmospheric corrosion-resistant steel that has been heat-treated. ISO standard 898-1, SAE J1199, and ASTM F568M are standards that specify metric thread-series fastener strength and their respective head markings (American Fastener, 2010). **Table 6-2** gives examples of ISO metric fastener strength information. Equivalent strength comparison between SAE and ISO standards is as follows:

SAE Grade	ISO Class
2	4.6 – 5.8
5	8.8
8	10.9

SAMPLE PROBLEM

Determine the grade specification of a sample bolt that has <u>A325</u> marked on the head.

SOLUTION

This bolt is not intended to meet the SAE grade specification because it is not marked with raised radial lines on the head. The markings on this bolt head include an alphanumeric symbol that is underlined. The underline indicates that the bolt meets the ASTM specification of A325 type 3, which is a stainless-steel alloy that has been heat-treated to increase the strength.

Material Strength

Previously, material strength was specified as fastener grade or listed by a strength class characteristic according to the respective standards organization. When reviewing the fastener strength class or grade tables, notice that each grade or strength level has an associated *ultimate* and *yield* strength. These two values are determined by tensile testing of the fastener materials. A tensile test is performed using a material test station like the one shown in **Figure 6-9**. This test station can apply an increasing compressive or tensile load and record the force (pounds or newtons) and displacement (stretch

TABLE 6-2

Comparison of ISO Grade and Strength Characteristics

Head Marking	Class of Material	Nominal size	Minimum Yield Strength (MPa)	Minimum Tensile Strength (MPa)
 8.8	Class 8.8 Medium-carbon steel, tempered by quenching	All sizes below 16 mm 16 mm–72 mm	640 660	800 830
10.9	Class 10.9 Medium-carbon boron steel or medium-carbon alloy steel, tempered by quenching	5 mm–100 mm	940	1,040
12.9	Class 12.9 Alloy steel, tempered by quenching	1.6 mm–100 mm	1,100	1,220

FIGURE 6-9 *Material test station*

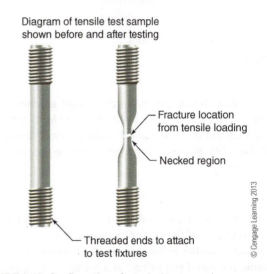

Diagram of tensile test sample shown before and after testing

Fracture location from tensile loading

Necked region

Threaded ends to attach to test fixtures

FIGURE 6-10 *Configuration of a typical test sample*

in inches or millimeters) of the sample during the test. **Figure 6-10** shows the configuration of a typical test sample. As the test is performed, the sample stretches as the force increases. This increase in length or elongation because of the test is relative to the strength of the material. For example, grade 2 (low-carbon steel) will stretch more than a grade 8 (heat-treated medium-carbon steel) fastener. Typically, the percent elongation at the failures of grades 2 and 8 are >18% and >10%, respectively. A material that stretches more than

5% before failure is known as a *ductile material*. Using the same example, a grade 2 fastener has an ultimate strength of 74 ksi (74,000 pounds/inch^2), and the grade 8 fastener has an ultimate strength of 150 ksi or nearly twice the strength of the grade 2 fastener. **Ultimate strength** is the maximum stress a material can withstand before failure. **Stress** (σ) is defined as the load (F) divided by the fastener cross-sectional area (A):

$$\sigma = \frac{F}{A}$$

These are great examples of the material strength, but how are these strength specifications related to bolting practices? The yield strength is another value listed in the fastener-grade or strength class-characteristic tables shown previously. This value is used to determine the fastener application

SAMPLE PROBLEM

Determine the stress on a grade 8.8, M20 × 2.5 fastener that is loaded in tension to 100 kilonewton (kN). Determine whether this load will create a permanent elongation or will return to its original length when the load is removed.

SOLUTION

The M20 × 2.5 fastener has a minor diameter measured at 16.6 mm (Figure 6-4). Fastener minor diameter is measured at the thread root as shown in Figure 6-1.

Calculate the cross-sectional area of the fastener and the corresponding stress created by the 100-kN load:

$$A = \pi r^2$$

$$= 3.141 \times \left(\frac{16.6 \text{ mm}}{2}\right)^2$$

$$= 216.4 \text{ mm}^2$$

$$\text{Stress }(\sigma) = \frac{\text{load}}{\text{area}}$$

$$= \frac{100,000 \text{ N}}{216.4 \text{ mm}^2}$$

$$= 462 \text{ N/mm}^2$$

$$1 \text{N/mm}^2 = 1 \text{MPa}$$

Therefore,

$$\sigma = 462 \text{ MPa}$$

Will the applied stress cause permanent elongation? The 100-kN load creates a stress of 462 MPa, which is less than the ISO grade 8.8 minimum yield strength of 640 MPa, so the fastener will not have permanent elongation.

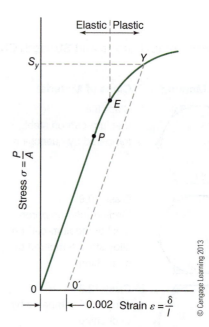

FIGURE 6-11 *Characteristic stress-versus-strain curve of a ductile metal*

and the torque or tension value to which the fastener will be tightened or loaded for the fastener assembly.

When a ductile metal is loaded in **tension** or in **compression**, the metal will behave as a spring up to a specified load. **Figure 6-11** shows the characteristic stress-versus-strain curve of a ductile metal. **Strain** (ε) is defined as the change in length (δ) divided by the original length (l): $\varepsilon = \delta/l$. The bottom left of the curve is the point of zero stress and strain. The curve moves up in a straight sloped line to the right as the load is applied to the metal. This slope ($E = \sigma/\varepsilon$) is characteristic for each ductile metal and is known as **Young's modulus** (or **modulus of elasticity**). The next point of interest along this line is the **proportional limit (P)**. Below this point, the material behaves according to the linear elastic theory or **Hooke's law**. This is also the point at which the line initially deviates from a straight line. The next point along the curve is the **elastic limit (E)**. Beyond this point, the metal will no longer return to its original length. It is said to have plastic deformation or a permanent offset when the load is removed. Yield strength is the next point of interest along

the stress-versus-strain curve. The **yield strength** value is determined by drawing a line that is parallel to the Young's modulus slope that intersects the stress-versus-strain curve and intersects the 0.002 strain value (0.2%) on the lower axis of the graph.

Continued loading of the sample will further increase the stress and strain until the ultimate strength is achieved. Beyond this point, the stress value will decrease until the sample finally reaches its **rupture strength**. This is the strain value where the material fails and breaks. **Figure 6-12** shows the entire stress-versus-strain curve for a typical ductile material. Noticeable differences between the low- and high-strength metal's stress-versus-strain curve would be the slope of the curve up to the proportional limit and the ultimate strain the metal could withstand before failure. The higher-strength metals would have a steeper slope and a lower strain value at failure.

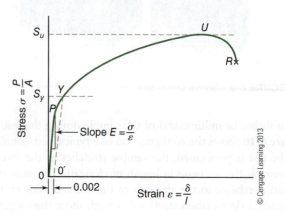

FIGURE 6-12 *Entire stress-versus-strain curve for a typical ductile material*

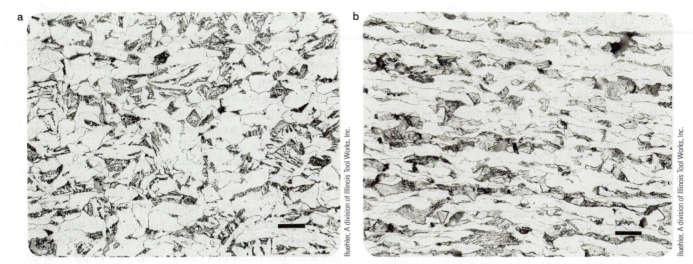

a   b

Buehler, A division of Illinois Tool Works, Inc.

Buehler, A division of Illinois Tool Works, Inc.

FIGURE 6-13 *Microsections of a carbon steel ball stud before (a) and after testing to failure (b)*

The appearance of the test sample shows the typical necking and failure point of the metal because of tensile testing, but what is the appearance of the metal structure because of the test? **Figure 6-13** shows microsections of a carbon steel ball stud before and after testing to failure. The microsection before failure shows a uniform microstructure; after failure shows that the microstructure is deformed along the axis of the tensile load. This deformation occurs throughout testing but becomes permanent after the metal exceeds the elastic limit. Why is this important? When you are in the field inspecting bolted assemblies, the torque or tension value used for the test is very important. If you exceed the manufacturer's specification for the applied torque, the bolt or stud may be compromised. Remember: the manufacturer is relying on the fastener to maintain a clamping force on the joint during any loading scenario that has been considered for the life of the fastener. When the fastener is exposed to a greater load, the permanent deformation of the fastener will reduce the clamping force on the joint. This would be the same as having a reduced torque value—or worse, a loose nut after the joint is exposed to vibration during operation of the wind turbine.

Consider the following. The blade bearing stud and nut assemblies are torqued to an excessive torque value with the reasoning that a higher torque value should be better than a lower value. The turbine is placed in service, and the joint is exposed to an extreme temperature cycle along with the normal loading and vibration. Depending on the amount of excessive torque, the nuts could loosen or the studs could crack and fail. This would make the front page of the local news! It would also be an expensive embarrassment to the company that provided the maintenance service to the wind turbine. Remember, more is not necessarily better.

The next section will apply information about loading scenarios, yield strength, and required torque or tension requirements of a joint.

Fastener Service Conditions

An appropriate fastener and joint design considers the loading scenario the equipment would typically see during operation. Wind turbines are exposed to several possible loading scenarios during operation, including temperature cycles, cyclic loading, and static loads. If you happen to be from the southern part of the United States or Europe, you may not realize the temperature extremes a wind turbine might encounter. The summer months expose the wind turbine to an upper temperature around 95 °F (35 °C), but in the winter months the turbine would be exposed to temperatures reaching down around −40 °F (−40 °C). If the turbine was assembled around 45 °F, this would be a shift of 50 °F during the summer. This temperature increase would stretch the bolts because of the metal expansion of the tower flange joints. The winter months will bring a 90 °F decrease in temperature that would reduce the tension on the bolts as the metals contract. This temperature cycling may loosen the bolted joint assembly if it did not have the proper torque or pretensioning value to maintain a minimum clamping force.

The blade-to-hub bearing joint along with tower joints would be a couple of examples where cyclic load fluctuations would be present during service. One load change on the blade-to-hub bearing joint would be created during rotation. The bolts would have an increase in tension as the blades rotate toward the ground and decrease as the blades reach top dead center of each rotation. Another load fluctuation on the blades would be the change in wind force during variable winds. This would increase the tension on the upwind blade bolts during a gust and decrease the tension on the bolts on the downwind side. This would be true for an upwind rotor wind turbine and be the opposite for a downwind rotor wind turbine.

The tower joints would be exposed to cyclic loading as the wind velocity changed. The bolts on the upwind side would have an increase in tension, and the downwind side

SAMPLE PROBLEM

A low-carbon steel flange joint is bolted together with a medium-carbon steel bolt and exposed to a 10 °C temperature increase. Determine the expansion of the flange joint and change in bolt tension because of the temperature change. The joint is made up of flanges 2–100 mm thick and is bolted together with an M25 bolt and nut assembly.

SOLUTION

Thickness of the 2–100 mm flange joint: 200 mm

Length between the head of the bolt and nut used to clamp the joint: 200 mm

Coefficient of thermal expansion: for low-carbon steel, 11.7×10^{-6} mm/(mm °C); for medium-carbon steel, 11.3×10^{-6} mm/(mm °C)

Expansion of the flange assembly: 200 mm $\times$ $(11.7 \times 10^{-6}/°C) \times 10$ °C = 0.0234 mm

Expansion of the bolt length: 200 mm $\times$ $(11.3 \times 10^{-6}/°C) \times 10$ °C = 0.0226 mm

Bolt expands 0.0008 mm less than the flange joint, so the joint will stretch the bolt.

What would the tension increase be on the bolt because of this stretch? The tension increase can be determined using the formula for change in elongation because of an applied force:

$$\delta = \frac{F(l)}{AE}$$

where F is the applied force (tension), l is the length in tension, A is the bolt cross-sectional area (490.9 mm² for an M25 bolt), and E is the modulus of elasticity for the bolt material. For medium-carbon steel,

$E = 207,000$ N/mm² (MPa) [1 N/mm² = 1 MPa]

Rearranging the equation to find force to create the 0.0008-mm elongation,

$$\delta = \frac{F(l)}{AE}$$

Therefore,

$$F = \frac{\delta AE}{l}$$

$$= \frac{0.0008 \text{ mm } (490.9 \text{ mm}^2)(207,000 \text{ N/mm}^2)}{200 \text{ mm}}$$

Force = 406.5 N

This is around 95 pounds force for those who think in American units.

would have a decrease. The wind is not always from the same direction, so the upwind and downwind sides will vary daily or with the season. This means the bolts in the tower joints and foundation will be constantly under cycling loads. The

next sample problem describes the variation in bolt tension due to differential temperature expansion between components in a bolted joint. The formula used for this sample problem is developed from the relationship between elongation and change in load $[\delta = F(l)/AE]$. With a cyclic loading scenario, there is a change in the amount of elongation with each change in the value of tension or compression. If an improperly tensioned fastener was exposed to cyclic loading, in time it would fracture and fail.

We now examine a bolted joint with a variation in loading from 0–5,000 lbs. Using a 1/2–13UNC stud that is 4 inches in length from the flange to the nut, the bolt cross-sectional area would be 0.129 in² and the modulus of elasticity (E) 30,000,000 lb/in². This fastener would experience 0.005 inches of change in length during each cycle. This does not seem like much of a change. This joint could cycle 20 times a minute, 1,200 an hour, and 10,512,000 a year. Get the picture? I'm sure you have had the chance to flex a piece of steel wire or strip before. Remember what happened to the metal as you continued to flex? It got hot and eventually fractured. The same could happen to the fastener in this example, not a desirable outcome for a $2 million piece of equipment. To prevent this from happening, ensure the fasteners are tensioned in excess of any loading that they may encounter during service. If this fastener had a pretension of 5,100 lbs, then it would not experience a cyclic change in length. The fastener would always remain at the initial elongation, and the joint would remain under a compressive load.

Fastener-Tensioning Considerations

These examples show that proper tensioning of fasteners is very important to ensure continuous service of equipment. How do we determine the proper tension or torque for a fastener? For maintenance applications, the information should be provided to the technician in the form of a maintenance procedure or manual. If these are not available, then consult reputable sources such as the SAE, ASTM, or ISO standards. These standards take into consideration grade, surface finish, and size.

Considerations

Fastener Grade. Proof load of a fastener is determined by a percent of the material's yield strength. If the joint is to be reused, the proof load is typically 75% of the material yield strength. For a permanent assembly, the proof load value is typically 90% of the material yield strength. For example, a medium-carbon steel fastener with a yield strength of 130 MPa would use a proof load stress of $S_p = 0.90(130 \text{ MPa}) = 117$ MPa for a permanent assembly. This stress value ensures a predetermined strain offset so the material will still act as a spring to provide a continuous clamping force. An item to remember in the field: never reuse fasteners. If the service activity requires components be disassembled, discard the fasteners—nut, washers, bolt, or stud—and replace them with new ones. These items have been deformed during the tensioning or torque process of assembly and will not perform the same. The possible failure of a reused fastener is not worth the perceived short-term savings.

Fastener Surface Finish. Fastener surface finish takes into account any coatings or lubricants that may be present that would change the friction characteristics of the thread surfaces. Torque values should be reduced for lubricated and plated fasteners as compared to bare, dry fasteners. Hot-dip galvanized surfaces require a higher torque value. Lubricants such as Never-Seez or other **antiseize lubricants** that are rich in molybdenum disulfide are used on fasteners to prevent galling or seizure of the threads. Stainless-steel bolts and nuts have a tendency to seize or lock together if there are any surface irregularities on the mating threads.

The addition of **metal coatings** such as zinc, nickel, chromium, and cadmium plating or the use of hot-dip **galvanizing** will prevent or slow down the normal corrosion process that occurs when carbon steel is exposed to the environment. The carbon steel components will quickly turn to the characteristic rust (iron oxide) appearance if they are not protected. The rough oxide finish will damage the exposed threads and make it difficult to take them apart in the future. The addition of these protective materials will also change the friction characteristics of the threads and must be considered when determining the required torque for a fastener.

Fastener Size. The proof load stress discussed previously can be used to calculate required tension for a specific fastener size. For example, an M12 medium-carbon steel fastener with a cross-sectional area around 113 mm² would have a tension of 13,221 N:

$$F = S_p \times A$$

where F is the fastener tension, S_p is the proof strength (MPa), A is the cross-sectional area withstanding the load (mm²):

$$F = (117 \text{ N/mm}^2) \times 113 \text{ mm}^2$$
$$= 13{,}221 \text{ N } (2{,}972 \text{ pounds-force})$$

Consult appropriate equipment manufacturer's specifications before proceeding with any maintenance activity.

Torque Calculation

Several Web sites present industry standards on proper tensioning and equivalent torque values that are required for commercially available fasteners. These Web sites use the formula $T = KDF$ for tightening torque in inch-pounds and $T = KDF/12$ for tightening torque in foot-pounds, where T is the calculated tightening torque, K is the dimensionless torque coefficient that encompasses variables such as surface finish and lubrication, D is the nominal diameter of the

SAMPLE PROBLEM

Determine the torque requirement for an assembly that uses an SAE grade 8, 1–8UNC hot-dip galvanized stud. The suggested clamp force for the joint is around 54,500 lbs.

SOLUTION

Fastener nominal diameter is 1.00 inch (D), torque coefficient for a hot-dip galvanized stud is 0.25 (K), and clamp force is 54,500 lbs (Morbark, 2011).

$$\text{Torque} = \frac{KDF}{12}$$
$$= \frac{0.25(1.00 \text{ inch})(54{,}500 \text{ lbs})}{\left(\frac{12 \text{ inches}}{1 \text{ foot}}\right)}$$
$$= 1{,}135 \text{ foot-lbs}$$

fastener in inches, and F is the required clamping force in pounds. Values of K include:

0.10 for a waxed surface,

0.15 for a lubricated surface,

0.18 for a dry zinc-plated surface,

0.20 for a slightly oily plain-steel surface, and

0.25 for a hot-dip galvanized surface.

This formula will work for metric units as well, with T as the desired tightening torque in newton meters, D as the bolt nominal diameter in meters, and F as the required clamping force in newtons. Appendix F shows examples of reference torque values for SAE and ASTM grade bolts.

TORQUE TOOLS

Torque tools come in a variety of configurations, sizes, and ranges depending on the application. A **torque wrench** can be set to a user defined torque value to match specific fastener requirements. When the user achieves the defined torque value, the wrench will indicate the value has been achieved by use of a numeric readout or a mechanical snap action. **Figure 6-14** shows examples of manual torque tools. Some applications may only require a hundred inch-pounds or less. A small torque wrench or torque screwdriver with a dial or digital readout would be a good choice for these applications. Some applications may require several hundred foot-pounds of torque. A three foot long torque wrench with a range of 100 to 600 foot-pounds may be a suitable choice for these applications. In another application, a technician may need to torque bolts connecting wind-turbine tower sections together during assembly or during a maintenance procedure. This bolting application may require a thousand foot-pounds of torque or more. An application like this would require a nine foot long wrench and a torque range of 100 to 2,000 foot-pounds.

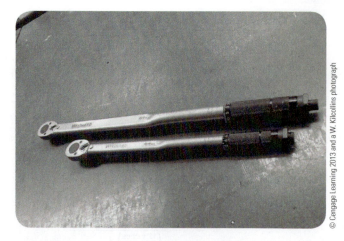

FIGURE 6-14 *Examples of manual torque tools*

This size wrench may be fine for an industrial setting, but very difficult to maneuver within a wind-turbine nacelle or tower section. **Figure 6-15** shows a variety of power torque tools. Technicians should choose a power-bolting tool that uses an energy source such as electricity, pneumatics, or hydraulics for this application. As you can see, there are as many variations in torque tools as there are fastener applications. This section will highlight considerations such as proper tool size, safety, operation, calibration, and care.

Tool Choice

Choosing the correct torque tool for the maintenance procedure is important. It would be inappropriate to use a torque wrench with a range of 10 to 125 foot-pounds to tighten a fastener requiring only 10 foot-pounds (120 inch-pounds). The wrench may even be physically too large and cumbersome to handle if the fastener was located in a control cabinet. A torque screwdriver or a small torque wrench with a working range of 50 to 250 inch-pounds would be a better choice. The general rule is to use a tool that allows the specified torque value to fall between 10% and 90% of its working range. This ensures the tool working range is a good match for the required procedure. The 120 inch-pounds specified for this procedure would fall within the tool's 50 to 250 inch-pound working range. Using a manual torque wrench to tighten a

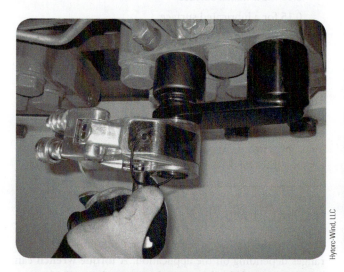

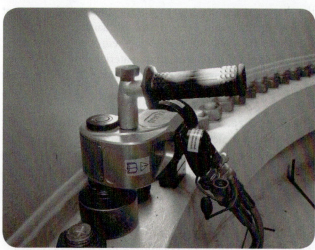

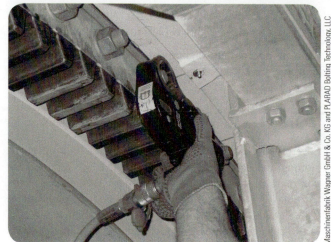

FIGURE 6-15 *A variety of power torque tools*

FIGURE 6-16 *Hydraulic torque tool with its reaction arm contacting an adjacent fastener*

nut with a 1,750 foot-pounds torque requirement would also be inappropriate unless you were using a **torque multiplier** to increase its mechanical advantage. Torque multipliers will be discussed later in this chapter. The appropriate tool for a procedure requiring 1,750 foot-pounds of torque would be a power-bolting tool using a power unit for the energy source. These tools are relatively compact in size and designed to apply high torque values without added physical stress on the operator. The tools are designed with a **reaction arm** that contacts an adjacent bolt head, nut, or structural component.

SAMPLE PROBLEM

Determine the force necessary to achieve 500 foot-pounds of torque to tighten a nut. The wrench for this application has an overall length of 70 inches and the handgrip location is 64 inches from the centerline of the nut.

SOLUTION

The torque requirement for this procedure is 500 foot-pounds or 500 pounds applied at a distance of 1 foot from the centerline of the nut. The technician has a handgrip location that is 64 inches from the centerline of the nut:

$$64 \text{ inches} \times \left(\frac{1 \text{ foot}}{12 \text{ inches}}\right) = 5.33 \text{ feet}$$

$$\frac{(500 \text{ foot-pounds})}{5.33 \text{ feet}} = 93.75 \text{ pounds}$$

The wrench operator has to apply around 94 pounds at the handle grip to achieve 500 foot-pounds of torque at the nut.

Applying this amount of torque to several fasteners during a maintenance procedure, a technician will gain an appreciation for power tools.

Figure 6-16 shows a hydraulic torque tool with the reaction arm contacting an adjacent fastener. The reaction force being applied to adjacent components reduces the operator's stress, but when not positioned properly it can create a safety hazard. Care must be taken when choosing your reaction point. The **reaction point** has an equal and opposite force to that being applied to the fastener. If the torque tool is applying 1,750 foot-pounds of torque to a 6-inch reaction arm, then the reaction point will be supporting 3,500 pounds of force. Example: 0.5 feet × 3,500 pounds = 1,750 foot-pounds. If the reaction point does not provide adequate support for the reaction arm, then the tool may break the reaction point or the tool may slide off the nut during operation. Either situation can cause serious injury to the operator or damage to the equipment.

CAUTION: **Ensure the tool operator has no body parts between the reaction arm and reaction point during tool operation. Most of these tools require a complete operation cycle before they can be released. If a body part becomes caught between the reaction arm and support structure, it will be crushed. Power tools make life easier, but the operator must have a thorough understanding of equipment safety practices and proper operation techniques. Required operation information and specific safety practices are listed in the tool manufacturer's manuals. Read and follow these guidelines to ensure safe and trouble-free tool operation.**

Safety

Safety is an important part of any job. Understanding what hazards exist and knowing how to prevent them are the first priority of any job. Placing job deadlines ahead of safety practices or not using appropriate personal protective equipment (PPE) is always unacceptable. When using manual or **power-bolting tools**, the operator should have a working knowledge of tool operation or be working under the direct supervision of a co-worker or supervisor who is providing instruction.

Safety Considerations

Daily safety discussions (i.e., tailgate meetings) before each job are a good practice. During the meeting, ensure the appropriate working procedures, technical manuals, and job-safety analysis (JSA) forms are available and reviewed. Once the team is up a tower or unloading equipment, it is difficult to maintain everyone's attention. Running a safety meeting after assignments are handed out provides reinforcement that safety is a priority.

Inspect equipment before heading to the job site. This is the time when problems can be caught and corrected. Finding a problem in the field causes delays and may create an attitude that the operator can cut corners to save time. This is the way to set up for an accident. Use preventative measures and plan ahead.

Equipment Inspection

Many different tool models and suppliers are available for bolting operations, so the inspection guidelines presented here are generic. For specific guidelines, consult the manufacturer's manual supplied with each tool.

Manual Tools

Manual tools are used every day on a job site. Proper selection, use, and care of these tools is important to complete a task safely and efficiently. Guidelines for using manual tools include the following:

- Ensure that the appropriate material safety data sheet (MSDS) is available and reviewed for each chemical to be used.
- Ensure appropriate PPE is available as prescribed in the MSDS and job procedures.
- Inspect for any defects that will create a safety hazard or cause the tool to not perform as designed.
- Ensure the right tool will be used for the job. Using a tool for the wrong application will damage equipment and create a safety hazard. Consult the job procedure for an approved tool list.
- Ensure all components are in the tool case. Not having all the components is the same as not having the tool available. Do not improvise. Plan ahead.
- Has the tool been calibrated? If the tool is out of calibration, then the quality of the job is compromised. Doing the job right the first time prevents having to do it over.
- Are required safety tags missing? If OSHA-required inspection tags are damaged or missing, the equipment is not available for use. Tag the equipment as "Nonconforming" or give it to a designated individual for disposal and request conforming equipment.

Power-Bolting Tools

Many items are the same as for manual tools, but the following are specific to most power tools:

- Ensure that hydraulic oil reservoirs are filled to appropriate levels and are not contaminated.
- Ensure that a backup supply of appropriate hydraulic fluid and any other lubrication oils are available.
- Ensure that spill cleanup supplies and appropriate disposal containers are available.
- Are any of the hydraulic lines, air hoses, or power cords damaged? Inspect for cracks, cuts, appropriate pressure ratings, damaged insulation, and exposed conductors. Tag these items as "Nonconforming" or give them to a designated individual for disposal and request conforming equipment.
- Is there damage to gauges? Missing lenses or visible damage to an assembly makes the equipment inoperable. It will be tough to read a gauge if the indicator needle or digital readout is damaged or missing.

- Is there damage to valves, fittings, or electronics? If the unit can be repaired, do so. If not, tag it as "Nonconforming" and retrieve conforming equipment. If no other unit is available, request another assignment.
- Ensure that safety guards are available and operate properly per the manufacturer's instructions.
- Ensure rated tool bits, adapters, and sockets are available. Improper tool components may break and create a safety hazard for the operator and damage the equipment.
- Ensure operation manuals are available and reviewed before operation.
- Ensure operator training is current for equipment to be used.

These are some considerations that apply to the use of manual and power-bolting tools. Consult the manufacturer's information, the safety coordinator, or the supervisor for specific details and available information. Proper operation is as important as the safety inspection. The tools may be safe but when used improperly create hazards to personnel and equipment.

Torque Tool Operation

There are many styles and sizes of torque tools available on the market so a general overview will be discussed. For specific tool requirements and operation, consult available manufacturer's manuals. Manual torque tools are available as wrench and screwdriver designs which one you choose will depend on the job requirement.

Manual Tools

Manual torque wrenches are available as beam-style, dial, and micrometer adjustments. Beam, dial, or digital readout styles are designed to read the applied torque on the gauge.

- Determine the torque requirement for the procedure.
- Choose the appropriately sized torque tool and socket or tool bit.
- Secure the socket or tool bit to the torque tool as recommended by manufacturer.
- Place the tool on the nut or screw head as required.
- Using a smooth application of force, tighten the fastener until the desired torque is achieved. Do not exceed the torque value, or you may damage the fastener, joint assembly, or torque tool. This is especially important for applications with copper, brass, or aluminum fasteners. These materials are relatively soft and break easily.

Micrometer-style torque wrenches or screwdrivers can be adjusted to a predetermined level of torque. When the predetermined torque level is achieved, the wrench will make a snap sound or produce a break away motion to indicate the torque value is achieved.

- Determine the torque requirement for the procedure.
- Unlock the micrometer barrel and turn the barrel to the required torque level. The lock is located on the end of

FIGURE 6-17 *Micrometer-adjustment torque wrench*

FIGURE 6-18 *Torque multiplier torquing tower flange bolts*

the barrel assembly. **Figure 6-17** shows a micrometer-adjustment torque wrench barrel assembly. The barrel is marked with the minor divisions of the wrench, and the shaft of the wrench handle has the major divisions. Always adjust up to the torque value to ensure a proper setting. If the specified value is exceeded, reduce the setting to well below the specified setting and work back up to the value.

- Example: Adjust a 10- to 125-foot-pound torque wrench to 95 foot-pounds. The barrel is marked with 1-foot-pound increments, and the shaft has 10-foot-pound increments. Rotate the barrel until the top edge lines up with the 90-foot-pound division on the shaft and the barrel increment 0 lines up with the shaft center line. Continue to turn the barrel past 1, 2, and so on until the 5 lines up with the shaft center line. The wrench is now set at 95 foot-pounds.
- Lock the barrel before proceeding to torque the fastener(s).
- Using a smooth application of force, tighten the fastener until the desired torque is achieved. The wrench will produce a snap sound when the preset torque value is achieved.
- When finished with the torque wrench, unlock and set the torque value back to the lowest value on the shaft. The torque adjustment spring may be damaged when stored under a load.

Torque nultipliers are used with a manual torque wrench to achieve higher values of torque. A torque multiplier is essentially a gear box. Recall that a gearbox is used to convert high-torque, low-RPM input to a low-torque, high-RPM output. **Figure 6-18** shows an image of a torque multiplier being used to torque tower flange bolts. The gear ratio for this model is 1:13.6, so for every 1 foot-pound of torque input the multiplier will supply 13.6 foot-pounds of output. The steps to use the torque wrench with the torque multiplier are the same as discussed previously. An additional step required for this procedure is to divide the required torque value by 13.6 to get the torque wrench setting.

Here is an example. The fastener requires a torque value of 1,360 foot-pounds. What is the torque wrench setting? 1,360 foot-pounds/13.6 = 100 foot-pounds. The ratio works

with metric units of torque as well as the American units. For example, an ISO-grade fastener torque requirement of 2,200 N-m would require 162 N-m set on the torque wrench:

$$\frac{2,200 \text{ N-m}}{13.6} = 161.76 \text{ N-m}$$

Always refer to the manufacturer's instructions for the specific ratio of each torque multiplier model used.

SAMPLE PROBLEM

Torque an SAE grade 8, 1–1/4 7UNC bolt assembly to 1,975 foot-pounds. Available torque wrench has a range of 10–125 foot-pounds, and the torque multiplier has a ratio of 1:13.6. Is this an appropriately sized torque wrench for the procedure?

SOLUTION

Determine the torque wrench setting. Torque multiplier has a ratio of 1:13.6:

$$\frac{1,975 \text{ foot-pounds}}{13.6} = 145.2 \text{ foot-pounds}$$

The torque wrench setting should be 145 foot-pounds. Is this a reasonable torque level for the available torque wrench? No. The maximum torque value for this wrench is 125 foot-pounds, and the recommended use would be the 10% to 90% working range from the previous discussion:

$$0.90(125 \text{ foot-pounds}) = 112.5 \text{ foot-pounds}$$

Exceeding the working range is pushing the limit of tool accuracy and exceeding the maximum torque limit will damage the wrench. A better choice for this procedure is a 20- to 250-foot-pound wrench.

$$0.90(250 \text{ foot-pounds}) = 225 \text{ foot-pounds}$$

Thus, 145 foot-pounds is less than the 225-foot-pound working value.

Power Tools

Using a large torque wrench that may be 5 feet long and weigh 50 pounds would be physically fatiguing after several hours. A 10% inspection for maintenance of a tube tower would require inspecting between 10 and 12 bolts for each tower section. These bolt assemblies are around waist height. How about bolts positioned over your head? Yaw bearing bolts in the top tower section are located above your head. Holding a 50-pound wrench in this posture becomes very difficult. An alternative to carrying a 50-pound wrench would be a power-bolting tool. A pneumatic or hydraulic power wrench may weigh around 8 pounds, depending on the style or rated capacity. The lighter weight and compact size of the power bolting tool make these jobs easier. These power wrenches use a **pump unit** to supply pressurized fluid through hoses to the tool. The **high-pressure fluid** does the work, so the technician only needs to hold the tool in place during the tool operation. The pump unit for some hydraulic torque wrenches is capable of supplying as much as 10,000 psi pressure [690 ($\times$100) KPa]. The previous sample problem discussed using a manual torque wrench and torque multiplier to produce 1,975 foot-pounds of torque. One of these power wrenches could produce 1,975 foot-pounds of torque with 6,200-psi pressurized hydraulic fluid. The following section shows a generic procedure to operate a fluid-powered wrench. Consult the specific manufacturer's procedure for the wrench used at your site. Examples of fluid-power wrench manufacturers are Norbar pneumatic tools and PLARAD hydraulic tools.

The following is the generic procedure for the use of a pneumatic or hydraulic wrench:

- Determine the required torque level for the fastener. Use the wind-turbine manufacturer's procedure or industry standards for recommended torque values for each size and grade of fastener.
- Choose the appropriate tool and socket for the procedure. **WARNING: Power-bolting tools require sockets and tool bits that would be appropriate for impact wrenches. Forged sockets that are used for manual socket sets will shatter under the loads applied by power wrenches.**
- Consult the pressure and torque conversion chart supplied by the tool manufacturer. Refer to Appendix G for a sample chart.
  - Connect the pressure lines to the tool and pump assembly. **CAUTION: Connections should be only finger-tight to ensure components are not damaged and may be disassembled after use.**
  - Mount the socket to the tool with an approved locking-pin assembly. Consult manufacturer's instructions for an approved locking pin.
  - Turn on the pump assembly and adjust the pressure to the appropriate level. The previous example used a pressure setting of 6,200 psi for a torque output of

1,993 foot-pounds (Hytorc 3MXT tool). If the tool does not index when the pump is activated, the hose connections may not be tight or there may be contamination in the connections. Turn off the pump, drain any line pressure, remove connections, inspect, clean if necessary, and reassemble.

- Set the torque wrench output. The tool will index each time the operator toggles the pump control. Place the wrench on a surface that will not allow anything to contact the tool bit or socket. Toggle and hold the pump control until the wrench turns to the internal hard stop. While holding the pump control on, unlock and adjust the pressure-control valve until the desired pressure is indicated on the gauge. If you exceed the required pressure value, decrease the pressure to below the target pressure value and readjust back up the target value. Lock the valve and release the pump-control toggle switch. Toggle and hold the pump-control switch and read the pressure gauge to ensure the setting did not change when the valve was locked. If it changed, unlock the valve and follow the previous step. This process prevents the wrench from exceeding the torque value required for the fastener.
- Place the wrench on the fastener for torque inspection.
- Ensure the reaction arm is positioned firmly against an adjacent bolt or a structure that is capable of withstanding the anticipated load. **WARNING: Do not place any part of your body between the reaction arm and the reaction point. An amputation or crushing injury will result. It is recommended that the operator control the wrench and pump unit to prevent injuries. Figure 6-19 shows a few reaction arm styles and supporting structures.**
- Toggle the pump control switch if necessary.
- Some wrenches will index to the internal hard stop if the fastener is not at the required torque value. Continue to toggle the pump-control switch until the fastener does not rotate to the internal hard stop. The fastener is now at the required torque value. Some power-tool manufacturers have an automatic setting on the control pendant that will continue to index the tool until the internal hard stop is not contacted.
- Move the wrench to the next fastener and continue until the job is completed.

Torque tools also come in an electric-drive–style that resembles a large electric drill. This style does not require a power unit like the pneumatic or hydraulic versions. The source of power is derived from an electric motor. The electric motor supplies high RPM and low torque to a gearbox assembly like the torque multiplier mentioned previously. The electric style also has a reaction arm attached to the tool to offset the torque output developed. One advantage of the electric-style tool is that it can supply a continuous rotation

FIGURE 6-19 *A few reaction arm styles and supporting structures*

FIGURE 6-20 *An example of an electric bolting tool*

to the tool bit to assemble a fastener. **Figure 6-20** shows an example of an electric bolting tool. The pneumatic and hydraulic versions may only turn a portion of a turn or step as they rotate. A drawback of an electric-style nut driver is its inability to supply a higher level of torque output compared to hydraulic designs.

Calibration

Manufacturers provide a chart that lists a pump pressure setting for each torque value output by tool model number. These settings are very close for a new tool and pump assembly. Most wind-turbine equipment manufacturers require a tool calibration each time the torque wrench is set up for use. This procedure eliminates any guesswork on the part of the technician. Typically, the torque wrench is calibrated at the start of each day or whenever a new torque value is required for a different tool bit. **Figure 6-21** shows a calibration stand and control unit.

The calibration procedure is similar to the generic operation process discussed previously, except that the pump pressure value would be adjusted to generate the desired output torque shown on the calibration stand and not the pressure value listed in the manufacturer's chart. For example, the pump pressure may be set to 6,100 psi or 6,300 psi instead of the manufacturer's listed 6,200 psi for an output torque of 1,993 foot-pounds.

Calibration is a good practice. A pump may be set to the required pressure value, but the setting may be incorrect for a particular tool. What if the pump pressure is set for tool A to get 1,993 foot-pounds output, but tool B is the tool in use? Tool B may produce 2,250 foot-pounds with this same pump pressure setting. This error may possibly break the bolts or equipment. Even worse, the fastener may be overtensioned and create a delayed failure because the yield strength was exceeded.

FIGURE 6-21 *A calibration stand and control unit readout*

TENSIONING TOOL APPLICATIONS

In some cases, fasteners are tightened directly with power **tension tools** instead of torque tools to set the tension. Typical locations for direct fastener tensioning would be on wind-tower foundation rods or **rock anchor rods**. **Tower foundation rods** extend up through the concrete foundation assembly and are used to secure the tower base section. These foundation rods are threaded into an embedment ring that is secured within the concrete foundation subframe

FIGURE 6-22 *Examples of tower foundation rods*

with a rebar assembly. This allows the foundation rods to be tensioned to produce a compressive load between the tower base flange and the concrete foundation. Typical foundation rods have a 60° V profile thread or rounded threads such as Williams's all-thread rebar. **Figure 6-22** shows examples of tower foundation rods. This author has seen a variety of all-thread rods used with several wind-turbine foundation designs. All-thread rod may be used in a variety of civil engineering applications, so it is no surprise that it is used as a structural connection between the tower and foundation assembly. Another use for the all-thread rebar is with foundation rock anchor assemblies.

Rock Anchors

The typical size of a wind-tower spread footing foundation may be reduced by the use of anchors that secure it to the subterrain rock formations. **Figure 6-23** shows a view of a rock anchor rods grouted into the rock formation and the

FIGURE 6-23 *View of the rock anchor rods before and after the foundation structure is assembled*

foundation assembly constructed before addition of concrete. Holes are drilled into the subterrain rock, and the anchor rods are secured to formations with a specialized grout. The tower foundation rebars and steel-frame assembly is then constructed around the **rock anchor rods**, which are protected from the tower foundation concrete by plastic sleeves that prevent them from becoming locked into the foundation. This construction technique allows the foundation to be stabilized by compressing subsoil between the foundation and the rock formation. Hex nuts and steel bearing plates positioned on the anchor rods are tightened against the tower foundation by tensioning the rods and applying enough torque to the hex nuts to seat them before releasing the tension of the anchor rods. The rock-anchor tensioning process is followed periodically to ensure the foundation maintains stability.

Tension Tools

Tension tools use a pump unit to supply high-pressure fluid to the tool in a similar manner as the torque tools. The tension-tool process is different because it uses the pressurized fluid to move a single piston within a **hydraulic cylinder** assembly to develop tension in the fastener. The torque tools use a cylinder to provide rotary motion in the wrench. The tension-tool assembly is constructed to fit around the fastener with a capturing nut to connect the fastener to the piston. When the high-pressure fluid enters the cylinder, it pushes up on the piston, which in turn applies a tension along the axis of the fastener. **Figure 6-24** shows examples of fastener tensioning tools. The procedure to use a tension tool is similar to the setup and use of a power torque tool. The operating procedure listed below is generic and should be used as a reference only. It is recommended to follow the specific tool manufacturer's instructions for safe and dependable operation. Fastener tension-tool suppliers include PLARAD, Enerpac, and Hytorc.

Operation

The purpose of a tensioning tool is to ensure the proper assembly of a mechanical joint. The tension tool may be used to tighten bolts for fluid pressure lines in an industrial setting or to tighten the foundation or rock anchor threaded rods of a wind-turbine application. Each manufacturer has developed tensioning tools to provide safe and consistent operation over the life of the tool. Always read and follow the provided safety and operation instructions for the tool in use. The generic operating procedure in this text is for reference discussion only.

- Determine the required tension level for the fastener assembly. The wind-turbine manufacturer will typically have a document for each maintenance procedure.
- Consult the pressure and tension conversion chart supplied by the tool manufacturer for each fastener size.
- Ensure that the pump, tension-tool, and pressure lines are rated for the job.
- Connect the pressure line to the tool and pump assembly. **CAUTION: Connections should be only finger-tight for threaded connections to ensure that components are not damaged and may be disassembled after use. For quick-disconnect types, ensure the coupler is seated and the locking collar is positioned correctly. Never disconnect pressure lines when they are under pressure. Always consult the manufacturer's instructions to ensure safe operation of the tool.**
- Connect a pressure line cap to the tension-tool output coupler if so equipped. This coupler is used for multiple tools connected in series.

WARNING: When working on the inside of the tower base around the down tower assembly, be sure to follow the site or wind-turbine manufacturer's safety procedure around power cables. Most manufacturers recommend using a LOTO procedure to deenergize and secure power cables

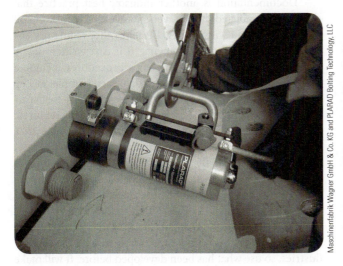

FIGURE 6-24 *Examples of fastener tensioning tools*

that may create a risk. It is always better to eliminate a risk than expose personnel to possible electrocution.

- Place the tool over the threaded rod and ensure the socket collar lines up with the hex nut.
- Thread the adapter on the threaded rod to be tensioned and ensure it is positioned firmly against the top of the tool. **WARNING: Ensure the proper length of threaded rod extends above the top nut. Consult the tool manufacturer's recommendation to prevent breakage of the tool or the threaded rod. Never stand over the tool during tensioning. If the tool or threaded rod breaks, then injury or death may result. Safety first!**
- Toggle the pump control and extend the piston to tension the threaded rod. Never exceed the fastener rated tension, tool pressure, or extension limit listed in the manufacturer's specifications. (Note that some tension-tool manufacturers include a load cell in the tension tool. The load cell output provides a direct measurement of the fastener tension force on an electronic display.)
- Use a torque wrench or manufacturer-provided tool to tighten the hex nut down against the washer and tower foundation flange. Some manufacturers only require seating the nut and washer against the foundation flange, whereas other tool manufacturers recommend setting the hex nut torque to around 100 foot-pounds. Follow the procedure recommended by the tool or wind-turbine manufacturer.
- Release the cylinder pressure and remove the top nut.
- Continue the tensioning procedure until the job is complete.

MAINTENANCE PRACTICES

General maintenance practices cover workmanship and industry best practices. This section will cover items that a maintenance technician should follow during any procedure. Workmanship addresses such questions as, Was the job done correctly, safely, and efficiently? Was the equipment cleaned? Were tools returned to their proper storage location? Previous sections highlighted the need to follow appropriate documentation along with choosing the proper tool and accessories for the procedure to be performed. This is very important and should never be underrated to ensure safety and the effective use of tools. Efficiency is the mark of understanding the maintenance process and proper use of the tools. This comes with proper training and practice. The last two questions address cleaning of the equipment and proper storage of the tools. Cleaning is an important part of any maintenance procedure, especially when using tools that contain hydraulic oil. A few drops of oil will spread over a large surface when not cleaned properly. This oil may degrade painted surfaces, create a slipping hazard, and collect dust between scheduled maintenance activities. As you can see, cleaning is an important step in any activity. This also holds true for the tools used for an activity. Clean and well-maintained tools will last a long time and save money for the technician and site operation.

Industry best practices are simple steps that make a procedure run more smoothly and improve overall efficiency. Best practices include using power tools instead of manual tools as discussed previously. It also includes tightening bolts in a specific order to ensure that the bolted joint is aligned and compressed evenly. Uneven clamping from one side of a circular bolt pattern to another can create mechanical stresses that may damage the structure over time. This uneven stress would be similar to not tightening fasteners to the proper tension as required by specification. If you have changed a tire on an automobile, then you may recall that the bolt-tightening procedure is a **star pattern**. For example, numbering the lugs as 1 in the top center and counting clockwise for a five-lug rim, the tightening sequence would be 1, 4, 2, 5, and 3. This would be true for any circular bolt pattern on a tube-tower structure or pipe flange connection. Refer to **Figure 6-25** for tightening patterns of a few circular bolt patterns. This information may be found in the equipment maintenance procedure or on a supplier's Web site for bolting tools.

Another industry best practice for bolt maintenance is marking the fasteners with a paint pen after each torque inspection. Keep two things in mind: use a designated color paint pen for each maintenance cycle and write the initials of the inspector along with the inspection date near the joint. Why is a single color important? This allows another technician or the farm owner to see the paint marking and quickly determine if the equipment has had its maintenance completed without pulling the maintenance log. **Figure 6-26** shows a tower flange joint with the hex nuts marked with a paint pen after an inspection. Typically construction technicians will use a black paint pen to indicate the fasteners have been tightened to the assembly specification. Maintenance cycles should use a color other than black, and the color for each cycle should be listed in a site maintenance plan or other controlled document.

Documentation is another industry best practice that should be completed during every maintenance cycle. If you have ever worked with OSHA, EPA, or other regulatory entity, you may recall the statement, "If it isn't written down, it didn't happen." Most maintenance procedures include a space for technicians to initial and date each completed step in the procedure. This is a great reference to ensure that critical steps are completed and in a specific order if required. This information will also assist with predictive maintenance and providing support documentation for warranty issues should they arise. If the bolt inspection and maintenance procedure does not include this information, then the site manager or lead technician should develop a checklist that will supplement the procedure. Other industry best practices will be covered throughout the text. Many bolting procedures you will encounter are similar to ones used in other industries, so use what has been developed before. It will make your work safe and improve efficiency.

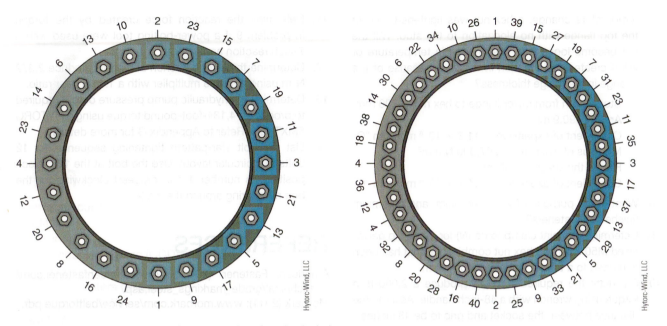

Hytorc-Wind, LLC

Hytorc-Wind, LLC

FIGURE 6-25 *Examples of tightening patterns for a few circular bolt patterns*

Cianbro Coporation

FIGURE 6-26 *Tower flange joint with hex nuts marked with paint pen after inspection*

SUMMARY

This chapter has covered many of the topics that are important when working with fasteners. Understanding fastener types, grades, and where to gather appropriate industry specifications will ensure proper joint assembly and inspection techniques. Tool discussions on proper choice should assist in size, style, and type when this information is not available in maintenance procedures. Another benefit of understanding what tools are available for a particular procedure would be requesting a tool that would make the job safer and more efficient. Previous examples pointed out that a large torque wrench can be replaced with a smaller torque

wrench and a torque multiplier or even a power-bolting tool. Each of these tools has their benefit, but reducing tool operator physical stress and improving safety will improve overall performance and team morale. Why reinvent the wheel? Knowing the proper tool for a job and using it is important.

Consistent use of written procedures and completing proper documentation ensures that bolting maintenance activities are recorded and available for reference in the future. Do not wait for a machine failure to decide what information was important. Determine this information before it is necessary and develop documentation that will make gathering it convenient. If it is not convenient to gather and maintain, then it usually does not happen.

REVIEW QUESTIONS AND EXERCISES

1. List several mechanical joining components.
2. Describe what the terms UNF and UNC mean with respect to fasteners, and describe the difference between them.
3. What organizations do the acronyms ANSI, ASTM, SAE, and ISO refer?
4. What does the term *grade* refer to with fasteners?
5. How is the yield strength determined for a material?
6. Determine the stress on a grade 2, 5/8–11 fastener that is loaded in tension to 15,000 pounds force (lbf). Determine whether this load will create a permanent elongation or return to its original length when the load is removed.
7. Determine the change in length of a stud and flange joint with an 85°C decrease in temperature. Before the

temperature change, a hex nut was tightened against the top flange with no elongation to the stud. Will the nut become loose with this change in temperature or will it create an elongation in the stud because of the change in the flange thickness?

> Stud length from lower flange to hex nut = 200 mm
> Area = 490.9 mm^2
> Coefficient of expansion = 11.3 × 10^{-6} mm/mm °C
> Modulus of elasticity = 207,000 N/mm^2
> Flange thickness = 200 mm
> Coefficient of expansion = 11.7 × 10^{-6} mm/mm °C

8. What is the purpose of zinc, cadmium, and chromium finishes on fasteners?
9. Determine the joint clamp force (*N*) for a hot-dip galvanized M50 stud and hex nut combination that has been tightened to a torque value of 12,511 N-m.
10. Calculate the required force to produce a 2,000-ft-lb torque using wrench with a 48-inch handle. Assume the distance between the socket and grip to be 48 inches.

11. Determine the reaction force created by the torque in problem 9 if a power-bolting tool were used with a 7-inch reaction arm.
12. Determine the torque wrench setting to produce 2,372 N-m using a torque multiplier with a 1:11.5 gear ratio.
13. Determine the hydraulic pump pressure output required to produce a 4,134-foot-pound torque using a HYTORC 5MXT tool. Refer to Appendix G for more details.
14. List the bolt star-pattern tightening sequence for 12 bolts in a circular layout. Use the bolt at the 12 o'clock position as number 1 and proceed clockwise for the bolt numbering around the circle.

REFERENCES

American Fastener (2010): www.americanfastener.com/technical/grade_markings_steel.asp.
Morbark (2011): www.morbark.com/service/belttorque.pdf.

7

Test Equipment

KEY TERMS

alternating current (AC)
amplitude
axial runout
charge-coupled device (CCD)
crest factor
digital multimeters (DMMs)
direct current (DC)
distortion
electrolytes
frequency
infrared
mega-ohm meter
nondestructive testing

peak-to-peak value
peak value
personal computer (PC) predictive maintenance
radial runout
resonance
root mean square (RMS)
sag
swells
sympathetic vibration
thermography
transients
uninterruptible power supply (UPS)
visible spectrum

OBJECTIVES

After reading this chapter and completing the review questions, you should be able to:

- Describe test equipment that may be used during a preventative-maintenance cycle to verify the status of wind-turbine systems.
- Describe the purpose of video inspection systems as part of a PM plan.
- Describe applications of thermal imaging that may be used to determine equipment condition.
- Describe how testing wire insulation can be used as an indicator of cabling, electrical motor, or generator condition.
- Describe how a load or impedance test can verify the condition of batteries.
- Describe the use of electrical test equipment as preventative-maintenance tools.
- Describe how vibration monitoring can be used to verify equipment condition.

INTRODUCTION

Preventative-maintenance (PM) activities may take on functions from mechanical to electrical, depending on scheduling and technician experience. Previous discussions focused on lubricants, bolting, and fluid-power component inspections. This chapter will focus on test equipment that may be used to determine the condition of equipment during a preventative-maintenance activity. Some test equipment described in this chapter will have data-storage features that enable the technician to track trends of electrical-system parameters over several minutes or days. This information will be invaluable in determining the overall condition of a wind turbine. Other test equipment will provide data that can be logged

in a computer database or on a written document to track and analyze trends over several PM cycles. Trend analysis is an effective method to determine changes that may indicate deterioration of components.

Electrical Test Instruments

Many electrical test instruments are available for determining the condition of control, power, and generation systems with wind turbines. This chapter is not intended to focus on basic test equipment such as multimeters and their functions of resistance, current, voltage, and so on. The focus will be on equipment that may be used to determine power variations

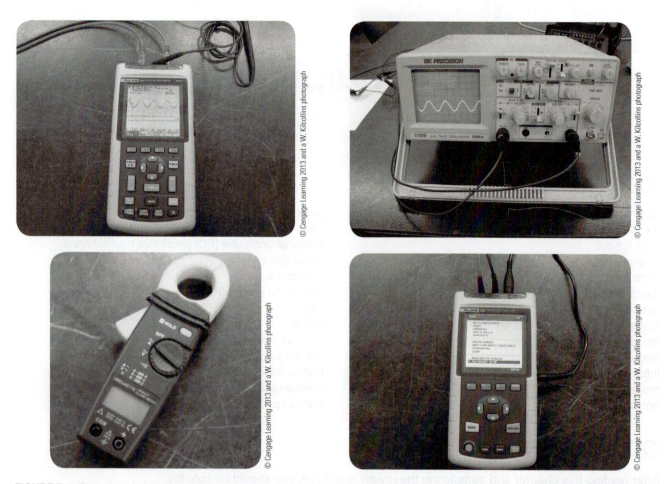

© Cengage Learning 2013 and a W. Kilcollins photograph

FIGURE 7-1 *Examples of various electrical instruments used to measure power and signal variations*

that can lead to problems with equipment. **Figure 7-1** shows examples of various electrical instruments that may be used to measure power and signal variations. The premise of this chapter will be that the technician has had experience with these types of devices and can use them safely in conjunction with other devices with advanced features.

Mechanical Test Instruments

Mechanical test instruments for use with PM activities include instruments for examining physical or thermal (temperature) variations created by abrasion or other load-related stress. Components such as gears, bearings, cable connections, and electrical devices may show noticeable variations that may be recorded as digital images and analyzed or filed for future comparisons. **Figure 7-2** shows examples of visual inspection and recording devices.

VISUAL INSPECTION TOOLS

Visual inspections are a quick way to determine the condition of a component. Changes in color; deterioration resulting from corrosion, abrasion, or cracks; and missing

components are easily seen and can be addressed with corrective measures. Some inspections are easily done by a walk around the equipment with a predefined checklist or notes from previous inspections. Other visual inspections may take a bit more time to complete because access panels or covers must be removed first. **Figure 7-3** shows examples of access panels or covers that may be encountered in a wind turbine. These inspections require a qualified technician who is familiar with the types of hazards that may be present behind the panels and covers. These inspections may require a lockout–tagout (LOTO) plan and special tools to ensure the safety and effectiveness of results. Even the removal of panels and covers may not provide the access a technician needs to conduct a thorough inspection. This author has seen many mirrors of various sizes and shapes and even mirrors attached to a telescopic wand to help the inspection process. **Figure 7-4** shows examples of simple inspection tools. The activity becomes interesting when the inspection requires a photograph or digital image of the area. True, the inspection can be done, but the use of a video-inspection system with a flexible wand (also known as a *borescope*) and digital file-storage capability sure improves the results over mirrors and a standard camera. These inspection systems may be available with removable viewers, different length wands, different wand diameters, removable data-storage devices,

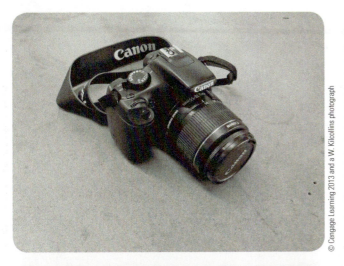

FIGURE 7-4 *Examples of simple inspection tools*

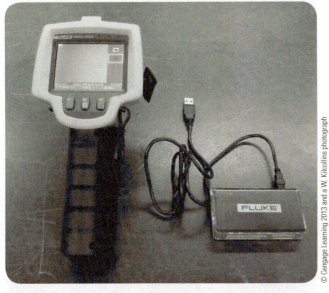

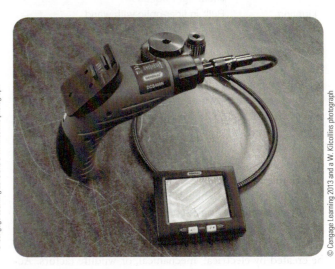

FIGURE 7-2 *Examples of visual inspection and recording devices*

FIGURE 7-5 *A typical video-inspection system*

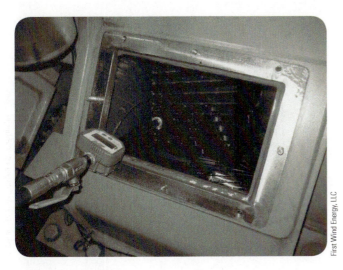

FIGURE 7-3 *Gearbox access panel that may be encountered in a wind turbine*

universal serial bus (USB) port, and still-imaging capability. The list goes on, depending on the features needed for the inspection process. **Figure 7-5** shows an image of a typical video-inspection system.

Equipment Summary

When purchasing an inspection system, consider the types of locations you may need to access. **Table 7-1** lists features available on a typical inspection system. A few features to consider may be a detachable viewer with wireless communication to the handheld unit, flexible wand, and data-transfer capability. A removable viewer allows easier viewing during an inspection. Having the wireless capability eliminates cables running around the inspection area and eliminates a tripping hazard. The inspection may require extended arms and placement of the viewer in an access panel.

TABLE 7-1

Typical Features Available with Vision Inspection Equipment	
Feature	**Specification**
Display	Camera viewfinder, multi-inch diagonal screen, or radio-frequency remote screen
Images	Still shot Video Color
Zoom	1× to 3×
Memory	Internal memory or plug-in micro secure digital (SD) card
Communication	USB adapter
Power	AC adapter or rechargeable batteries or both
Wand	Multiple lengths, and some models have liquid-tight wands
Lighting	Built-in fiber-optic with wand

© Cengage Learning 2013

It may be physically uncomfortable for the technician to see the viewer if it is attached to the unit. Being able to position the viewer at eye level and at a comfortable viewing distance makes an easier, more thorough inspection. If the technician is not comfortable, then the task will be a rushed activity. Compound this with a cold working environment, and the quality of the inspection will depend on the physical stress the technician can endure.

A flexible wand enables an inspection to take place in areas that would be considered a blind reach if the only way to view the area was with a small mirror and an extended arm. A blind reach is not a safe practice. Unseen hazards may be created by electrical, mechanical, or wildlife factors. Accidentally placing an arm into an exposed electrical circuit can cause a serious injury. Play it safe. Understand the work activity and use tools that improve safety.

Using a sealed or liquid-tight wand assembly also enables use in areas that may have pooled moisture or oil. This wand feature will prevent contamination and extend the life of the instrument.

The use of a data-storage card or USB port with an inspection system enables images and videos to be e-mailed or transferred to maintenance reports to improve clarity of information. File creation and storage is quite simple with most of these systems. The viewer screen has either a button on the bezel or a touch screen area for image or video capture. With the image on the screen, depressing the capture button creates a file. **Figures 7-6 through 7-9** show some of the capture and play features on a system. Transferring image files to a laptop or **personal computer (PC)** can be accomplished by removing the memory card from the system and plugging it into a reader port or adapter on a computer. Most computers will recognize the addition of the memory card in the reader and ask to play or save the files. **Figures 7-10**

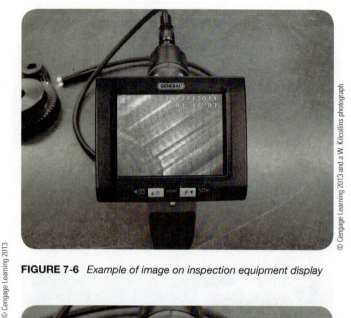

FIGURE 7-6 *Example of image on inspection equipment display*

© Cengage Learning 2013 and a W. Kilcollins photograph

FIGURE 7-7 *Close-up of image capture buttons on display for still shots and video*

© Cengage Learning 2013 and a W. Kilcollins photograph

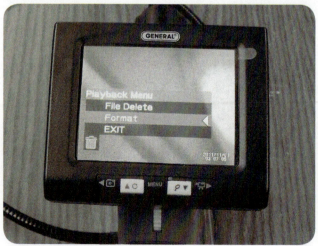

FIGURE 7-8 *View of vision system option menu*

© Cengage Learning 2013 and a W. Kilcollins photograph

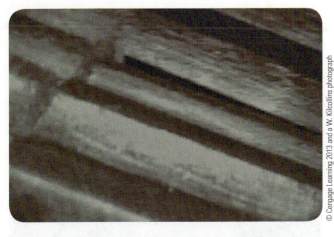

FIGURE 7-9 *Display of stored image file on computer monitor*

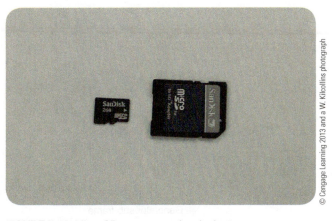

FIGURE 7-10 *Micro SD memory card and adapter*

through 7-12 show examples of memory cards, readers, and USB data connections.

Each video-inspection system available has many standard features and options. The features discussed in this section are worth considering when purchasing a video-inspection system. Consider the features and options of several systems and choose those that will make your inspection activity safe and effective.

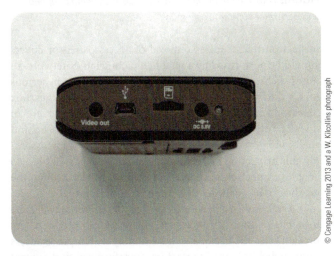

FIGURE 7-11 *Example of a USB data port*

Applications

Typical applications for video-inspection equipment include inspections of gearboxes, generators, control cabinets, and structural assemblies. As previously mentioned, any

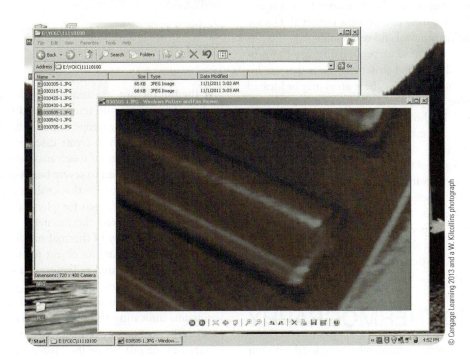

FIGURE 7-12 *Screen shot of PC with data device recognized and stored image*

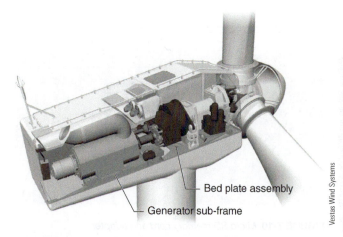

FIGURE 7-13 *Layout of wind-turbine nacelle showing bedplate and generator subframe assembly*

FIGURE 7-14 *Examples of welded joints that may be inspected during a PM activity*

inspection that requires removal of access panels or covers should be performed by a qualified technician. Safety procedures such as LOTO should be considered to prevent serious injury during these inspection activities. Opening a gearbox access cover allows inspection of components that are in one's line of sight, but inspecting components that are behind shafts and gear assemblies is difficult. A flexible wand assembly aids inspecting these blind areas. The wand assembly also enables inspection of cables that may be routed through the gearbox main shaft that feed to the rotor assembly.

Areas within a generator or control cabinet may also be inspected with a video-inspection system and proper safety precautions. The use of a wand assembly will improve access to areas behind large contactors, transformers, and control assemblies. Some of these areas may only be available for inspection after components have been removed. An inspection system would save time and prevent potential damage to components during removal and reassembly.

Other areas that may need inspection during a PM process included welded or bolted structural joints that are located in areas with limited access. These areas may be located on the wind-turbine bedplate and the generator subframe. **Figures 7-13 and 7-14** show an example of a bedplate and generator subframe assembly along with inspection areas such as welded joints.

These are only some of the inspection activities that may be captured with a video-inspection system. Other inspection activities will be discussed in later chapters as PM activities are highlighted for the various areas within a wind turbine. Sometimes, inspecting components with normal video equipment does not show the entire picture. The use of imaging equipment that detects variations our eyes cannot see may be necessary.

INFRARED THERMOGRAPHY

Visual inspection of components can be completed with our eyes, cameras, or video equipment as described in the previous section. There are times when our sense of sight does not

give the full picture. The sense of touch adds valuable feedback for a situation with temperature differences. Our sense of vision does not give temperature feedback unless the component is hot enough to create discoloration or flames that we can see. Our sense of touch may tell a difference, but it can expose a technician to severe burns or electrical hazards.

An inspection process that enables the feedback of temperature without the need for physical contact is through the use of an **infrared** thermometer or infrared imaging equipment. The study of thermal imaging or using emitted infrared radiation from an object for imaging is known as **thermography**. Infrared radiation is a portion of the electromagnetic spectrum located below the range of visible light. **Figure 7-15** shows a diagram of the electromagnetic spectrum, indicating the **visible spectrum**, or visible light region. Thermal energy radiated by warm objects falls in the infrared range just below the visible light region. Our eyes may not detect this thermal energy, but we can feel the temperature difference with our sense of touch. Infrared imaging

Electromagnetic spectrum

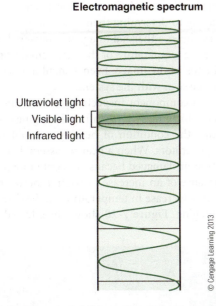

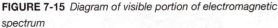

Ultraviolet light
Visible light
Infrared light

© Cengage Learning 2013

FIGURE 7-15 *Diagram of visible portion of electromagnetic spectrum*

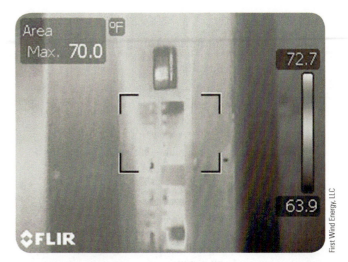

First Wind Energy, LLC

FIGURE 7-16 *Image captured with infrared imaging equipment note the temperature range scale provided on the right side of the display*

© Cengage Learning 2013 and a W. Kilcollins photograph

FIGURE 7-17 *Infrared thermometer*

equipment is designed to detect infrared frequencies of the electromagnetic spectrum and display them as visual images. An infrared imaging system uses a **charge-coupled device (CCD)** to detect photons and convert their energy into a signal that can be displayed as an image similar to a digital camera or video recorder. The system display can be set up to show an image with a color scale that corresponds to temperature values. Objects that appear white, red, or yellow in color on the display screen would be warmer than those that appear blue, violet, or black. White would be the warmest temperature, and black would be the coolest. **Figure 7-16** shows an example of an image that has been captured with infrared imaging equipment. Note the color variation of the objects that are warmer compared to the background using the temperature scale on the equipment display.

Equipment Summary

Several different infrared measuring instruments are available for determining the temperature of an object without physical contact. The simplest instrument is an infrared thermometer: it can be used to quickly determine the temperature of a component during an inspection to determine if there is an issue. An infrared thermometer has a similar configuration to a gun with a handgrip and detector to improve pointing. **Figure 7-17** shows an example of an infrared thermometer. Some models have a laser pointer that improves aiming the detector. An infrared thermometer has a liquid crystal display (LCD) that shows an object's temperature in either degrees Fahrenheit (°F) or Celsius (°C). Some models have features that will enable capturing maximum and minimum temperature readings along with other statistical information.

When an issue is found with a component during an inspection, it is convenient to be able to see the device's temperature variation and not just a temperature value. This is where infrared imaging is valuable to determine the extent of the problem. Infrared imaging equipment can display the thermal image on a screen to be analyzed as a photograph or the use of a video function can be used to show the increase in temperature over time. **Table 7-2** lists features that may be found on infrared imaging equipment. **Figure 7-18** shows views of an infrared imaging instrument to highlight its features.

Test Applications

Infrared thermometers or imaging equipment may be used to determine issues with cable crimps or terminal connections. If a crimp is not compressed to create a low-resistance connection, it will create heat because of higher resistance. Inspecting power cable crimps while a system is under load

TABLE 7-2

List of Features for Infrared Imaging Equipment

Feature	Specification
Display	Camera viewfinder, multi-inch diagonal screen, and backlighting
Images	Still shot Color Temperature-scale adjustment (°C or °F) Infrared fusion level
Zoom	1× to 3×
Memory	Internal memory External adapter for multiple memory card formats: microdrive (MD), secure digital (SD), SmartMedia (SM), Memory Stick (MS)
Communication	USB adapter
Power	AC adapter or rechargeable batteries or both

© Cengage Learning 2013

Circuit breakers and fuses can be inspected with infrared imaging to determine their load-carrying capability. **Figure 7-19** shows a thermal image of a power-distribution panel. Note the temperature of the main circuit breaker located at the top of the image. Even a small temperature difference can be seen with the system.

Electrical component inspections are not the only use of an infrared imaging system. Infrared imaging can be used to determine the condition of bearing assemblies used in motors or generators. When a bearing assembly is deprived of lubrication or damaged because of contamination, it will heat up because of an increase in friction between moving elements. This increase in temperature can be detected with infrared imaging. **Figure 7-20** shows an infrared image of

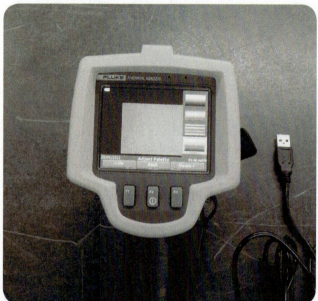

© Cengage Learning 2013 and a W. Kilcollins photograph

will show a variation in crimp connections because of the quality of the crimped splice. This inspection method can show a problem crimp before it causes damage to equipment or create downtime.

Infrared imaging can also be used to determine the condition of terminal lug connections on relays, motor starters, and other switching devices.

Circuit board components can also be viewed with infrared imaging to determine if discrete components such as resistors, diodes, transformers, and other devices are being subjected to abnormally high levels of current that may cause failure of the device and ultimately the control board.

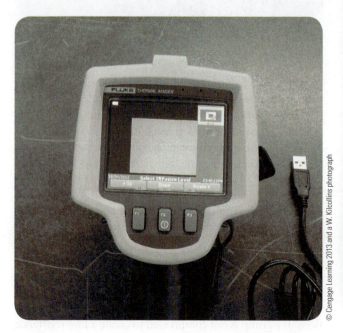

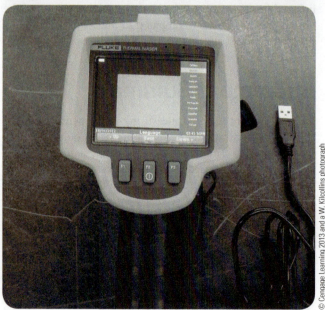

© Cengage Learning 2013 and a W. Kilcollins photograph

FIGURE 7-18 *Infrared imaging instrument features*

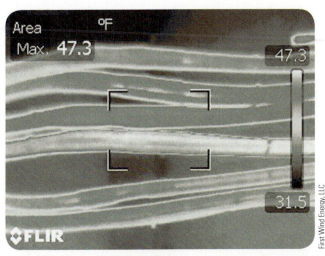

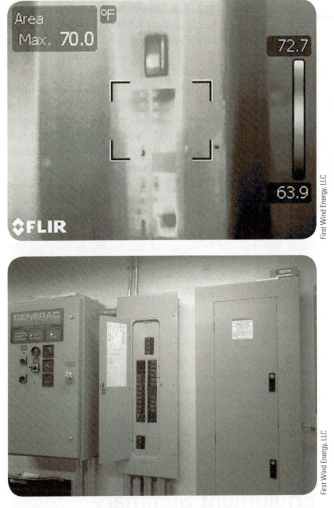

FIGURE 7-19 *Infrared image of power-distribution panel*

FIGURE 7-20 *Infrared image of power cables inside a tower*

power cables running up the inside of a tower. Discovering a faulty phase cable or crimp connection during a maintenance cycle can prevent downtime and reduce the cost of damage to other equipment.

Infrared imaging equipment goes beyond our ability to inspect components and systems with our eyes. These tools are extremely useful in determining issues before they can become costly. Other inspection methods for determining issues can use basic electrical principles to quantify the condition of a component. This can be seen with equipment that is used to determine the condition of insulation covering electrical conductors.

WIRE-INSULATION RESISTANCE

Wire-insulation measurements have been used in industry for some time to determine the insulation value of wire coatings used with motors and generators. Wires wound on the stator or rotor of a motor or generator are considered windings. **Figure 7-21** shows an example of an AC motor stator and rotor assembly. The wire used for the stator assembly is coated with a thin layer of varnish or lacquer that separates the wires and prevents shorting between the layers. The varnish insulation on the wires will deteriorate because of heat, moisture, dust, and lubricant contaminants during the life of the assembly. A good method to determine the integrity of the insulation is through the use of an insulation tester, also known as a **mega-ohm meter** or *megger*. Insulation testing with this type of equipment is considered nondestructive. **Nondestructive testing** means that the insulation will not be damaged during the procedure. After completing the test, the equipment may be put back in service if no issues were discovered. A mega-ohm meter functions on the principle of Ohm's law:

$$V = IR$$

where V is voltage, I is current, and R is resistance. Leakage current that passes through the wire insulation from a test conductor to an adjacent conductor can be used to calculate insulation resistance.

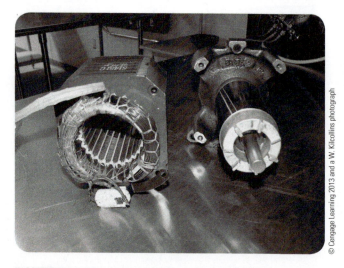

FIGURE 7-21 *An AC motor stator and rotor assembly*

A mega-ohm meter supplies a fixed voltage level to a conductor being tested and measures the leakage current available from adjacent coil windings, cables, or a ground circuit. Typical test voltages can be set from 500 V to 2,500 V, AC or DC, depending on the instrument and test requirements. Consult the manufacturer's instructions for the test instrument being used to ensure safe reliable operation.

Other things to consider during insulation testing are ambient temperature and humidity. Insulation resistance decreases with increased temperature and elevated humidity. Note ambient temperature, test sample temperature, and dew point with each test measurement. Equipment manufacturers recommend taking insulation-resistance measurements at approximately the same temperature each time to reduce

variations from temperature. Insulation-resistance measurements should also be taken over a period of time such as a minute or several minutes to ensure time for the current to saturate the insulation and stabilize any capacitive charging that may occur between other conductors and structural components. Most insulation test meters have a time function available, so refer to the manufacturer's instructions for proper use. Each consideration will ensure consistent insulation resistance results over time. Tracking these results will make this testing a valuable tool in predicting equipment issues before they become problems. Many insulation test instrument models have data-storage features that enable storage and retrieval from the instrument with data cables, removable memory devices, and Bluetooth connectivity. "Bluetooth" is the registered trademark of Bluetooth SIG Inc. These features enable storage of test results so they may be downloaded to a computer for later analysis.

Testing wire-insulation resistance may also be applied to power cables used within a wind turbine. To use a mega-ohm meter to determine power cable insulation, the cable ends have to be disconnected from bus bars or terminals on both ends and isolated to prevent shorting the applied voltage to ground, other equipment, or personnel. **SAFETY FIRST: Remember: deenergize, isolate, and LOTO power sources before attempting to disconnect any power cables. This type of work should only be attempted by qualified technicians.** Applying the test voltage to anything other than the cable under test can damage equipment or cause serious injury to co-workers.

Equipment Summary

The basic purpose of a Mega-Ohm meter is to charge a conductor and measure the leakage current transmitted to adjacent conductors. Meters available may have multiple voltage-level settings along with **alternating current (AC)** and **direct current (DC)** modes. AC voltage varies or alters with time, and DC voltage remains constant. **Figure 7-22** shows an example of an AC voltage signal. Some meters have insulation resistance as one of many functions available. Choose a test instrument that meets the needs of the farm or service center. Multifunction instruments that have features that enable them to function as scope meters or power-quality analyzers are useful. These instrument features will be discussed briefly in later sections.

Test Applications

Wind turbines use different motors for functions such as yaw, pitch, and multiple cooling fans. Cooling fans of various sizes are used for the power-conversion cabinet, generator assembly, gearbox oil, or hydraulic systems. These motor assemblies could be tested on a schedule to determine the insulation condition of their windings. Collected data may be tracked and compared to the manufacturer's recommended minimum values as a means of **predictive maintenance** analysis.

SAMPLE PROBLEM

Determine the insulation resistance of the natural rubber jacket on a power cable. The test instrument has been set at 1,000 V_{DC}, and the detected current from an adjacent cable is measured at 0.001 amps. What would be the insulation resistance between the conductors?

SOLUTION

From Ohm's law,

$$V = IR$$

so

$$R = \frac{V}{I}$$

Then

$$R = \frac{1,000\ V_{DC}}{0.001}$$

Resistance = 1,000,000 ohms

or 1 mega-ohm of insulation resistance.

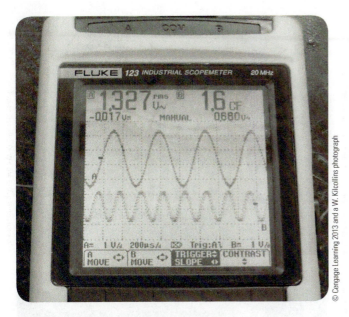

© Cengage Learning 2013 and a W. Kilcollins photograph

FIGURE 7-22 *Example of an AC voltage signals*

These motors range in cost from several hundred dollars to thousands of dollars, depending on the application. The other common factor for these motors is that the turbine will be out of service if any of them are inoperative. Predictive maintenance of the winding assemblies may be well worth the expense to prevent downtime.

Inspecting the insulation resistance of windings in the generator assembly is equally as important in preventing unscheduled downtime as the motors. Time required to perform a proper LOTO procedure and test the rotor and stator windings will pay dividends over the life of the equipment. Always consult the equipment manufacturer for test specifications and safety procedures.

Power cables are another important area to inspect insulation resistance. Vibration, temperature changes, and tension on the cables because of hanging can cause the insulation to chafe, stretch, and distort. Variation in cable splices between cables may also generate excessive heating around the splices. This heat may deteriorate the insulation adjacent to the splice or shrink sleeves around these splices. Isolating and inspecting the cables can determine insulation changes and predict whether there are problems. An unwanted short between adjacent phase cables and ground could create tens of thousands of dollars in equipment damage and many days of lost power generation waiting for replacement components and repair.

Inspecting transformer coils for insulation condition is another valuable area to remember when performing preventative maintenance. Wind-turbine systems have transformers that are necessary to reduce the voltage supplied to the tower for use in control circuits and power circuits for operating lighting, motors, and other systems. A short circuit in one of these transformers will remove the turbine from service because of the loss of system power. The short

could also damage equipment being fed by the transformer. As with any insulation-testing procedure, follow the equipment manufacturer's recommendations and only perform the activity if qualified. Several of these power systems operate at medium voltage levels or higher and can cause serious injuries if work activities are not performed properly.

Data gathering and analysis of insulation resistance for these tests is very important to watch for trends that may occur during the life of the equipment. Predictive maintenance practices can save money and time if a component is repaired or replaced before a failure occurs.

ELECTRICAL MEASUREMENTS

Electrical test measurements are not limited to insulation resistance. Test equipment for spot sample readings along with trend analysis can be used to check voltage, current, and frequency measurements for system output along with individual control and power circuits. Units for this type of testing include **digital multimeters (DMMs)**, oscilloscopes, scope meters, and the more-sophisticated power-quality analyzers. Most technicians have had some experience with basic multimeters, so discussions in the following sections will focus on the testing features of the other test instruments.

Oscilloscopes are a great tool for gathering data on AC signals used for control and power circuits. These devices can show multiple signals on their display along with enabling adjustments for voltage levels and time values. **CAUTION: Make sure the test instrument is rated for the voltage level to be measured. Consult the manufacturer's operation manual for details on safety features and test setup requirements.** Figure 7-23 shows a typical two-channel oscilloscope model with test probes. Many oscilloscope models have data-storage capabilities so that a trend analysis may be used to determine variations in signal quality over several minutes or hours. This trend capability is very useful in determining issues that are infrequent but may shut down a system each time they occur.

Scope meters are also available for simple analysis of control and power-circuit signal quality. These devices resemble a typical multimeter but have the capability of multichannel input and display of electrical signals like larger lab oscilloscope models. The smaller size of scope meters makes them a great choice for use by a technician in many test applications. They may be stored conveniently in most technicians' tool bags and made available for easy access. **Figure 7-24** shows an example of a scope meter.

Another test instrument that may be used is a power-quality analyzer. These instruments have several test options that make them a good choice for many signal-quality issues. These devices are also similar in size to scope meters or larger digital multimeters, so they are a good size for most

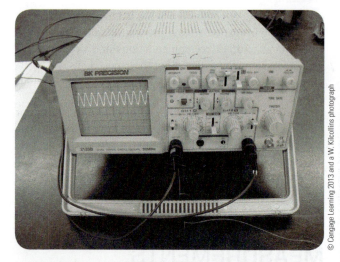

© Cengage Learning 2013 and a W. Kilcollins photograph

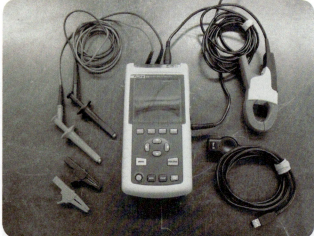

© Cengage Learning 2013 and a W. Kilcollins photograph

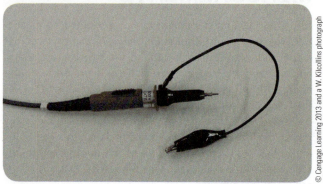

© Cengage Learning 2013 and a W. Kilcollins photograph

FIGURE 7-23 *Typical two-channel oscilloscope model with test probes*

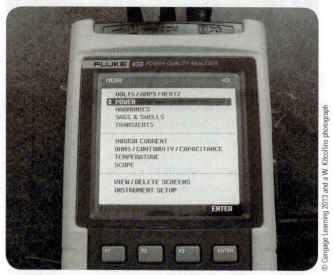

© Cengage Learning 2013 and a W. Kilcollins photograph

FIGURE 7-25 *A power-quality analyzer*

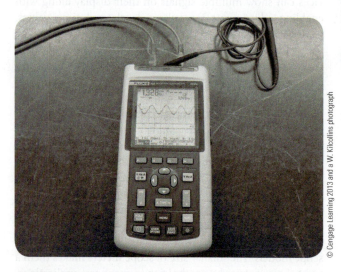

© Cengage Learning 2013 and a W. Kilcollins photograph

FIGURE 7-24 *A two-channel scope meter*

tool bags. **Figure 7-25** shows an example of a power-quality analyzer. Power-quality analyzer models are configured for single-phase testing or may be designed for use with three-phase applications. Three-phase models are supplied with four channels for attaching current and voltage probes for simultaneous measurements of each phase and ground.

Equipment Summary

Oscilloscopes have several features that make them useful in testing electrical signals. They may be set up to display AC signals for circuit-board components, control-system circuits, and power circuits provided they are used with appropriate test leads and proper settings and performed by qualified technicians. Do not attempt any work on live electrical circuits if you do not have proper technical or safety training for the activity. Consult the equipment manufacturer's documentation before any testing activities to determine voltage and current levels that may be encountered during testing. It is always a good practice to reference the company's safety policy or NFPA 70E electrical workplace standards for appropriate PPE before performing electrical test activities. **Figure 7-26** shows the face panel on a typical oscilloscope. Note the instrument's channel controls, signal settings, time adjustment, and other features. These may be used to determine the shape, voltage amplitude, and frequency of a signal.

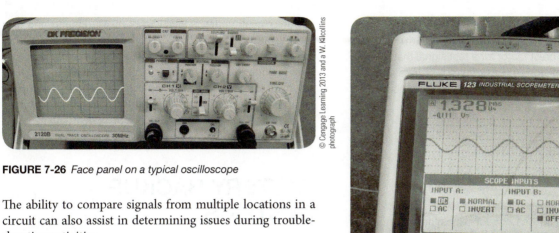

FIGURE 7-26 *Face panel on a typical oscilloscope*

The ability to compare signals from multiple locations in a circuit can also assist in determining issues during troubleshooting activities.

Scope meters can be used in the same fashion as an oscilloscope. Benefits of the scope meter are its compact size and reduced weight. The instrument can display and store signal data from a variety of sources just like the oscilloscope. **Figure 7-27** shows the display of a typical scope meter.

Power-quality analyzers are also useful test instruments for checking electrical signals. These devices are rated for higher voltage and current levels and typically are a lot more expensive than scope meters. Depending on the instrument

SAMPLE PROBLEM

Determine the peak, **peak-to-peak value**, and frequency of the signal shown on the oscilloscope display with the channel voltage set at 2 V/division, time set at 5 mS/division, and test probe set on 1×.

SOLUTION

The signal extends 5.1 divisions from peak to peak, and the distance for one complete cycle (alteration) is 2.5 divisions:

Voltage peak to peak = 5.1 × 2 V/division

= 10.2 V

Peak value would be one half the peak-to-peak value or

$$\frac{1}{2}(10.2) = 5.1 \text{ V}$$

If the test probe was set on the 10× position, the measured value would be 1/10 that displayed. The 10× position amplifies the signal strength by a factor of 10.

$$\text{Frequency} = \frac{1}{T}$$

where T is the time for one complete signal alteration. T for the complete signal alteration is

2.5 divisions × 5 mS/division = 12.5 mS

So the frequency,

$$f = \frac{1}{12.5 \text{ mS}} = 80 \text{ Hz}$$

FIGURE 7-27 *Display of a typical scope meter*

model, it may be used to measure multiple phases of a system for voltage and current. Models may be configured for single- or three-phase power-quality testing. Tests that may be performed include voltage and current levels, frequency, phase imbalance, improper phase sequence, determining signal distortion (crest factor), **root mean square (RMS)** readings, and data-logging features. When an electrical signal is considered to be distorted, it would not appear to be a true sine wave when viewed on an oscilloscope or scope meter. A true sine wave would have a **crest factor** = 1/0.707, or 1.414. Anything other than this would be considered a distorted signal. Increased distortion in an electrical signal can create issues with equipment relying on the signal for smooth operation. Signal distortion can also create issues to determine the true RMS value of a signal when using a multimeter for voltage measurements. AC signal **distortion** can be determined using the relationship between the voltage

SAMPLE PROBLEM

Determine the crest factor for a test signal that is suspected to be distorted. Peak voltage is measured at 25.6 V_P, and the RMS voltage value is measured to be 10.25 V_{RMS}.

SOLUTION

Crest factor is calculated using the formula

$$CF = \frac{V_p}{V_{RMS}} = \frac{25.6\,V_p}{10.25\,V_{RMS}} = 2.498$$

This value is greater than 1.414, so the suspicion of a distorted signal was correct. This type of distortion could be created by harmonics being introduced to the line from a load.

peak value of the alternating signal (sine wave) and the voltage RMS value. If distortion is suspected with the circuit being tested, then ensure that the test equipment is rated for the highest crest factor available. One suggestion is to have a test instrument with a crest factor rating of 3 for full scale and 6 for half scale to ensure proper measurements. Deenergize circuits before applying current and voltage test probes when practical. Once the test probes are attached, the circuit may be reenergized and measurements taken. **If deenergizing the circuit is not an option, then use appropriate approach boundary considerations and PPE to complete the testing. Always follow approved safety practices! Contacting live conductors or shorting phases together can cause severe burns or death.**

Another benefit for many of these test instruments is the ability to store data for analysis. Many electrical faults may be captured as they occur during the operation of a system. The presence of voltage **sags** or **swells** during normal system operation may be created by large motors or other heavy load devices switching on or off. This information along with supervisory control and data acquisition (SCADA) data may enable understanding of a premature device failure or the cause of a nuisance circuit-breaker trip. Understanding failure mechanisms can lead to system improvements that will prevent future occurrences.

Test Functions

Common test measurements that can be obtained by electrical test instruments include the voltage sags and swells mentioned previously along with other signal variations that may be created by line variations such as transients. **Transients** may be introduced to a line through switching of large electrical loads or through a lightning strike on or near a transmission line.

Electrical test instruments such as DMMs, oscilloscopes, scope meters, and some of the more sophisticated units such as power-quality meters enable spot measurements and data logging of electrical circuits to determine system status. Other electrical-testing activities for preventative maintenance may include the analysis of battery systems. Wind-turbine emergency-power systems may include the **uninterruptible power supply (UPS)** for the PC or the programmable logic controller (PLC) processor along with sealed battery packs required for some manufacturers' blade-pitch systems.

BATTERY BACKUP SYSTEMS

Wind-turbine control system battery backups are used in a variety of different roles such as the UPS for the interface computer or PC, communication hub, or PLC backup power. UPS devices supply AC voltage in the absence of grid power. These devices use a battery or batteries to supply DC power to an inverter circuit to produce AC power to appliances. These battery backup systems may be checked for system voltage or charging voltage through SCADA data registers or through a spot check of the battery pack voltage with a DMM. Determining the battery condition may require a check of the individual cell's internal resistance or voltage or the application of a predetermined load to evaluate the response of the battery. Specialized battery testers are designed to test the charge status and endurance of sealed batteries.

Battery testing for emergency power backup systems is very important to ensure that the wind-turbine control system can function long enough after a grid event to safely shut down the equipment. A safe shutdown process may include opening the main contactors that connect the power-conversion system to the grid, pitching the blade assembly to the feathered or 90° position, engaging the parking brake, and shutting down vital control functions. If any of these safety functions do not occur, there is a possibility of a catastrophic system failure.

Equipment Summary

Lead–acid storage batteries come in a couple of different designs. Some designs may be serviced such as by testing **electrolytes**—solutions of acid and water—for specific gravity to determine the condition of the cells. If the electrolyte needs adjustment, then more may be added during the PM procedure. These batteries include deep cycle designs that enable a slow discharge over a period of time. These batteries may also be load tested with equipment that is designed to draw a high load for a short interval to determine battery integrity. **Figure 7-28** shows an example of a battery tester designed for a load test. Other lead–acid storage batteries are designed in a sealed case for use as portable power systems or for stationary UPS applications. These batteries are

FIGURE 7-28 *Battery tester designed for load test*

FIGURE 7-29 *Test connections for battery-load tester setup*

not designed for electrolyte testing or replenishing. Many of these battery systems may also require testing while in operation because shutting the power system down may cause other issues.

These battery cells require a test method for determining the internal resistance of the cell along with the voltage. Fresh sealed lead–acid storage batteries have an internal resistance in the micro-ohm ($\mu\Omega$) range that increases with the number of charge and discharge cycles. As these batteries near the end of their usable life, the increase in internal resistance will have an associated increase in charging time and a decrease in discharge capacity. Charge and discharge parameters may give feedback to the power system to indicate when a battery should be replaced or when it has failed, but they would not assist in predicting usable life. Performing a battery test after installation and using that information as a baseline with the manufacturer's recommendations would help predict usable life. This would enable replacing the battery or battery pack during a scheduled maintenance procedure rather than after a failure occurred.

Operation

It is important to remember safety precautions when working around batteries. Shorting battery terminals can cause an explosion of the hydrogen gas that is produced during the discharge cycle of the lead–acid storage battery. This explosion will also spray the acid electrolyte that is contained within the battery. Always wear safety goggles, a face shield,

chemical-resistant gloves, and clothing that will minimize exposure of skin to battery acid.

Testing of serviceable lead–acid storage batteries should be completed per the manufacturer's recommendations. Compare the measured electrolyte sample's specific gravity to the battery's specifications. Add acid or distilled water as necessary to bring the electrolyte back into the specified range. When using a battery-load tester, ensure the proper lead is connected to the correct battery terminal. Make the connection to the positive terminal before attaching to system ground. This will eliminate any sparking around the battery and minimize the risk of an explosion. **Figure 7-29** shows test connections for the battery-load tester setup.

CAUTION: Always wear safety glasses, a face shield, and clothing that will protect your body in case of a battery explosion. Read the operations manual and familiarize yourself with the battery-load tester before beginning a battery test. Follow the test instrument manufacturer's recommendations for the duration of the test and interpretation of measured results. Use extreme caution around the resistor side of the test equipment. The load resistor assembly will become extremely hot during the test and will remain hot for several minutes after the test. Severe burns to skin or fire to clothing and other combustible materials can result in contact with the load assembly.

Testing a sealed lead–acid storage battery may be accomplished using a battery tester with functions for resistance, voltage, and temperature. Some battery test instruments enable entering test parameters of internal resistance and voltage so that the test may be simplified to display a pass, marginal, or fail outcome on the LCD screen or a series of three light-emitting diodes (LEDs) of green, yellow, or red. This enables the contact of the test probes to the individual battery and then interpreting the LED light readout for battery status. These instruments enable a safe and accurate method to determine the status of backup batteries. Some battery test instrument models have a data-collection and data-storage feature that enables test results from multiple wind turbines to be captured and downloaded to a spreadsheet to aid in predictive analysis of each backup power system.

Test results from either of these systems should be documented and used for future reference on the status of backup power systems. Accumulating this data in a spreadsheet software file will also enable trend analysis to determine when battery packs will need to be replaced before a failure occurs. This type of trend analysis is very useful for predictive analysis and improving farm management activities. Analysis of the data from multiple wind turbines may show which activities can wait for future PM activities and which will need to be scheduled earlier. The data will also enable the purchase and delivery of several batteries for a major change-out activity. Bulk purchase of batteries may help reduce component cost and shipping charges, not to mention eliminate unscheduled equipment downtime that can reduce equipment availability.

VIBRATION ANALYSIS

Vibration analysis is a useful maintenance tool for monitoring changes in equipment operating characteristics of a machine throughout its service life. Like oil analysis or wire-insulation resistance analysis, vibration analysis is a tool that can be used to improve overall performance of equipment beyond the PM activity. Each of these tools allows tracking of equipment characteristics that can aid in predicting failures. Understanding equipment condition and predicting failures can be used to schedule repairs before they create downtime and possibly cause catastrophic failures. Effective use of vibration monitoring requires multiple measurements over time at several locations on the equipment to ensure adequate data points for analysis. Data collected during vibration-monitoring activities will include items such as frequency, amplitude, and equipment speed (RPM). **Frequency** is the number of signal repetitions over a period of time. Typical units for frequency are cycles per second (cps) (or hertz) and cycles per minute (cpm). One important frequency level to note for a piece of equipment is the natural or resonance frequency. The **resonance** frequency level is the frequency when the vibration amplitude is at a maximum because of its mechanical configuration. Operating a piece of equipment for an extended period of time at this frequency will cause premature failure and possibly catastrophic failure.

Amplitude is the measured value of signal strength at any given time or frequency level. Amplitude readings for vibration-monitoring activities would be taken in terms of displacement, velocity, and acceleration. **Table 7-3** shows examples of amplitude readings and units.

Collecting initial measurements on new equipment or using historical data supplied by the equipment manufacturer can provide a suitable baseline for vibration analysis. Without a baseline, there would be nothing to compare subsequent measurements to determine whether conditions have deteriorated. Deciding what equipment should be monitored for vibration analysis is up to design engineers and experienced technicians within an organization. Any piece of equipment that slides or rotates during operation may be a candidate for vibration monitoring. Primary causes of vibration in rotating machines are the result of misalignment and unbalance. Rotating equipment found in a wind turbine include components such as motors, generators, pumps, fans, and gearboxes. Component wear or lubrication contamination within bearing elements, couplings, gears, and pump impellers can produce noticeable changes in vibration. Accumulation of contaminants such as lubrication and dust on fan blades can also create unbalance and subsequent increases in vibration levels. An increase in vibration levels can lead to damage and premature failure of seals, couplings, and other machine elements. Factors used to determine equipment candidates for vibration monitoring include component cost, labor requirements, and how critical the component is to safe operation. Components in a wind turbine that meet these criteria may be the generator, gearbox, and main bearing assembly. A failure in any of these components would create extended downtime and a safety hazard, as well as require specialized equipment and technicians to complete repairs.

Equipment Summary

Vibration-monitoring equipment can be summarized in a couple of categories: vibration measurement and vibration analysis. Vibration measurement equipment can measure, display, and store the data values. Vibration analysis equipment not only can measure, display, and store data values but also display the information in a graphical form to aid in interpreting the results. Measurement display options

TABLE 7-3

Units Associated with Vibration Amplitude Measurements		
	Units	
Amplitude	**U.S. Customary Units**	**Metric**
Displacement	Inches (in.)	Millimeters (mm)
Velocity	Inches per second (ips)	Millimeters per second (mm/sec)
Acceleration	Inches per second squared (in./s^2)	Millimeters per second squared (mm/s^2)

TABLE 7-4

Vibration-Monitoring Functions	
Test Function	**Display**
Temperature	°F or °C Requires optional temperature probe
Speed	RPM Requires optional tachometer probe
Filters	2 Hz, 100 Hz, ISO Used to eliminate frequency ranges that may interfere with test-measurement accuracy. Consult the manufacturer's test-instrument documentation and site procedure for details on use of filter options.
Velocity	Ips or mm/s
Acceleration	gPK (peak) or gSE (spike energy)

© Cengage Learning 2013

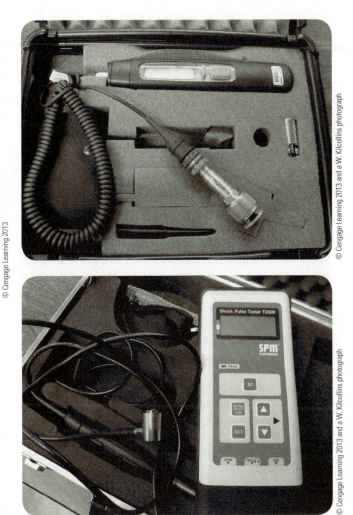

© Cengage Learning 2013 and a W. Kilcollins photograph

FIGURE 7-30 *Examples of vibration test instruments and attachments*

with a typical vibration-monitoring instrument are shown in **Table 7-4**. The display layout, functions, and units are dependent on the manufacturer's specifications and included test accessories. **Figure 7-30** shows examples of vibration test instruments and attachments.

Operation

Vibration-monitoring equipment is used to determine the extent that a moving assembly has become misaligned, unbalanced, or to assess the amount of wear since the last monitoring activity. Wear of moving bearing elements such as rollers or balls enable the inner race assembly to slide relative to the axis of the bearing assembly. The inner race of a bearing assembly is used to support the drive shaft so that movement of the inner race enables axial movement of the shaft. Movement or displacement of the shaft within the equipment is known as **axial runout**. This condition can cause coupling damage, misaligned gears, or issues with other power-transmission components. This displacement can be measured with the equipment not running using an approved LOTO procedure and a dial indicator with a test stand. Forcing the shaft assembly along the axis of the equipment will show up as a linear measurement on the dial indicator. This increase in motion can also be seen as a change in the vibration characteristics along the same equipment axis as the shaft.

Wear in the bearing assembly can also cause a horizontal or vertical displacement of the shaft as viewed looking at the bearing face. Movement in this plane is known as **radial runout**. This displacement can also be measured with a dial indicator test stand mounted perpendicular to the shaft. Vibration measurements measured with the test probe located perpendicular to the horizontal and vertical planes at each end of the assembly will also show this change in vibration characteristics. **Figure 7-31** shows the test positions for these measurements. Vibration monitoring in the horizontal and vertical planes can also show bearing contamination.

Contaminated lubricant within a bearing may cause the moving elements to slide or shift in the race instead of following normal rolling action. This abnormal motion will show up as an acceleration spike in the vibration reading. Other issues that will show up as changes in vibration characteristics because of unbalance include damage to gears or other rotating components, buildup of contaminants, and missing components.

Test measurements for vibration monitoring should be taken as close to bearing locations as practical for best results.

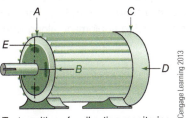

Test positions for vibration monitoring

© Cengage Learning 2013

FIGURE 7-31 *Test positions for vibration monitoring*

TABLE 7-5

Table for Collecting Generator or Motor-Vibration Test Data					
		Test Locations			
Speed Level RPM	A	B	C	D	E
1					
2					
3					
4					
5					
6					

© Cengage Learning 2013

In the case of a motor or generator assembly, the bearing assemblies are located within the end plates. **Table 7-5** shows an example of a data-collection sheet for vibration monitoring of a generator or motor. Collect test data as recommended by the preventative-maintenance procedure and document findings for further analysis. Compare the test results to the vibration test equipment's manufacturer recommendations to determine appropriate action. When a problem is found, schedule corrective action to repair or replace the component that is causing the excess vibration. Operating equipment with excessive vibration not only may create unscheduled downtime for that component to be replaced but also damage other components in the equipment because of induced vibration in other components known as **sympathetic vibration**.

Data collection at different equipment speed levels may help determine whether resonance may be an issue.

SUMMARY

Test equipment used for preventative-maintenance activities can include visual aids to improve the inspection of many electrical and mechanical systems in a wind turbine. These instruments include digital photography and video devices that can document the condition of equipment with easy access. Other instruments are available to aid inspections in hard-to-reach locations such as in gearboxes or in generator assemblies. Infrared thermography can aid visual inspection as an enhanced set of eyes that can assess the status of components that may have an elevated temperature signature because of wear or poor electrical connections. Available electrical test equipment can be used for a variety of tests such as assessing motor assemblies, generator components, and power-conversion equipment, along with power cables

that are used to deliver electricity from the turbine to the utility grid. These tests include insulation resistance, evaluation of electrical signals, and battery backup power equipment.

The use of test measurements can be enhanced even further through documentation of test results gathered during PM activities and using the data for evaluation of how farm equipment is performing over time. This evaluation process can be very useful in predicting an equipment failure so that repairs can be scheduled before downtime occurs.

REVIEW QUESTIONS AND EXERCISES

1. List equipment that may be used to aid in a visual inspection activity.
2. Describe the relationship between IR frequencies and the visual portion of the electromagnetic spectrum.
3. What advantage does thermography have in some PM activities?
4. What are several safety precautions that should be considered when doing wire-insulation resistance measurements?
5. Describe how Ohm's law can be applied to insulation-resistance–measuring instruments.
6. Name a few PM activities that require the use of a mega-ohm meter.
7. How can a mega-ohm meter be useful in determining the useful life of a motor?
8. Name environmental factors that can deteriorate the wire insulation in a motor assembly.
9. Calculate the value of the power cable insulation resistance if 1000 V_{DC} is applied between two adjacent power cables and the current between them is measured at 0.002 amps.
10. What are some of the safety requirements to consider when performing electrical-system tests?
11. Calculate the time for one cycle of an alternating wave if the frequency is measured at 120 Hz.
12. Calculate the crest factor for an electrical signal given the following: $V_P = 163 \ V_{AC}$, and $V_{RMS} = 115 \ V_{AC}$. Is this a sine wave without distortion?
13. What are several safety precautions that should be considered when doing battery-load testing?
14. What is the relationship between battery internal resistance and useful life?
15. Why is it important to track and analyze data from PM activities?

Component Alignment

KEY TERMS

angular misalignment
axial misalignment
bedplate
compression coupling
flexible coupling
gearbox ratio
gear ratio (GR)
horizontal axis wind turbine (HAWT)
laser
main bearing

main shaft
parallel misalignment
pitch diameter (PD)
prime mover
radial misalignment
rotor assembly
slip ring
soft foot
synchronous speed

OBJECTIVES

After reading this chapter and completing the review questions, you should be able to:

- Describe the different components that make up a wind-turbine drive system.
- Describe the difference between drive systems with a gearbox and those that are direct drive.
- Describe the different types of shaft misalignment.
- Describe some of the issues created by a misaligned generator.

- Describe the process to align components with a laser system.
- Describe the issues that may create a soft foot condition.
- Describe the process of correcting a soft foot condition.

INTRODUCTION

The typical wind-turbine drive system is made of several components such as a rotor assembly, bearings, a gearbox, shafts, couplings, and a generator. Many of the newer wind-turbine designs have been developed without a gearbox in the drive system. These direct-drive systems use a large generator assembly that has hundreds of magnetic poles per phase to enable them to produce the required output AC frequency at much lower shaft speeds. Utility-sized generators produce three-phase power to match the utility grid. Typical gearbox drive systems run the generator at 1,500 to 1,800 revolutions per minute (RPM), depending on the utility grid frequency. Typical European and North African grids use a frequency of 50 Hz, whereas North American and most South American countries use 60 Hz. **Figure 8-1** shows the typical arrangement of a wind-turbine drive system with a gearbox.

To minimize mechanical issues with drive systems, close tolerances are placed on component shaft alignment and a

coupling system is used. A **flexible coupling** is a coupling design used to transfer power between two shafts with a flexible structure to reduce shaft stress created by minor misalignment. Power transfer between shafts is a function of speed and torque. The formula for power can be represented as

$$P = \frac{(T \times S)}{9,549}$$

where P is power (kW), T is torque (N-m), and S is shaft speed (RPM); 9,549 is a constant used with these measurement units. This relationship is important because it shows the forces exerted on coupling components as a result of the power transfer from the rotor assembly through the gearbox to the generator. The torsion forces acting on the main drive components are very high during normal operation. These forces in combination with any bending forces resulting from misalignment can cause premature failure of drive-system components. **Figure 8-2** shows a typical flexible coupling that may be found connecting the gearbox high-speed shaft to the generator. Note the electrical isolation section of the coupling,

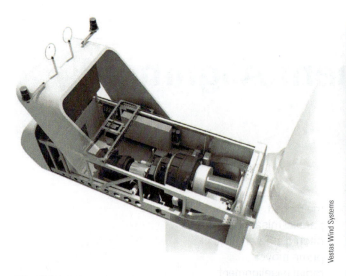

FIGURE 8-1 *Typical arrangement of a wind-turbine drive system with a gearbox*

Vestas Wind Systems

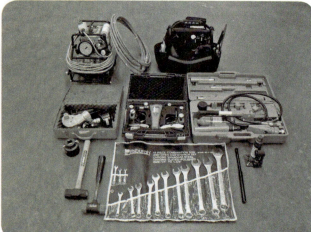

FIGURE 8-3 *Typical laser-alignment toolkit*

© Cengage Learning 2013 and a W. Kilcollins photograph

FIGURE 8-2 *Flexible coupling that may be found connecting a gearbox high-speed shaft to a generator*

University of Maine at Presque Isle

which may be made of fiberglass or other ridged nonconductive material. This isolation section prevents current flow through the drive system should there be an electrical fault or loss of generator grounding. Having a flexible coupling in the system helps with minor misalignments but still requires an acceptable alignment between the gearbox and the generator. Poorly aligned shafts between the generator and the gearbox may cause issues such as reduced efficiency, shaft fatigue, excessive wear of the coupling, vibration, and bearing damage.

Aligning the generator properly with the gearbox shaft should be carried out on final field assembly of the wind turbine and verified during preventative-maintenance (PM) cycles. The initial configuration of the generator to the gearbox is typically mounted so that it appears higher than necessary at the factory because the bedplate flexes when the **rotor assembly** is mounted on the main shaft. The **bedplate** is the common frame assembly that is used to mount the drive

system. This assembly is also the frame that connects the nacelle components to the yaw bearing and tower. The weight of the rotor assembly will deflect the main bearing end of the bedplate down, which in turn can raise the back of the gearbox. Note of caution: whenever a service project requires the removal of the rotor assembly from the wind turbine, refer to the manufacturer's recommendations for removing the flex coupler. Removal of the rotor assembly will return the bedplate back to its unloaded state. This change in position of the drive components may damage the flex coupler.

Aligning the generator to the gearbox can be accomplished with a measuring device such as a large caliper or a dial indicator with a bit of math or with laser-alignment equipment. The laser-alignment equipment can do the math for you. A typical wind-turbine generator-alignment kit includes the laser equipment, a selection of spacers or shims, a portable fluid-power unit, pancake hydraulic cylinders, and an assortment of hand tools. A typical laser-alignment kit is shown in **Figure 8-3**. Laser-alignment equipment, tools, and safety information will be discussed later in the chapter. Our focus now will be on explaining the different drive components and their functions.

DRIVE-SYSTEM COMPONENT IDENTIFICATION

The drive system of a **horizontal axis wind turbine (HAWT)** is made up of several components that are typical to most commercial systems today. As mentioned previously, the use of a direct-drive system in the three-plus megawatt units being developed eliminates the need for a gearbox. The discussion in this section will focus around the components that would be found while performing PM activities in the wind-turbine models with a gearbox. The higher shaft speed

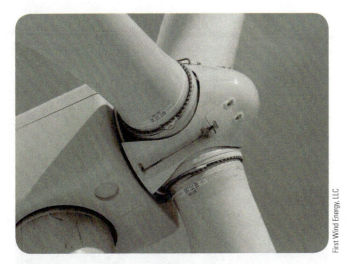

First Wind Energy, LLC

FIGURE 8-4 *A HAWT rotor assembly*

FIGURE 8-5 *A hub adapter*

First Wind Energy, LLC

necessary to drive these systems requires very close shaft-alignment tolerances between the gearbox high-speed shaft and the generator.

Rotor Assembly

The hub assembly is the portion of the drive system that serves to attach the blades. The hub assembly with the blades attached is considered the rotor. **Figure 8-4** shows an example of a HAWT rotor assembly. This assembly may contain the control-system components necessary to provide pitch control and emergency backup power to feather the blades in the case of grid power loss. This assembly may weigh in excess of 40 tons, depending on the model's power-output rating. For example, a 1,500-kW (1.5-MW) wind-turbine model with a 250-foot (80-m) rotor diameter has three 5.5-ton blades and a cast-steel hub assembly that may weigh around 15 tons. With this type of weight, it is no wonder the addition of the rotor assembly can cause the bedplate assembly to flex.

Hub Adapter

The hub adapter is a large flange assembly that allows the connection of the rotor assembly to the main shaft. **Figure 8-5** shows an example of a hub adapter. This connection is made through several studs threaded into the hub and held to the adapter through the use of large nuts. These nuts are inspected for integrity through the use of hydraulic bolting equipment set to a specified torque value during each PM cycle. Details on this activity will be discussed in a later chapter.

Main Bearing

The **main bearing** assembly is used to support a large portion of the rotor assembly's weight. The combination of the main bearing and the gearbox mounting pillow blocks trans-

fer the weight of the rotor assembly, main shaft, and gearbox to the bedplate assembly.

Main Shaft

The **main shaft** assembly is used to transfer the wind power captured by the rotor assembly to the gearbox. The main shaft of typical utility-scale wind turbines is a large-diameter steel shaft with a bore through the center to run control cables and electrical power cables or hydraulic line, depending on the pitch system design. This shaft assembly may be a few hundred millimeters (around 12") in diameter or larger, depending on the wind-turbine model.

Compression Coupling

A **compression coupling** is a collar that connects the main shaft to the gearbox input shaft. A compression coupling uses a series of bolts located around a collar to compress a tapered bushing onto the main shaft. This type of connection is used in many industrial applications to attach power-transmission devices to motors or generators. This particular design uses a compression coupling assembly that has an outer flange diameter around 1 meter (3.25 feet) or larger. **Figure 8-6** shows an example of a compression coupling.

Gearbox

The gearbox in a wind-turbine generator is used to increase the input speed of the prime mover (rotor assembly) to the synchronous speed of the generator. A **prime mover** is a device used to provide mechanical power to equipment such as a generator. A common example of a prime mover is a gasoline engine. A gasoline engine is a device that converts energy from the chemical reaction of burning fuel to a rotational

SAMPLE PROBLEM

Determine the torque transferred by the main shaft to the gearbox of a utility-sized wind turbine. The generator for this wind turbine has a rated capacity of 3,000 kW (3-MW) at 1,500 RPM for 50-Hz output. The gearbox input shaft speed for a 1,500-RPM gearbox output speed is 25 RPM. Consider the efficiency (η) of the gearbox to be 90% for the calculation.

SOLUTION

Use the power formula provided previously in the chapter,

$$P\ (kW) = \frac{[T(\text{N-m}) \times S(\text{RPM})]}{9,549}$$

for the calculation. Given P = 3,000 kW and S = 1,500 RPM,

$$P = \frac{(T \times S)}{9,549}$$

Rearranging the formula to calculate for T yields

$$T = \frac{(P \times 9,549)}{S}$$

So,

$$T = \frac{(3,000\ \text{kW} \times 9,549)}{1,500}$$
$$= 19,098.6\ \text{N-m}$$

This would be the required torque input to the generator for the rated output of 3,000 kW if the system were 100% efficient. With a gearbox efficiency rating (η) of 90%, the rotor assembly has to provide more input power to the gearbox to get the 3,000-kW output at the generator. So the actual power-input requirement may be calculated as

$$3,000\ \text{kW} = P_I \times \eta$$
$$P_I = \frac{3,000\ \text{kW}}{0.90}$$
$$= 3,333\ \text{kW}$$

where P_I is the input power requirement. Calculating the main shaft torque to provide the 3,333-kW input power:

$$T = \frac{(3,333\ \text{kW} \times 9,549)}{25\ \text{RPM}}$$
$$= 1,273,113\ \text{N-m}$$

This calculation shows the relationship between torque and shaft speed. A large decrease in shaft speed (RPM) made a large increase in required torque. With this amount of torque, it is no wonder the main shaft is so large in diameter compared to the generator shaft.

output. Coupling a gasoline engine to a small electrical generator provides a source of electricity. Typical prime movers in the power-generation industry include natural gas, steam, and water turbine assemblies. These devices are used to

FIGURE 8-6 *A compression coupling*

transfer energy from a moving fluid to a rotational output for an electrical generator. The **synchronous speed** of a generator is the RPM value that is required to achieve the line or grid frequency. A previous discussion mentioned that the line frequency in Europe was 50 Hz and that of North America was 60 Hz. To achieve an output frequency that matches the utility grid, a generator has to rotate at a speed (RPM) that depends on its design. For instance, a generator designed with four poles per phase (two north and two south magnetic poles) has to rotate at 1,500 RPM to achieve a 50-Hz output:

$$S\ (\text{RPM}) = \frac{120[f(\text{Hz})]}{N\ (\text{poles})}$$

where S is generator synchronous speed, f is utility line frequency, N is the number of magnetic poles, and 120 is a constant that relates the variables. Remember from basic electricity that generating an electrical output requires three things: (1) a conductor coil connected to a load, (2) a magnetic field, and (3) the relative motion between the coil and the magnet field. The prime mover provides power (torque and rotational speed) to the gearbox, which in turn increases the speed to the generator synchronous speed. This is why wind turbines have been developed around a gearbox for many years. The basic design of a generator has been to use a small number of magnetic pole sets and a high shaft speed to produce electrical power at utility line frequency. Direct-drive wind turbines shift this paradigm on its head by using a low shaft speed and a large number of magnetic pole sets to provide the utility line frequency.

A simple gearbox design is made up of an input shaft, gears, frame, and output shaft. The gearbox assembly may be designed with multiple gear sets on parallel shafts or as a planetary system. A planetary gear system has a central gear

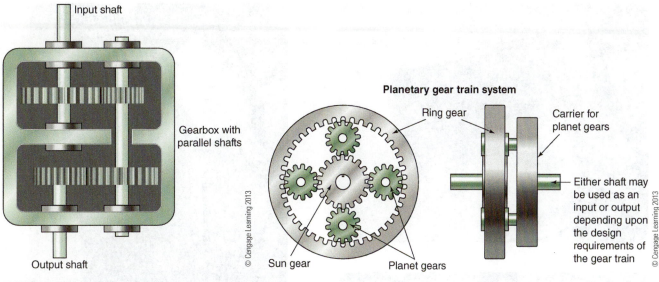

Input shaft

Gearbox with parallel shafts

Output shaft

Planetary gear train system

Ring gear

Carrier for planet gears

Either shaft may be used as an input or output depending upon the design requirements of the gear train

Sun gear

Planet gears

© Cengage Learning 2013

FIGURE 8-7 *Gearbox designs*

or *sun* gear, outer fixed ring gear, and pinion gears or planet gears located between them mounted to a carrier assembly. **Figure 8-7** shows a couple of gearbox designs. Good terms to remember when working with gearboxes are *gearbox ratio* and *gear ratio*. **Gearbox ratio** is the ratio between input shaft speed and output shaft speed. This value may seem incorrect with a wind turbine because the gearbox is actually used in reverse compared to standard industrial applications. Normally, the output shaft of industrial applications is low speed and the input shaft is high speed. This is not the case for a wind turbine. Another term is **gear ratio (GR)**, which is expressed as the ratio associated with the gear pitch diameters or the number of teeth on each gear in a set. **Pitch diameter (PD)** is the diameter that passes through the midpoint of the teeth on a gear. For example, two gears meshed together include one with a PD of 40 and one with a PD of 20. Say, for example, that the PD 40 gear drives the PD 20 gear. The gear ratio is determined by the formula

$$GR = \frac{PD \text{ driver}}{PD \text{ driven}}$$

So,

$$GR = \frac{40}{20} = 2 \text{ or } 2{:}1$$

A GR of 2:1 would double the output shaft speed compared to the input shaft. Our previous sample problem listed a gearbox that had an output shaft speed of 1,500 RPM with an input shaft speed of 25 RPM. What would the GR be for the gearbox in this problem? Knowing the change in speed from the input to the output gives

$$GR = \frac{1{,}500 \text{ RPM}}{25 \text{ RPM}}$$
$$= 60$$

or a GR of 60:1. There may be several gears in the gearbox, but the relationship of all of the gears gives a GR of 60:1. The larger number listed first in the ratio indicates that the output

shaft turns faster than the input shaft. The GR for a particular wind-turbine drive system would be found on the gearbox nameplate.

Another thing to consider with a gearbox is the rotation relationship between the input shaft and the output shaft. Even numbers of gears give an output shaft direction that is opposite to the input. If a gearbox with an even number of gears was turned clockwise (CW) on the input shaft, the output shaft would turn counterclockwise (CCW). Odd numbers of gears would yield the same rotation direction for the output and input shafts. Our previous sample problem did not specify the shaft rotations, but if it had stated that the input shaft rotated CW and the output shaft also rotated CW, then we may conclude the gearbox had an odd number of gear sets in the assembly. If you find this interesting, one specialty track for wind technicians is gearbox inspection and service. This specialty track develops a technician's skills to inspect and diagnose gearbox issues. If the gearbox can be serviced, it is done so in the field. **Figure 8-8** shows a view into a gearbox assembly used in a wind turbine. Complicated repairs or teardowns would require removing the gearbox and returning it to the factory for service. The gearbox service technician would coordinate this replacement project with the site manager and the gearbox manufacturer.

Other items that would be encountered when working around a wind-turbine gearbox are the parking brake, slip-ring assembly, and ancillary lubrication equipment. **Figures 8-9 and 10** show examples of a parking brake and slip-ring assembly. The **slip-ring** assembly enables the transfer of electrical power and control signals to the hub control equipment through a series of rotating rings and stationary brushes. A stationary brush in contact with a ring attached to the rotating gearbox main shaft can complete the electrical circuit from the stationary nacelle to the hub. This type of device has been used in many other industrial applications when a power or control circuit has to be used with a moving system.

FIGURE 8-8 *View of gearbox assembly with access panel removed*

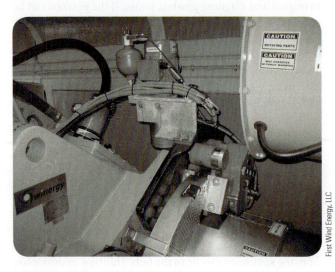

FIGURE 8-9 *Parking-brake assembly*

FIGURE 8-10 *Slip-ring assembly*

Flexible Coupling

A coupling assembly is used to connect the output shaft of the gearbox to the generator. Couplings can be made in many design configurations, depending on power-transmission requirements of the drive system and the amount of misalignment that can be tolerated. Many motor drive systems used with fluid and ventilation systems use a three-part coupling that includes two hubs that interlock with a flexible component. The flexible component can be as simple as a rubber "spider" that transfers the power between the hub components or as complex as a series of metal spring components. The typical coupling connecting drive motors to pumps and gear drives in a wind turbine is a simple hub-and-spider design. **Figure 8-11** shows an example of a hub-and-spider coupling assembly. The coupling assembly used to transfer power between the gearbox and generator is a much larger design with flexible components such as flat metal springs and a central electrical isolation component. **Figure 8-12** shows an example of a coupling used between the gearbox and generator.

Generator

The generator assembly is used to produce electrical output from the mechanical input captured from moving air (wind). The design used with a gearbox, as mentioned previously, has high speed and a lower number of magnetic poles as compared to the newer direct-drive wind-turbine models. It may be relatively smaller than direct-drive generators, but they

FIGURE 8-11 *A hub-and-spider coupling assembly*

FIGURE 8-12 *Example of a flexible coupling used between gearbox and generator*

FIGURE 8-13 *Generator with view of mounting feet*

are still massive in size. These generators are not moved easily for service work unless a hydraulic jack or hydraulic cylinder and power unit are used. A discussion on moving the generator for alignment will be covered later in the chapter. **Figure 8-13** shows the generator with a view of the mounting feet. An acceptable alignment of the generator to the gearbox is recommended by the manufacturer of each drive component: gearbox, coupling, and generator. If care is not taken with the original assembly alignment and subsequent PM alignment activities, then the components can be damaged. A couple of hours spent with each PM cycle on alignment can save thousands of dollars in component-replacement cost and equipment availability.

IDENTIFICATION OF MISALIGNMENT PROBLEMS

Shaft misalignment problems can appear in several ways during operation of the wind turbine. Misalignment can cause reduced system efficiency and excess vibration, along with premature shaft, coupling, and bearing failure. Reduction in system efficiency can be created because of binding of the coupling components during rotation. This binding creates friction between the moving elements of the coupling and can be seen as excessive heat buildup in the assembly. This may be seen as a discoloration of the flexible elements or the collars attached to the shafts. If the issue is not corrected, then the friction and excess loading in the coupling can cause the flexible components or shafts to fracture and break. A failure of any of these components during operation would be catastrophic. This type of coupling failure would damage any electrical, hydraulic, or structural components in the path of any ejected components.

Alignment issues do not have to be catastrophic to be observed with turbine operation. A turbine may show anomalies with its SCADA data as compared to other units with an acceptable alignment. The data may show reduced electrical output, higher levels of vibration, or an elevated temperature of the gearbox and generator bearings closest to the coupling connection. Excess vibration can create issues with sensors, printed circuit (PC) boards, or electronic components such as programmable logic controller modules located in the nacelle or rotor assembly. Excess vibration can also loosen wire connections that are vital to communication, control, and power circuits. Elevated temperatures and the associated stress on bearing assemblies can cause the premature failure of these components. Anytime these flags are observed, the system should be shut down and investigated to determine the root cause of the problem.

Other signs of misalignment can be seen during a walk-around inspection of the generator, coupling, and gearbox. Any debris that may have been chafed or abraded from components should be investigated. Observing the flexible elements of the coupling will indicate whether the elements are in a relaxed position or whether they are distorted because of misalignment of the assembly. The types of misalignment that may be seen in the drive system will be discussed in the following sections.

Types of Misalignment

Shaft misalignment between the gearbox and generator may be of three types: axial, angular, or parallel. Each of these misalignments is caused by the orientation of the generator shaft with respect to gearbox output shaft. As you recall, the gearbox is mounted rigidly to the bedplate as part of the structural support for the rotor assembly. The generator is not structurally connected to the gearbox but only connected through a flexible coupling assembly that allows the transfer of power. It would require a crane to move the gearbox because of the weight it supports on the input shaft and mounting blocks. Alignment of a wind-turbine generator is not as easy as moving a motor connected to a pump in many industrial settings, but it can be accomplished with proper training and the use of appropriate tools.

Axial

Axial misalignment is created by improper spacing of the gearbox and generator. The two components can be out of tolerance by spacing that is too wide or too close. If this shift occurs after initial installation of the coupling, then it will cause the coupling flexible components to be compressed or tensioned, depending on the condition. Either condition will cause excess loading of the coupling assembly along with excess force applied to the gearbox, generator shafts, and associated bearings.

Angular

Angular misalignment is caused by a shift in the angle of the generator with respect to the gearbox. This misalignment

FIGURE 8-14 *Example of angular misalignment*

type may be seen in both the horizontal and vertical planes. If the generator angle is out in either plane, then it will create a binding or cyclic side loading of both shafts and the coupling assembly. Cyclic loading will create a bending stress in the associated shafts and may ultimately cause a fatigue failure. A fatigue failure would start as a surface crack on the component that would propagate until failure occurred. Correcting this may be done by adjusting the generator height, back or front, or by shifting the generator from right to left to correct the angular misalignment. **Figure 8-14** shows an example of angular misalignment.

Parallel

Parallel misalignment or **radial misalignment** of the generator to the gearbox is caused by the generator being parallel to the gearbox axis but positioned higher, lower, left, or right of the axis. This alignment issue can create side loading on the shafts and bearing assemblies. Extended operation of the drive system in this condition can cause fatigue and failure of the shafts just as with the angular misalignment. Side-loading bearings can distort the assembly, deteriorate the surface of moving elements, and generate excess heat. Each condition will substantially reduce the operational life of bearings. Correcting this can be done by repositioning the generator back on axis with the gearbox. **Figure 8-15** shows an example of parallel misalignment.

ALIGNMENT PRACTICES

Alignment of motor assemblies to pumps and other rotating equipment has been a standard practice in the industrial setting for many years. The main purpose of component alignment has been to reduce power consumption, vibration, and shaft fatigue. Alignment practices have also been implemented with large drive systems in power-generation

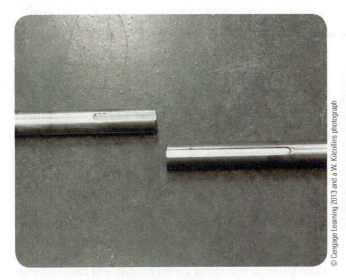

FIGURE 8-15 *Example of parallel misalignment*

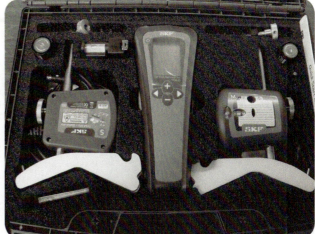

FIGURE 8-16 *Typical laser-alignment kit*

facilities with the goal of improving system efficiency and eliminate other mechanical issues created by misalignment. Alignment procedures have incorporated the use of dial calipers, dial indicators, and other measuring devices to determine offset of the coupling hubs and then use a bit of geometry to determine appropriate corrective action. The wind-power industry has adopted the same PM philosophy to ensure that the generator is aligned with the gearbox to eliminate alignment issues.

For the purpose of our discussion, we will look at the use of laser equipment to perform the generator alignment. Some technicians may prefer to use a dial indicator and dial calipers over a laser system because of their previous experience. The use of these tools may be used to compare techniques during our discussion. Make no mistake: these measuring devices are equally capable of providing an acceptable alignment in the hands of a qualified technician. A laser system removes the mental gymnastics required for geometry in an alignment activity. These aspects may be removed, but a technician should still understand how the math relates to the activity so they do not blindly follow the numbers of a malfunctioning unit. Laser equipment is only one component of the tools needed to perform an acceptable generator alignment. The next section will discuss this device and other tools necessary to perform an alignment.

ALIGNMENT TOOLS

Laser-alignment equipment uses three main components to measure and give feedback on an alignment process: a control unit, a transmitter, and a receiver. **Figure 8-16** shows a typical laser-alignment kit. The term **laser** stands for *light amplification by stimulated emission of radiation*. The laser source in this application is used to produce a fine coherent beam of light that can be transmitted and received between two devices to determine a distance. Other laser system

designs may be used for communication, cutting, welding, or medical applications. The National Aeronautics and Space Administration has used laser devices to accurately measure long distances between objects such as Earth and the moon since the 1960s. If this type of system can measure these large distances accurately, then it is more than capable of measuring the distance between coupling hubs on a drive system. Laser-alignment equipment may be designed in several variations: two heads that both transmit and receive the laser signal, one head that transmits and a second to receive the signal, and one head to transmit and receive the signal with a second head to act as a reflector for the signal. The control unit is used to coordinate the signal between the two measuring heads and display distance values to the operator. **Figure 8-17** shows an example of the display on a laser-alignment control unit. More information on these components and the meaning of the display values will be discussed in the alignment procedure section to follow.

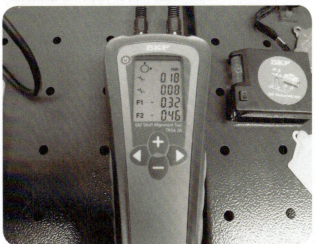

FIGURE 8-17 *Display on a laser-alignment control unit*

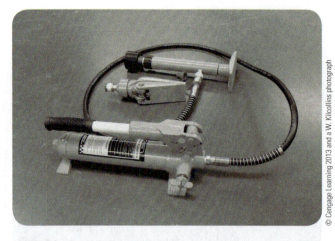

FIGURE 8-18 *Typical hydraulic power components*

The task of moving a generator assembly weighing several tons is not easy. To do this safely and efficiently, technicians need to use tools to carefully lift and move the generator. The use of a hydraulic jack or a hydraulic power unit and small hydraulic cylinders is necessary. These devices can lift and lower heavy objects easily and safely. Placing a jack or hydraulic cylinder on its side makes it useful in pushing the generator from side to side to aid in the horizontal positioning process. **Figure 8-18** shows typical hydraulic power components that may be used to move the generator during an alignment procedure.

The use of specialty tools such as jack screws may also be used to move the generator from side to side as part of the alignment procedure. These specialty tools have been designed to lock onto the mounting stud and apply a side force to the generator feet for positioning.

Once the generator has been lifted to adjust the height of the assembly, a series of shims may be added under the feet to maintain the proper height. **Figure 8-19** shows an assortment of shims that may be used to adjust the height of a generator. These shims may be purchased in a variety of thicknesses from 0.001" up to larger values such as 0.50". The shims may

also be purchased in a variety of metric thickness as well, depending on the measuring system that is used for the activity.

ALIGNMENT PROCEDURE

The specific generator-alignment procedure used will depend on the wind-turbine model. Different manufacturers use different system layouts, depending on their design philosophy and the components used. Refer to the alignment practices of the original equipment manufacturer (OEM) for your wind-turbine model. This text will cover a generic alignment procedure that a technician may see. Completing a generator alignment, like any activity, requires practice to become proficient, so do not get discouraged when your colleagues boast about their time to complete an alignment. The main things to remember are do it safely and correctly. Proficiency will come with practice.

Safety Practices

Safety is a key component of any activity, including generator alignment. Always ensure the control system is set to manual operation and appropriate LOTO procedures are implemented. Remember that the gearbox input shaft is connected to the rotor assembly and will spin freely if the parking brake and a portable drive system are not used to control shaft rotation. Allowing the gearbox output shaft to spin freely will damage the alignment equipment and create a possible pinch or entanglement hazard with an unsuspecting technician. During normal operation of the wind turbine, a guard assembly is located over the coupling and parking-brake assembly to prevent service technicians from coming into contact with the parking-brake caliper assembly, brake disc, and the spinning coupler. **Figure 8-20** shows a view of the parking brake moving components along with the exposed coupler.

FIGURE 8-19 *Assortment of shims used with alignment process*

FIGURE 8-20 *Parking brake components along with exposed shaft coupler*

Wind speed and direction should also be observed during the generator-alignment process. Typically, the equipment manufacturer will specify the maximum wind speed allowed for a successful alignment. This information may be listed in the maintenance manual or other service reference material supplied with the wind turbine. Values for maximum wind speed may range from 5 to 9 meters per second (m/s) [around 10 to 20 miles per hour (mph)]. This number is important because it minimizes the flexing of the bedplate assembly and reduces variation in the numbers that appear on the laser control. Higher wind speed will also increase the loading on the portable equipment used to rotate the gearbox output shaft during the alignment procedure. Another item to keep in mind is direction of the wind with respect to the rotor assembly. Turn the rotor assembly into the wind to reduce side loading on the assembly, which may increase variation in the values seen on the alignment control-unit display. Chasing a moving target will increase the time to complete the procedure and decrease the precision of the final outcome.

Before starting the process, place the wind turbine in manual operation, disable remote access, and notify the farm operator, operations department, and any remote-monitoring service of the work to be performed. This will ensure that no one can access the tower control system and create a safety hazard. Turning the rotor assembly into the wind not only will reduce side loading on the assembly but also aids in rescue should it become necessary. Positioning the rotor upwind will push items away from the blades and tower when they are being raised or lowered. These steps should be completed down tower before making the climb to the nacelle.

Alignment Setup

This section points out many of the major steps that are necessary to complete a generator-alignment procedure. Consult the manufacturer's recommended procedure or approved site procedure where you work and use this text as a supplemental guide.

Once the equipment and personnel are located in the nacelle, place the parking brake in manual operation and activate it to eliminate the gearbox output shaft from rotating during the alignment setup procedure. Engage the manual locking device on the parking-brake disc, if equipped, and then remove the safety guards located over the brake disc and coupling assembly. Place these components and any mounting hardware in a safe location out of the way to eliminate a tripping hazard and prevent loss of items. If the brake disc is not equipped with a manual locking device, then use an approved means to prevent accidental rotation.

Inspect the parking-brake assembly, pads, and disc to ensure proper function. Refer to manufacturer's specifications for acceptable brake wear and pad adjustments. The parking brake is used during the alignment process so this inspection will ensure the equipment is functioning properly. It may be the only time a visual inspection is possible during the PM

process, so take advantage of the situation. Brake pads that are out of adjustment will also side load the gearbox output shaft and create a noticeable change in the alignment values each time the brake is applied during the procedure.

Mount the portable drive equipment supplied by the wind-turbine manufacturer to the gearbox output shaft and ensure proper operation and control. Follow the manufacturer's recommendations or approved site procedure to complete this activity. **SAFETY: Improper control of the parking brake or pitch position of the blades can allow the rotor assembly to pinwheel and create an entanglement hazard. Placing body parts between the brake pads and brake disc may also create a pinching hazard. Once the drive equipment is installed, it can be used to move the gearbox output shaft to the required positions necessary for an accurate alignment activity.**

Alignment Equipment

Layout the laser-alignment components and tools to ensure everything is ready for use. The equipment should have been inspected and the calibration verified before it was loaded in the service truck at the operations and maintenance facility. This step ensures that everything is accounted for and nothing has been left down tower or damaged during transit or lift.

Assemble the alignment mounting hardware and mount it securely at the 12 o'clock position on the shafts at each end of the coupler assembly. **Figure 8-21** shows a typical setup of the laser-alignment heads and control-unit connections. Position the heads on the mounting hardware as follows: the head assembly indicated as the "stationary" unit should be mounted on the coupler side toward the gearbox, and the head indicated as "movable" should be mounted on the generator side of the coupler. The heads should be orientated so the laser light passes between them. The stationary unit is considered because it is connected to the gearbox. The gearbox is not moved during the alignment process for reasons mentioned previously in our discussion. The movable unit is connected to the generator side because it moves along with the generator during the alignment process. Now connect the two heads to the control unit, matching the appropriate cables to the jacks on the controller. Stationary cable should connect to the stationary jack and movable cable to the movable jack as indicated by markings on the controller. Energize the laser equipment and set the units of measure as needed for your procedure. Refer to the laser equipment manufacturer's material for specifics on operation and adjustments. Align the heads so that the laser light is aimed as necessary at the detector windows for the procedure and use the included torpedo level with the set for correct orientation. **CAUTION: Do not aim the laser light into your eye. Typically, the laser light emitted from an alignment system is low power, but it is always a good practice to follow this recommendation.**

Next, measure the distance between the laser heads and enter this value in the control unit as the A reference value. Ensure that the measurement units entered match the system

FIGURE 8-21 *Typical setup of laser-alignment heads and control-unit connections*

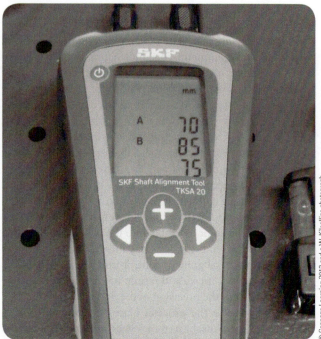

FIGURE 8-22 *Control-unit display*

being used. Inputting a value in millimeters when the system is set up for inches would create an issue when using the display values for the appropriate adjustments later. Measure the distance between the laser head closest to the generator

and the generator front foot slot centerline. Input this value in the control unit as B. Finally, measure the distance between the foot slot centerlines and enter this in the control unit as C. These values will be used by the control unit to determine the extent the generator has to be moved to get an acceptable alignment. **Figure 8-22** shows an example of the control-unit display for the input of the reference values A, B, and C.

Now check the setup for alignment of the laser light into the detector windows. Disengage the parking brake and use the portable drive unit to rotate the shaft and heads to the 9 o'clock position back to 12 and then to the 3 o'clock position and ensure that the laser light remains in the detector window. Use the spirit levels located in the two head assemblies to determine when they are level to the horizontal at both the 3 and 9 o'clock positions. Make adjustments as necessary and repeat the process to ensure the adjustments are correct. Move the shaft back to the 12 o'clock position and engage the parking brake and prepare for the initial alignment reading on the generator.

Initial Alignment

Checking the generator for initial alignment is accomplished by disengaging the parking brake and rotating the shaft and heads to the 9 o'clock position as viewed from the back of the generator. Use the spirit levels located in the two head assemblies to ensure that the horizontal position is achieved and then enter the location as directed by the display menu. Rotate the shaft and heads to the 3 o'clock

FIGURE 8-23 *Control-unit back cover*

position and enter the value as indicated on the control-unit display. The values displayed on the control unit as F1 and F2 will indicate the alignment in the horizontal plane. F1 indicates the adjustment of the front feet, and F2 indicates the adjustment of the back feet. Refer to the equipment instructions or the back of the control unit for the orientation information of left and right with respect to a negative (−) value on the display. **Figure 8-23** shows a view of the control-unit back cover. Write these numbers on the alignment inspection report for initial horizontal alignment. Now rotate the shaft to the 12 o'clock position and enter the position on the display as indicated. The F1 and F2 values displayed on the control unit now indicate the generator alignment in the vertical plane. Refer to the equipment instructions for the adjustment with respect to a negative F1 or F2 value. Write these values in the inspection report under the section for initial vertical alignment. Appendix H shows an example of a generator-alignment inspection report. After the initial alignment is documented, compare these values to the manufacturer's requirements or **Table 8-1** for suggested maximum alignment tolerances with respect to equipment operation speed (RPM). The initial alignment process is now complete. If the initial alignment process shows all values within published specifications, there is no need to

proceed further. If any of the values exceed published specifications, a complete alignment procedure will need to be preformed. Part of the alignment procedure will include a check for a soft foot condition.

Soft Foot Correction

In a **soft foot** condition, one generator mounting foot is located above the plane of the other three feet. This condition can be created by variations in manufacturing of the generator frame, damage to the foot during shipment, or improper spacing of the foot during assembly. Remember from geometry class that three points determine a plane. If you do not, perhaps you have seen the concept in practice and did not realize it. If one of the four legs on a table or chair is shorter than the others, then it will rock between two legs of equal length, so the shorter leg can touch the floor by lifting the third leg of equal length. You probably attached a spacer or folded paper under the shorter leg to prevent the annoying motion. The same problem can occur with a generator assembly. If the soft foot condition is not corrected properly, it will create stress in the "shorter" foot assembly. This stress can cause issues with the generator frame or mounting hardware over time.

Determining a soft foot condition using the laser-alignment setup from the initial alignment procedure is as follows. With the heads located in the 12 o'clock position, change the operation mode to the soft foot mode. Refer to the alignment equipment operation manual for details. With the display in soft foot mode, a technician can loosen one back-foot mounting nut and determine the shift of the generator. Compare this shift with the manufacturer's specifications. If the generator motion does not exceed specifications, tighten the nut securely and proceed to the other foot in the back of the generator. Check the motion of the foot when the mounting nut is loosened and compare this to the specification. If it is within specification, then you can tighten the nut and move to the front feet. If it is out, then this foot may have a soft foot condition or the front foot diagonal to this foot may have the condition. For sake of discussion, we will say this foot is out of specification. Tighten the foot and move to the front foot that is diagonal to this foot. Loosen the mounting nut for this foot and compare the value to the specification. This foot is also

TABLE 8-1

Suggested Maximum Shaft-Misalignment Values				
RPM	**Angular (mm/100 mm*)**	**Parallel (mm)**	**Angular (0.001"/1"**)**	**Parallel (0.001")**
0–1,000	0.10	0.13	1.0	5.1
1,000–2,000	0.08	0.10	0.8	3.9
2,000–3,000	0.07	0.07	0.7	2.8
3,000–4,000	0.06	0.05	0.6	2.0
4,000–6,000	0.05	0.03	0.5	1.2

*Gap at coupling edge per 100 mm coupling diameter

**Gap at coupling edge per 1" coupling diameter

out but by a larger value than the back foot. The larger value indicates this is probably the soft foot. Using the control-unit display correction value indicated for the front (F1), add the recommended thickness of shims under this foot. You may have to lift the generator foot with the hydraulic jack to slide the shims under the foot. Only loosen the mounting nut and lift the generator enough to clear the stack of shims. Tighten the mounting nut and go back to the back foot diagonal to the one that was corrected and loosen it again. It should be within tolerance after adjusting the front foot. If not, then repeat the process until the four feet are within specification. Now return the control-unit display back to the main menu to continue the alignment process if necessary.

Alternate Method for Soft Foot Correction

Another method to correct for a soft foot condition is by using a dial indicator mounted on a magnetic stand. The process to determine a soft foot is similar to the one with the laser-alignment equipment except the measurement is read off the dial indicator. Mount the magnetic stand with the dial indicator on the generator mounting structure next to the foot to be inspected. Adjust the indicator so that the movable shaft is perpendicular to the foot top surface and the dial reads zero. Now loosen the mounting nut holding the foot and read the value shown on the indicator. **Figure 8-24** shows the dial indicator test setup. The foot will spring up if it is being forced down by the mounting nut. How does this value compare to the specification given previously? If it is within specification, tighten the nut and set up the same way on the next back foot. Loosen the nut and check this reading. Follow the same steps used with the laser-alignment process. If a foot is found to have movement out of the required specification, then tighten the nut and check the foot on the diagonal. Whichever foot has the largest deflection value will be the soft foot. Add the required thickness of shims under the foot to make the necessary correction. Repeat the process as necessary to ensure any movement is within specification.

© Cengage Learning 2013 and a W. Kilcollins photograph

FIGURE 8-24 *Dial indicator test setup*

Alignment Process

The generator-alignment process follows the same steps as listed previously for the initial alignment process. Use the F1 and F2 values displayed on the control unit for repositioning the generator. The rule of thumb is to make any vertical adjustments before completing the horizontal adjustment. If horizontal adjustments were completed first, then any subsequent adjustments in the vertical plane could change them. This would be twice the effort. A good phrase to remember for this activity is "Work smarter, not harder." Deactivate the parking brake and position the shaft and heads back to the 9 o'clock position and initiate a new alignment process. Refer to the equipment instructions for this procedure. Input the 9 o'clock position as before and follow the display instructions and then move to the 3 o'clock position. Input the 3 o'clock position, and the values on the display screen will indicate the horizontal misalignment. Remember—we are starting with the vertical plane. Move the shaft to the 12 o'clock position and input the location and reactivate the secondary brake. If the brake pads are adjusted properly, you should not see a change in the F1 and F2 values. The display screen now shows the vertical misalignment. Refer to the equipment instructions or the back of the control unit for what direction of movement is required to correct the misalignment. For example, the display may indicate the generator is misaligned by both parallel and angular. Correct for the angular first and then for the parallel. The angular values may show the generator front feet are low by 1.5 mm and the back feet are low by 1.9 mm. Most laser-alignment equipment displays live readings on the screen, so any change will be seen as feet are adjusted. Start with the back feet first, loosen each back nut and lift one foot at a time to add the recommended 1.9-mm shim thickness. Use the hydraulic jack or cylinder to lift the foot only enough to slide the shims into place. **CAUTION: Removal of the shims may create a space large enough to allow for a pinch hazard if the generator is lowered before the shims are replaced.** Lower the generator and retighten the mounting nut after adding the required shims. Check the control-unit display for the alignment progress. The display should now show that the front feet are still low and that the angular direction has probably shifted from the original value. The front of the generator is probably shown as tipping down. This makes sense because the back was raised by 1.9 mm and the front has not been adjusted. Now loosen the mounting nuts holding each front foot, add the required shim thickness, and retighten the nuts. Use the lifting tools as necessary to place the shims under each foot. The display should now show a corrected angular alignment, but the parallel misalignment may have shifted from the original values. The parallel misalignment may show the generator is low by 1.2 mm. Add the 1.2-mm thickness under each foot one at a time and make sure the mounting nuts are tightened after each stack of shims is in position. The display will now show the progress toward the proper alignment. If the alignment is still out of the acceptable range, then repeat these steps until both F1 and F2 fall within the acceptable range. Use the OEM recommendations for the acceptable tolerance range.

After the vertical misalignment is corrected, deactivate the parking brake and reposition the shaft back to the 3 o'clock position and observe the horizontal angular and parallel misalignment values. Reactivate the parking brake before proceeding to the next step. Correct the angular misalignment first and then the parallel. The adjustment for the horizontal misalignment is done by shifting the generator side to side. Use the hydraulic jack or hydraulic cylinder placed on its side or a jack screw attached to the mounting stud to aid the process. Follow the F1 and F2 values on the display screen to check progress. When shifting the generator for angular misalignment, one nut should be tight to act as a pivot and the others snug but not tight. If the nuts are too loose, then the generator may shift very easily, and it will travel past the proper location before it is realized. The object is

to move the generator a small amount at a time and check for progress. The alignment process for a PM realignment activity should not require a considerable amount of movement. If it does, then take note and see if the same amount of effort is required on the next PM activity. The mounting nuts may have not been tightened properly or there may be some other structural issue creating the problem. Ensure all of the mounting nuts for the generator are tightened to prevent shifting during operation. Write the new alignment values from the display screen in the alignment report under the final alignment section. Remember: documentation is a very important part of the PM process. This will enable a reference to equipment history to help track adjustments and variations because of normal wear.

Turn off all alignment equipment, remove it from the shafts, and place in its case. Remove the electric drive unit from the gearbox output shaft and stow in its case. Return all tools to their respective cases and inspect the area for any tools or material that may have been left in or on the coupler. Stray tools left in moving components can cause considerable damage if they bind or jam the equipment. Replace the coupling guards and secure them as specified in the maintenance manual or alignment procedure. Return the parking-brake system to automatic mode and remove any LOTO components that were necessary to control system energy. Once equipment, tools, and personnel have been moved down tower, notify the required groups that the system will be returned to service.

SAMPLE PROBLEM

For example, 300-mm coupling is running at a speed of 1,800 RPM. The measured coupling offsets are as follows.

Horizontal offset: 0.08 mm
Vertical offset: 0.10 mm
Horizontal angular offset: 0.25 mm
Vertical angular offset: 0.20 mm

SOLUTION

Determining the suggested values for each measurement can be found with Table 8-1. Suggested values for horizontal and vertical offset are 0.10 mm for a shaft speed range of 1,000–2,000 RPM. The measured offsets of 0.08 mm and 0.10 mm fall within the suggested value, so the shaft offset is considered acceptable.

Suggested angular offset values are determined with respect to the application coupling diameter. The suggested acceptable angular offset value for the 1,000–2,000 RPM range is 0.08 mm/100mm of diameter. To calculate the acceptable angular offset value for the above measured values,

$$\text{Acceptable angular offset value} = 0.08 \text{ mm} \times \frac{300}{100}$$
$$= 0.24 \text{ mm}$$

The measured vertical angular offset value of 0.20 falls under the acceptable value of 0.24 mm, whereas the horizontal angular offset value of 0.25 mm exceeds the acceptable values. This gap should be corrected during the preventative-maintenance activity to eliminate possible issues.

SUMMARY

The drive system of a wind turbine is made of several large components that transform wind power into an electrical output to supplement the utility grid. Periodic inspections during a PM program can ensure normal operation and an acceptable component life expectancy. One critical activity in ensuring the drive-system functions normally is through the periodic alignment of the generator to other drive components. This process is even more critical with wind-turbine systems that use a gearbox to increase shaft speed to the required generator synchronous speed. These drive systems operate at shaft speeds of 1,500 to 1,800 RPM, depending on the output line frequency. Higher shaft speeds require a tighter tolerance on maximum misalignment variation to ensure proper operation of the coupling components. Exceeding the recommended alignment tolerances can damage equipment and reduce its availability. Laser-alignment equipment can help ensure the generator is aligned properly and consistently to the gearbox without complicated math calculations that may be necessary for other manual alignment practices. No matter what method is used for generator alignment, technicians need to follow safety practices to eliminate hazards and ensure that all required documentation is followed for an acceptable project outcome.

REVIEW QUESTIONS AND EXERCISES

1. What is the purpose of the gearbox in a wind-turbine drive system?
2. What is the relationship between generator magnetic poles and synchronous speed?
3. Calculate the synchronous speed of a 50-Hz generator that has eight magnetic poles per phase.
4. Determine the input torque requirement of a 1,800-kW generator that has a synchronous speed of 1,200 RPM. Assume 100% efficiency.
5. What issues may be created by an unacceptable generator-alignment condition?
6. What are the three types of shaft misalignment?
7. Can manual measuring devices such as a dial indicator or dial caliper enable an acceptable generator alignment?

8. Name the three main components of a laser-alignment system.
9. What is the purpose of the transmitter and receiver heads for a laser-alignment system?
10. What is the acceptable alignment value for angular and parallel offset of a system that operates at 3,600 RPM and has a coupling hub diameter of 12 inches.
11. What is meant by the term *soft foot*?
12. How can a dial indicator be used to determine the correction of a soft foot?
13. What is the clock position of the laser transmitter and receiver heads during the vertical-adjustment phase of the alignment procedure?
14. What do the F1 and F2 values indicate on the laser-alignment control-unit display during the horizontal-adjustment process?

CHAPTER 9

Down Tower Assembly

KEY TERMS

bonding
disc defragmenter
down tower assembly (DTA)
fiber-optics
foundation bolts
foundation studs
grounding
heat exchanger
hot–cold–hot test
hydrometer
insulated gate bipolar transistor (IGBT)
ladder safety system
light-emitting diode (LED)

modules
operating system
personal computer (PC)
programmable logic controller (PLC)
retina
rock-anchor system
silicon-controlled rectifier (SCR)
specific gravity
supervisory control and data acquisition (SCADA)
terminal strip
top box
uninterruptable power supply (UPS)
universal serial bus (USB)

OBJECTIVES

After reading this chapter and completing the review questions, you should be able to:

- List several DTA maintenance activities necessary in a typical utility-sized wind turbine.
- Describe the purpose of inspecting wire connections on PLC modules and wire terminal strips.
- Describe some of the maintenance activities necessary for a tower PC.
- Describe the purpose of cleaning dust from cabinets and replacing air filters.
- Describe the purpose of verifying the safety switch function with the SCADA system.

- Describe the need for a LOTO program when inspecting electrical connections.
- Describe the purpose of inspecting power cables and terminations.
- Describe the maintenance activities necessary with ladder and associated safety systems.
- Describe the purpose of periodic tensioning of tower foundation bolts.
- Describe the purpose of periodic tensioning rock-anchor bolts.

INTRODUCTION

The chapters leading up to this section covered many of the basic topics necessary to complete preventative-maintenance (PM) activities. Topics such as bolting, lubrication, and alignment developed information necessary to see what tools are available for PM activities along with their safe operation. Information on hydraulic concepts will be useful in understanding the operation of fluid systems so they may be safely operated and maintained. The test-equipment information will be useful in understanding what equipment is available and how it may be used to verify normal system operation. Understanding normal system operation is important in

determining the root cause of a system problem. Knowing what tools are available for an activity will help a technician choose the correct ones to complete an activity. "The right tool for the job" is more than just a phrase. It is an important concept in ensuring that the job is done correctly and without creating other issues. Knowing what tools can be used together for a task is also important. For example, a 250-foot-pound torque wrench cannot be used to inspect tower bolts, but when used with a torque multiplier you may achieve a torque value in excess of 2,000 foot-pounds. This is important because there may be times when power-bolting equipment is not available. A positioning lanyard used with a full-body harness can free up hands and make an activity easier when working off a ladder or structure. Knowledge is the key to making an activity more efficient and for reducing hazards. Taking advantage of others' experience will save you time and ultimately money.

The next section of this text will discuss typical activities that may be seen when performing PM. The location of these systems or components may vary from one wind-turbine model to another, but the activities and safety precautions will be similar. For instance, the power-conversion system in a General Electric 1.5 SLE is located at the base of the tower assembly, whereas the conversion system in a Vestas V80 is located in the nacelle. The lowest level of the tower assembly that can be accessed by the entrance door is typically considered part of the **down tower assembly (DTA)**. Some manufacturers use an area below this entrance level for their power-cable entrance and termination, which is considered a basement.

Other items that may vary from one manufacturer to another include the quantities of components. For example, the Clipper Liberty wind turbine uses four induction generators attached to a single gearbox, whereas models from other manufacturers may only have a single generator. The specifics of the system you will be working on should be covered in a manufacturer's PM manual. Use the information supplied in this text as a supplement to that literature. The goal of this text is to familiarize a technician or manager with some of the topics and activities that may be encountered during a PM activity. This information should help jump-start your training process and make the formal training provided by an original equipment manufacturer (OEM) easier to grasp. The chapters in this section will be divided into areas where you may find some of the typical PM activities. The first chapter is developed around many of the activities that may be encountered in the first level of a utility-scale wind turbine. These areas may have names such as down tower assembly (DTA) or basement, depending on the OEM naming scheme for a particular model. Some experienced technicians may use other names, depending on their previous experience with other systems. The area designations mentioned here will be good for most applications.

One more item worth mentioning again is the importance of documentation during the PM process. Having a procedure or list of items to cover with each PM activity is important. The use of a checklist to mark off activities as they are completed will prevent overlooking something that may create serious trouble later. A properly documented checklist will provide proof that required tasks were completed and ensure warranty items are covered in the future.

TOWER FOUNDATION AND DTA OVERVIEW

When you approach the tower of a utility-scale wind turbine, one of the first things that come to mind is its size. A tube 14 feet (4.3 meters) in diameter rises a few hundred feet

into the air with a nacelle mounted on the top that is the size of a bus. On the front of the nacelle, a rotor assembly with three long blades extends out from the center. Each blade may be longer than a basketball court, and the weight of the rotor assembly may be in excess of 30 tons (60,000 pounds or 27,215 kg). It is amazing that something this massive can be moved by the wind, not to mention that it can extract energy from the wind.

The tower is mounted to a foundation assembly made of tons of concrete and steel to provide support for the structure. The foundation may be one of two designs: spread footing or rock anchor. A spread-footing or ballast-type design is constructed of a large mass of concrete and a rebar and steel frame to provide structural support and prevent toppling. The mass provided by this design is used like the weight in a sailboat's keel. A rock-anchor design is much smaller and, as the term implies, uses the natural geology of the area to support the tower. The foundation is held in place with several large threaded rebars that are locked into the sedimentary rock or granite. The foundation is then constructed around the threaded rebars so they may be used to hold the structure in place. **Figure 9-1** shows a cutaway view of a rock-anchor foundation design.

The foundation is also constructed with steel embedment rings locked into the rebar assembly with two rows of threaded rebars to hold the lower tower assembly. These threaded rebars are known as **foundation bolts** or **foundation studs**. Foundation bolts will be discussed in a later section as part of a PM activity. **Figure 9-2** shows the foundation bolts ready for a Vestas V47 wind turbine.

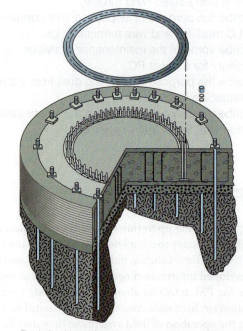

• Rock Anchored foundations are more suitable for sites with good, competent rock beneath turbines (Most in New England)

FIGURE 9-1 *Cutaway view of rock-anchor foundation design*

FIGURE 9-2 *Foundation bolts for vestas V47 wind turbine*

The entrance to the tower is made with a bulkhead door that looks like something used for a ship. It is elliptical in shape with a filter assembly constructed in the door that enables airflow for cooling but prevents dust and other contaminants from entering the tower. Depending on the tower DTA, this door may be a meter above the grade or several meters above to provide space for the basement area. Some turbine models use a basement area below the DTA for high-voltage cable connections. **Figure 9-3** shows an example of a wind-turbine tower entrance.

FIGURE 9-3 *Entrance area for a wind-turbine tower*

Once you enter the tower, you see cabinets used for control, communication, and power connections along with access equipment for ascending the tower to the nacelle. These cabinets, support structures, and access equipment will be the topic of discussion for the DTA PM activities. We will start the PM activities with the control and communication cabinets.

SAFETY NOTES 9-1—PC and PLC Cabinets

Safety hazards

- Never open a cabinet that you are not authorized to perform work activities.
- Voltage levels within these cabinets may range from 24 V_{DC} to more than 19,000 V_{AC}, depending on power-converter controls or contactors that may be co-located in each console.
- Always determine voltage levels, test-equipment requirements, and conductor and disconnect locations before performing electrical work activities.
- Refer to the site safety plan for required lockout–tagout (LOTO) procedures along with appropriate PPE necessary for the hazard risk category (HRC) in the work area. Use OEM supplied electrical schematics and procedures to verify isolation.
- Automatic equipment may be present in these consoles for heating and ventilation. Ensure that any automatic equipment is disconnected from the control system or isolated from the power source to prevent possible energizing.

INSPECTION OF PC AND PLC COMPONENTS

The control and communication cabinets may house the **personal computer (PC), programmable logic controller (PLC)**, and communication router. The PLC is a control system device that reads input signals from the wind-turbine equipment, processes the information using a predefined program, and sends output signals to devices to control normal operation. The PLC system includes a power supply, a processor unit, and a variety of **modules** that are used for control and communication functions. The communication modules enable the PLC processor to communicate with the tower PC and farm network, along with collecting data for processing. This PLC system may also connect to another PLC system in the nacelle control console. The nacelle control console is also known as the **top box**. The PC in the DTA console is connected to the PLC system through an Ethernet cable to enable communication with the **supervisor control and data acquisition (SCADA)** system. SCADA is the system that is used to monitor and control the wind farm. It can be used to monitor an individual wind turbine or the entire system of turbines that is collectively known as a *power plant*. It can store and

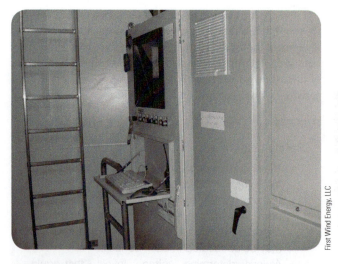

FIGURE 9-4 *DTA control-console setup for PC, PLC, and communication equipment*

FIGURE 9-5 *Wiring devices attached to a DIN rail and associated wire trays*

process data that enables a snapshot of system performance or track long-term trends. The system also enables remote access to the farm for remote control of individual turbines or the power plant. As versatile as this system may be, it is useless if a communication cable comes loose during operation. **Figure 9-4** shows a DTA control console that has been set up for the PC, PLC, and communication equipment.

SAFETY NOTES 9-2—Power-Distribution Panels

Safety hazards

- Never open a cabinet for which you are not authorized to perform work activities.
- Voltage levels within this panel may range from 24 V_{DC} to more than 600 V_{AC}, depending on transformers or circuit breakers that may be present.
- Always determine voltage levels and test-equipment requirements, along with conductor and disconnect locations before performing electrical work activities.
- Refer to the site safety plan for required LOTO procedures along with appropriate PPE necessary for the HRC in the cabinet. Use OEM supplied electrical schematics and procedures to verify isolation.
- Automatic equipment may be present in this cabinet for heating and ventilation. Ensure that any automatic equipment is disconnected from the control system or is isolated from the power source to prevent possible energizing.

Inspecting the Control Consoles

Inspecting the control console requires an initial visual of the cabinet, connections, wiring devices, insulation, fans, PLC components, and PC components. Any items that may have

shifted or come loose will need to be noted before you proceed. Before work starts on wire connections and terminal strips, the PC and PLC will need to be powered down and the power sources to the console will have to be isolated per approved LOTO procedure for the site. **Terminal strips** are wiring devices that allow the termination of multiple wires between control devices and cables. These strips are typically attached to the back panel of the control console with a standardized rail system or directly to the panel with machine screws. The standardized rail (or *DIN rail**) system is a shaped metal rail system that enables easy attachment and removal of terminal strips, circuit breakers, relays, and other electrical devices. **Figure 9-5** shows wiring devices attached to a DIN rail and associated wire trays.

Some DTA control consoles have 120 V_{AC} or higher supplied to power circuits, so LOTO is required to protect personnel and prevent damage from a short when inspecting wire connections. Some of these power circuits may be located in a distribution panel to the side of the control console. This panel may have several circuit breakers lined up and grouped by control or power circuit, depending on their function. For example, some of the circuit breakers protect tower lighting, tower receptacles, an **uninterruptable power supply (UPS)**, the cooling system, PLC components, or the PC. Always determine which circuit to isolate using the manufacturer's electrical schematic and panel markings before beginning an activity.

Once the electrical circuit is isolated, test it to verify a zero-energy state with a **hot–cold–hot test**. This ensures that the multimeter is functioning properly and the circuit is at zero potential. For the hot–cold–hot test, place the multimeter in the appropriate voltage mode, check a known live circuit, verify the isolated circuit, and then check the same live circuit again to confirm that the isolated circuit is deenergized. There may also be a transformer located in the console that drops the utility supply voltage to the required

---

*The name is taken from the initials of the Deutsche Institut für Normung, the German standards organization.

tower subsystem voltages. Refer to the manufacturer's documentation for approved isolation of the transformer and the required maintenance.

Checking wire connections is very important. Many visits to the tower for service are caused by loose wire connections. Time spent checking connections now will save on downtime later. Wire and cable connection inspections should include PC cables, communication cables, PLC module connections, power cables, cabinet grounding, and fiber-optic connections. **Fiber-optics** refers to a system of communication that uses light signals to pass information between devices by a glass or plastic fiber. The light source for the communication signal may be from a laser diode or a **light-emitting diode (LED)**. **WARNING: Never shine the light from a fiber connector or transmitter into your eye. The emitted light may cause permanent damage to the retina, the back area of the eye that contains nerves and other structures that are vital to vision.**

Ensure that each item is properly seated and that any fasteners that may have worked loose are tightened securely. Refer to the OEM's maintenance manual for required torque specifications or Appendix F. While the console is powered down, wipe off any dust or other debris that has accumulated in the cabinet. Dust will settle on electronic devices and reduce the cooling effect of air flowing through the console. Dust can also create a shorting hazard if moisture saturates a dust bridge that has formed between wires or component leads. This activity should include cleaning fan assemblies and replacing filters. Reduced airflow from a blocked air filter will create an overheating issue with the electronic components. Refer to the OEM's procedure for approved cleaning solutions and only spray the solution on the cleaning cloth and not directly on components.

Once the wire connections and components have been inspected and the cabinet has been cleaned, ensure that all wire tray covers have been secured and any foam insulation mounted to the inside of the cabinet is secure. The tray covers may have been inadvertently left off during a previous service activity or they may have come loose during operation.

Uninterruptable Power Supply

A UPS has a battery pack that is used to supply backup power during the loss of utility power to the wind turbine. The UPS typically can supply the control system (PLC, PC, and communication components) for several hours during a power outage. While the system is down for inspection, run a battery test to determine the status of each battery in the pack. If it shows a high internal resistance or higher value than a previous PM test, the battery may need to be replaced. Consult the UPS or battery manufacturer's information for acceptable readings. Document your findings and add information to the checklist of items that may need replacing on the next service visit. Once the inspection and cleaning are complete, remove the LOTO and restore power to the console.

PC MAINTENANCE ACTIVITIES

The PC system in a wind turbine is a vital link between local users for troubleshooting and remote communication with the SCADA system. PCs use an operating system such as Windows or other software-based systems. An **operating system** provides the interface between the user and computer system to process data storage, data retrieval, and data control between software applications. Most likely you have been working with a PC and have seen a pop-up window that indicated the system needed an upgrade or that a new software release was available. The PC in the wind turbine is no different. It may not be connected directly to the Internet so that it can get these updates or releases automatically, so a service or maintenance technician may be responsible for loading these operating system updates. The engineering group working with the farm should have an approved list of updates available or the updates may be supplied on a USB flash drive or CD-ROM. A **universal serial bus (USB)** is a communication port on the PC or laptop computer that is used to transfer files from external data-storage devices or to communicate with peripheral devices such as a modem or a router. Always follow site procedures for installation and documentation of changes to the PC operating system.

One PC system maintenance activity that is useful for speeding up a PC's operation is a **disc defragmenter** program. A PC operating for a long period of time will store data wherever it can find space on the hard drive. For this reason, all related data may not be stored in the same place. Retrieving data later for system operations may take extra time as the processor searches for the related data locations in the registry. The disc defragmenter utility tool finds all related data and places them in a group. Running this utility will speed up the operation of the PC.

SCADA INFORMATION VERIFICATION

Verification of system safety switches should be performed during each PM cycle. This verification ensures that all required Emergency Stop switches, limit switches, and other critical interlocks are functioning. A malfunction of a safety loop circuit component could be hazardous to a technician working on the system or catastrophic for the wind turbine if the system does not detect a critical fault and shut down. **Table 9-1** lists examples of the safety circuit sensors and switches. Consult the OEM documentation for location, type, and approved procedure for testing safety loop devices. This procedure may require two technicians to perform; a technician to monitor SCADA information and reset the system while another is used to activate the emergency devices. There should be constant communication between everyone working in the tower during this

TABLE 9-1

Examples of Safety Loop Circuit Devices and Possible Locations

Device	Location(s)
Emergency stop switches	DTA, nacelle, and hub
Pressure switches	Parking-brake hydraulics, accumulator circuit pressure for emergency blade-pitch, and critical equipment coolant and lubrication systems
Limit switches	Blade-pitch position and drip-loop cable twist
Temperature switches	Power-converter cabinets and critical equipment coolant and lubrication systems
Voltage-level sensors	Control system UPS and backup battery power for emergency blade pitch
Vibration sensor	Nacelle
Motion-level sensor	Nacelle

© Cengage Learning 2013

activity to ensure that an actual emergency situation is not ignored. A checklist indicating each device tested should be filed with each PM activity documentation package. The checklist will provide proof that the safety system has been verified.

Another important use of time during a PM activity may be the follow-up of punch list notes that may have been left in the SCADA system notepad. The notepad feature in some SCADA software allows technicians and operations personnel to place and retrieve notes in the system for later follow-up in an easily accessed electronic text format. These items may include replacement of noncritical items such as a missing screw on a console door, a failed lightbulb, a damaged door gasket, or a missing grease fitting. A hard copy of the punch list may have been attached to the documentation package handed out during the morning tailgate or safety briefing to clarify the day's activities. Make good use of downtime between activities to follow up on these items.

SAFETY NOTES 9-3 — Power-Conversion System Cabinets

Safety hazards

- Never open a cabinet for which you are not authorized to perform work activities. Voltage levels within this cabinet or panel may range from 24 V_{DC} to 1,000 V_{DC} and 120 V_{AC} to 600 V_{AC}, depending on transformers, capacitors, and switching equipment that may be present.
- Always determine voltage levels, test-equipment requirements, and conductor and disconnect locations before performing electrical work activities.

- Always wait the required time for capacitors to dissipate charge before verifying zero-energy level. Connect grounding cables to the capacitor assembly and other power-switching equipment to prevent charging during electrical work activities.
- Refer to the site safety plan for required LOTO procedures along with appropriate PPE necessary for the HRC in the work area. Refer to OEM electrical schematics and procedures to verify isolation.
- Automatic equipment may be present in this cabinet for heating and ventilation. Ensure that any automatic equipment is disconnected from the control system or isolated from the power source to prevent possible energizing.

POWER-CONVERSION SYSTEM MAINTENANCE

The power-conversion system of a wind turbine is an important subsystem that ensures the power output meets a utility's grid frequency and voltage requirements. PM activities with this system should be performed by qualified technicians who have been trained in the operation and maintenance requirements. This system may be located in the DTA or in the nacelle, depending on the manufacturer or the wind-turbine model. This text will cover the PM activities as part of the DTA.

The power-conversion system has large solid-state switching devices, such as an **insulated gate bipolar transistor (IGBT)** or a **silicon-controlled rectifier (SCR)** that can switch on and off quickly to control the output power signal. An IGBT is a solid-state device that is used to form an approximate AC sine-wave power output from a DC power source. An SCR is another device that may be used to control a power-circuit output signal. The size and type of devices used will depend on the switching speed and power requirements of the circuit. Switching schemes will not be covered in this text because of the proprietary nature of the system designs. The typical PM and safety recommendations may be similar between systems. These systems have voltage levels that run from the generator potential up to that of the utility grid or farm transmission infrastructure. There may also be charged capacitor assemblies within the conversion system that may charge to an elevated DC voltage level as part of normal operation. Isolation and LOTO of this cabinet is done through one of the primary circuit breakers in the tower. Refer to the OEM's documentation before proceeding with this activity.

Only a qualified technician should work on a power-conversion system. A qualified technician is trained on voltage levels, locations, and disconnects of all sources that feed this cabinet. **CAUTION: Ensure that the DC charge on the capacitor assembly has adequate time to discharge before performing any maintenance.**

Inspection of Electrical Connections

Once the power-conversion cabinet has been isolated and LOTO has been performed according to the site safety plan, the wire and cable connections that feed the system can be inspected. Visually examine all terminal connections on control components and circuit-board assemblies to ensure they are seated and that fasteners are secure. Carefully ensure that all fasteners are tightened as necessary to eliminate loose connections that may occur during the vibration of the components during normal operation. Refer to the OEM's procedure requirements. Do not overtighten the connections. Some of the connections are made with delicate brass screws that can easily break or strip the terminal assemblies. Such damage will create downtime as you wait for the replacement component to arrive.

Visually inspect the larger components located in the cabinet for discoloration because of abnormal heating and for any loose cable connections. Note any issues on the checklist that may need to be addressed during a later service activity. Ensure that all power-cable connections are secure to the bus bars and terminal assemblies. These fasteners may be larger than the ones located on the control components, but some of them may be mounted in fragile electrical isolators, so use caution. Deviating from the OEM's maximum torque values can damage the fasteners, components, or isolators.

Inspection of Cooling System

The cooling systems in power-conversion cabinets may be of two types: forced air over heat sink assemblies to remove excess heat or liquid coolant (glycol and water solution) flowing through plumbing attached to the power-conversion components. Some manufacturers may use a combination of the two methods to improve the system's overall cooling efficiency. Inspection of a forced air cooling system includes wiping down components and the cabinet interior of any dust that may have built up during operation. Use caution when wiping down components that may be susceptible to static-discharge damage. Refer to the OEM's procedure and use appropriate static grounding straps to ensure that components are not damaged. If the manufacture recommends a cleaning solution to remove contaminates, then spray the solution only on the cleaning cloth and not directly on the components. This will avoid leaving excess cleaning solution behind that may cause issues especially, if it is water based. Clean the fan assembly and replace air filters as necessary.

Liquid-coolant systems are made up of a several basic components: pump, expansion reservoir, **heat exchanger**, plumbing, and cooling jackets for the electronic components. There may also be ancillary devices in the system such as sensors, gauges, valves, and heaters, but this PM discussion will be limited to the basic components. Liquid-cooling systems should be visually inspected for leaks. Excessive heating of flexible coolant lines can create cracks and possibly swell the lines around connections. Note any suspect hoses on the checklist so they may be inspected for further deterioration during a future service activity. If a leak is found, clean any residue that may be present and tighten the hose connector to eliminate further leaking. If the connection cannot be tightened, replace the hose or connection or list it on the punch list for a future service activity.

Check the coolant system expansion reservoir for fluid level and add approved coolant to bring it up to the recommended level. Checking the coolant solution for proper mixture should also be performed. This will eliminate the possibility of boiling the solution in the system components when they are operating under load or allowing the solution to freeze when the system is setting in standby on a cold day. The coolant ratio of glycol to water (distilled water) should be 50:50. This should protect the solution from temperature extremes seen during normal operation of the wind turbine. Use of a hydrometer to check for specific gravity will give a good indication of the solution's properties. The **hydrometer** is a device that is calibrated to float at a specific level according to the solution's specific gravity. **Specific gravity** is the ratio of a fluid's density with respect to the density of freshwater. Water's specific gravity is the standard and is defined as a specific gravity of 1. A hydrometer for checking automotive antifreeze should work fine for this application. **Figure 9-6** shows an example of a hydrometer used to check coolant working temperatures. If the coolant does not pass the specific gravity test, then the system should be drained and flushed in accordance with the OEM's requirements. Replenish the system

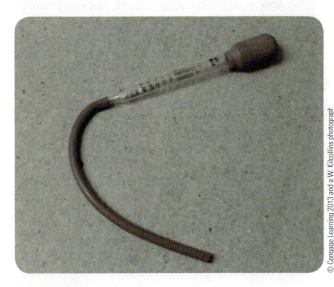

FIGURE 9-6 *Hydrometer used to check coolant working temperatures*

with approved coolant and bleed any air from the lines that can cause an overheating problem later.

The coolant heat-exchanger assembly will be located on the side or back of the power-conversion cabinet in close proximity to a fan assembly. The coolant system functions through a process of picking up heat (thermal energy) as it passes through the power electronic components and cooling the liquid as it passes through the air-cooled heat exchanger. A fan assembly is used to force air through the heat-exchanger assembly to increase the amount of heat removed from the coolant. PM activities for a liquid-cooled power-conversion cabinet should include the cleaning of the heat exchanger. Forcing air through the heat exchanger during normal operation will cause dust and other contaminants to be trapped in the fins of the assembly. As dust accumulates, it will reduce the cooling effect of the heat exchanger. To restore the efficiency of the heat exchanger, the surfaces of the fins should be wiped down or cleaned with a vacuum cleaner. Any air filters used with the heat exchangers should also be replaced at this time. This will reduce buildup on the heat exchanger because of drawing dust from a contaminated filter. Once inspections and cleaning activities are completed, LOTO can be removed from the power-conversion cabinet so it may be returned to service.

SAFETY NOTES 9-4 — Power and Grounding Cable Inspections

Safety hazards

- Inspection of cables may require opening cabinets to verify terminal or bus-bar connections.
- Never open a cabinet for which you are not authorized to perform work activities.
- Voltage levels within some cabinets may range from 120 V_{AC} to 600 V_{AC}, depending on circuits that may be present.
- Always determine voltage levels, test-equipment requirements, and conductor and disconnect locations before performing electrical work activities.
- Refer to the site safety plan for required LOTO procedures along with appropriate PPE necessary for the HRC of the isolation verification process.
- Power-cable insulation-resistance testing requires disconnecting both ends of the cables before performing the test.
- Power cables must be deenergized and isolated before ends are removed for testing.
- Ensure that cable ends are covered to prevent unauthorized or accidental contact with personnel during the testing activity. Accidental contact with the energized cable by employees may cause a shock hazard. Accidental contact with electronic equipment during the test may cause damage.

INSPECTION OF POWER CABLES, GROUNDING, AND BONDING CONNECTIONS

Once the cabinet wire and cable connections have been verified, the PM task moves to the cables running between the DTA, basement, and the nacelle assembly. These cables include power connections from the nacelle to the main tower contactor, tower grounding cables, and bonding cables between cabinets, the DTA structure, and the tower foundation. The main tower contactor is the large switch located between the utility grid and the wind-turbine power circuit. This switch is often referred to as CB1 for circuit breaker 1. Refer to the OEM's procedure for isolation, approved LOTO, and maintenance activities. This contactor switch is the final circuit connection engaged during normal operation after the turbine control system determines that output voltage amplitude and frequency are in sync with the utility grid. Because this contactor is located between the wind-turbine power circuit and the grid, it requires isolation and LOTO of the pad-mount transformer or the main fuse connections located on the grid utility pole. The isolation operation should only be completed by qualified OEM high-voltage specialists or qualified local utility personnel. If possible, a verification of the CB1 isolation process should be completed before removing panels to the contactor cabinet. This verification process may require the use of appropriate HRC PPE and testing equipment to minimize the exposure risk of an accidental arc flash. **Figure 9-7** shows an example of a 40 cal/cm^2 arc flash suit being used to isolate the pad mount transformer from the wind turbine main switch.

A visual and torque inspection of the cable terminations, bus-bar assembly, contactor assembly, and wiring devices will ensure that the connections are secure and will not generate excessive heat during operation. A visual inspection of all control devices associated with the contactor should be completed during this activity to ensure these devices are not showing signs of deterioration or damage. Another inspection that may be completed at this time is a cable insulation-resistance test.

Cable Insulation Resistance

Cable insulation-resistance testing requires the disconnection and isolation of each power-cable termination in the contactor cabinet along with the connections at the transformer or generator in the nacelle. Cable insulation testing will verify the integrity of the cable coatings, crimp connection insulation, and any other insulation components located in the power system. This process is completed one cable at a time to determine the integrity of each cable assembly. Once the test is complete, ensure that all cable connections

First Wind Energy, LLC

First Wind Energy, LLC

FIGURE 9-7 *A 40 cal/cm² arc flash suit*

are assembled in the proper phase sequence and tightened to the appropriate torque value listed in the OEM's procedure. Inform all personnel that testing is complete and that the system will be reenergized.

Infrared Inspection

Another power-cable inspection that can be completed without isolation and LOTO is done by infrared thermography. The use of infrared inspection equipment for power cables, splices, and cabinets while the system is under load or just after shutdown will enable a technician to see areas that exhibit higher temperature signatures. These areas may be created by poor cable crimp connections, loose cable terminations, or damaged clips that are used to secure fuse assemblies. Finding these areas before a PM activity shutdown may enable the components to be visually inspected to determine whether they can be repaired or need to be replaced. Document suspect areas and include the information in the PM package for later service work as necessary.

Bonding and Grounding Cable Inspections

As you visually inspect the DTA and basement area of the tower, you will see cables connected to the interior structures, cabinets, and tower tube. These cables are used to ensure that each structure is maintained at the same potential. This connection process is known as **bonding**.

Figure 9-8 shows examples of bonding connections. Bonding ensures that any static charge that may accumulate because of airflow around the tower or rotor assembly along with any capacitive charge that may accumulate between the power cables and the tower is bled off to prevent an uncontrolled electrical discharge. An uncontrolled electrical discharge could create a hazard for technicians working around the equipment or damage sensitive control circuits. Another step to protect the sensitive control circuits from electrical discharge during operation or from a lightning storm would be overvoltage protection modules. These modules will open during an overload condition and shut down the circuit to protect the components. Overvoltage modules are also linked together as part of a control feedback circuit to a PLC input module. The loss of this input signal to the PLC provides information to SCADA that an overvoltage fault has occurred. The connections to these modules would be inspected and tightened as part of the various control-console PM inspections.

Each bonding termination should be inspected to ensure the connections are tight and that no damage to the cables or terminations is visually apparent. Another part of this cable-inspection process is to ensure that tower grounding hardware is intact and secure. The grounding system assembly is integral to the tower foundation construction process. The foundation assembly is wired together and connected to remote grounding rods positioned around the tower. The purpose of **grounding** is to minimize the resistance of any path from the tower assembly to earth potential. Some OEMs may require a ground-resistance measurement be taken as part of the PM process. Refer to the OEM's grounding requirements for appropriate values and recommended equipment. Some insulation-resistance meters are also equipped with a ground-resistance measurement function. **Figure 9-9** shows an example of a grounding connection between the tower and grounding assembly buried around the tower.

First Wind Energy, LLC

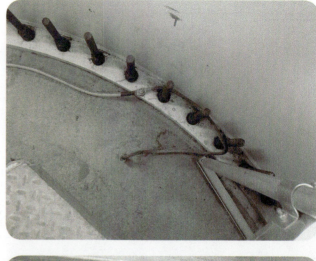

First Wind Energy, LLC

First Wind Energy, LLC

FIGURE 9-9 *Grounding connection between tower and grounding assembly*

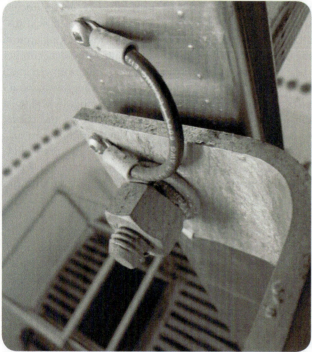

First Wind Energy, LLC

FIGURE 9-8 *Bonding connections*

INSPECTION OF LADDER, DECKING, AND CABINETS

A required safety inspection for the PM cycle is a check of the access equipment used to ascend the tower. Inspect the ladder assembly to ensure that all mounting hardware

is secure and that required safety devices are available. Access ladders may be set up with a **ladder safety system** that is used as a fall-arrest system. Permanently installed ladder safety systems must meet government safety requirements such as those regulated by OSHA in the United States. These requirements were discussed in Chapter 2 for working at heights. The ladder safety assembly must be checked to ensure that the attachment component (cable or rail) and mounting hardware are not damaged or missing. Items may come loose and drop because of vibration during system operation. Inspections of the ladder and ladder safety system should include a check of all attachment components located from the base to the top of the ladder. The first technician up the ladder should perform this bottom-to-top inspection to ensure that it is safe for others to use. This inspection should also include a check of the manufacturer's tags, labels, and placards that are present to ensure they meet current industry standards. These inspection results should be logged with the turbine safety documents and

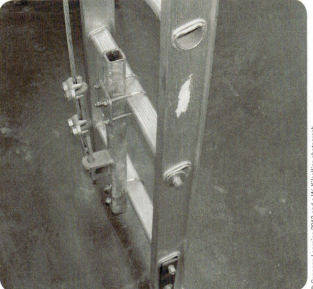

FIGURE 9-10 *Ladder safety system with a cable assembly*

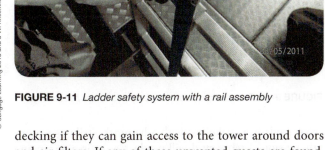

FIGURE 9-11 *Ladder safety system with a rail assembly*

any discrepancies corrected. **Figures 9-10 and 9-11** show examples of ladder safety systems. The first technician up the tower should also inspect the hoist assembly located at the yaw deck or in the nacelle before it is used to lift supplies and tools for PM operations in the upper levels of the wind turbine.

Other inspection items in the DTA include a review of decking, cable trays, cabinet supports, and mounting hardware. Ensure all items are secure and cleaned of dust and other containments that may have accumulated since the last PM cycle. A couple of notes of caution: depending on the geographic location of the wind farm and the season, there may be wildlife hazards associated with this inspection. Be cautious of areas that may have insects or reptiles present. Snakes love warm locations and will gather in accessible areas under steps and decking. Bees, hornets, and wasps may also build nests under cabinets and

decking if they can gain access to the tower around doors and air filters. If any of these unwanted guests are found, use caution to remove them from the tower. Professional exterminators may need to be called in if the guests cannot be removed safely.

Another safety hazard to consider when inspecting decking or areas below cabinets is electrical connections. Many of the power cables enter below the DTA decking and are routed up into the cabinets. These cables should be isolated and LOTO performed before inspecting these areas. Tower assemblies that use a basement for the high-voltage cable entrance should also be accessed with proper precautions to eliminate a shock or electrocution hazard. It is best to perform this inspection when the system is isolated during the main contactor inspection and maintenance activity to reduce the time the tower is isolated. Interior foundation bolt tensioning is another PM activity that should be performed when the tower is isolated. This activity will also require working under the decking and cabinets to access the bolts located on the inner flange of the tower.

TOWER FOUNDATION FASTENER TENSIONING

Tower foundation bolts or studs are used to secure the tower tube to the foundation assembly. The foundation studs are set up in two concentric circles that connect to the inner and outer flange of the tower. **Figure 9-12** shows the foundation studs connected to the tower base flange. The foundation studs are all-thread rebar secured in the foundation through large steel embedment rings. The tower tube flange assembly sets over these studs and is secured with nuts. The assembly is visually inspected, and the studs are tensioned according to OEM specifications or the manufacturer's recommendations. One test that can be used to determine the tension on

FIGURE 9-12 *Foundation studs connected to tower base flange*

rebar studs is a strike test with a hammer. Striking a properly tensioned rebar stud will produce a distinctive ringing sound or tone. A rebar stud that is not properly tensioned will produce a thud without a ringing sound. This test is subjective and only provides feedback that the rebar stud is under tension—but not necessarily the correct level of tension. Tensioning equipment will provide a quantitative value of the rebar stud tension. **Figure 9-13** shows an example of

tensioning equipment. The tension inspection is typically completed on a 10% sample of the rebar studs. If the tension findings on the sample are found to be below specification, then all of the studs will require tensioning. The rock-anchor fasteners around the foundation will be another area that will require tensioning as part of the PM cycle.

ROCK-ANCHOR FASTENER TENSIONING

The **rock-anchor system** consists of several threaded rebars that secure the tower foundation to the underlying rock formation. Using this construction design reduces the required volume of the foundation without compromising the strength. **Figure 9-14** shows a view of a typical rock-anchor foundation. To ensure that the foundation remains stable, it should have the rock-anchor threaded rebar tensioned annually. Tensioning of the threaded rebar maintains a consistent compression of the soil layers between the foundation and the underlying rock formation. The soil below the foundation may shift during the year because of changes in water content and temperature. Heavy rains or spring runoff of melting snow can shift soil along with the seasonal freeze and thaw cycles. Including tensioning of the rock-anchor threaded

FIGURE 9-13 *Tensioning equipment*

FIGURE 9-14 *Typical rock-anchor foundation threaded rebar*

FIGURE 9-15 *Rock-anchor tensioning system*

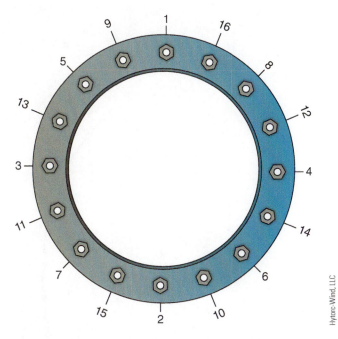

FIGURE 9-16 *Typical star pattern used with inspection of rock-anchor system*

rebar ensures that the foundation remains stable. Tensioning the large diameter threaded rebar requires a large hydraulic system to create a sufficient tension force. **Figure 9-15** shows an example of a rock-anchor tensioning system. The rock-anchor threaded Threaded rebar should also be tensioned in a star pattern to ensure that the foundation does not shift to one side during the activity. **Figure 9-16** shows a typical star pattern that may be used with a rock-anchor system.

Tensioning the rock-anchor assembly is a very specialized activity that requires qualified technicians, specialized hydraulic equipment, and a service truck to move the equipment. A wind-farm tool inventory does not typically include the specialized equipment for doing this activity. In most cases, subcontracting this activity to a construction organization that specializes in this type of foundation work makes financial sense. Most organizations have limited resources, so investing limited capital in specialized equipment reduces funds available for other required expenditures.

SUMMARY

Preventative-maintenance activities in the DTA and basement area of a wind turbine include both mechanical and electrical activities to ensure the system will function efficiently. Visual inspection and securing electrical connections ensure that the system will not shut down because electrical signals are lost with the control and feedback of the PLC. PC operating system and cable connection maintenance will ensure proper function of the file-management system along with communication with the farm SCADA system. The PC and PLC play vital roles in turbine system control and communication coordination for local users and remote groups to determine equipment status.

Maintenance activities performed on the power-control systems require proper electrical isolation and approved LOTO procedures to ensure safety of technicians performing activities. Proper electrical isolation will also prevent damage to equipment resulting from an accidental short between electrical devices during activities. Proper maintenance of system cooling equipment will ensure months of trouble-free operation.

Periodic inspection and maintenance of tower access equipment will ensure compliance with national regulations and provide a safe environment for technicians and operations managers who need to access the wind turbine for work activities.

REVIEW QUESTIONS AND EXERCISES

1. List several maintenance activities that are performed in the lower level of a utility-size wind turbine.
2. What is the purpose of inspecting wire connections on PLC modules and other wire devices?

3. List a couple of maintenance activities necessary for a tower PC and why each is important.

4. What is the purpose of cleaning dust and other contaminants from components in a control console during a PM cycle?

5. Why is it necessary to replace air filters on control cabinets?

6. What is the purpose of verifying the safety switch function for the SCADA system?

7. Why is it necessary to have an approved LOTO program for electrical maintenance?

8. List a few pieces of equipment that can be used to inspect power cables and terminations.

9. Why is it necessary to verify the specific gravity of a coolant solution?

10. What are two connection components used with ladder safety systems?

11. Why is it necessary to inspect safety equipment?

12. What are the foundation bolting systems that require periodic tensioning?

10

Tower

KEY TERMS

cable keepers
drip loop
flange
grease manifold
ground-fault circuit interrupt (GFCI)
impact-grade sockets
lift bag

saddle deck
star pattern
torque multiplier
vertical lifeline
yaw-bearing assembly
yaw deck

OBJECTIVES

After reading this chapter and completing the review questions, you should be able to:

- Describe the different maintenance activities performed in the tower section of a utility-sized wind turbine.
- List inspection items for the ladder assembly.
- Describe the purpose of inspecting the torque level of tower flange bolts.
- List winch-assembly components that should be inspected.
- Describe the purpose of the cable assembly draped below the yaw deck.

- Describe the purpose of inspecting the torque level of yaw-bearing bolts.
- Describe the purpose of inspecting the power-cable support system.
- List inspection items for tower lighting and electrical receptacles.
- List safety items associated with working with power-bolting tools.
- List safety items associated with working at heights.

INTRODUCTION

The tower assembly of the wind turbine is made up of several cylindrical components that are stacked and bolted together during the construction process. These cylindrical components or tubes have reduced construction time, decreased tower cost, and improved tower quality. Each tube is made of several steel plates that are rolled and welded together. Several sections of these rolled plates are welded together with staggered vertical joints to form a length of approximately 82 feet (25 meters). Each tube has a **flange** welded to each end that will be used to join the sections of the tower. The flanges may have more than 100 holes drilled through them to enable bolting the sections together. This construction process allows the assembly of a 300-foot tower in less than a day. This process also enables the complete structural assembly of a wind turbine from the foundation up within a day with two cranes. A large crane (~400 tons) is used to lift the major components, and a smaller one

(~90 tons) is used to assist in moving and balancing components. The smaller crane is also used to assist in the assembly and disassembly process of the larger crane. **Figure 10-1** shows the assembly of tower sections for a Vestas V47 (660-kW) wind turbine. These two cranes may be larger or smaller, depending on the size of the wind turbine being constructed.

During the tower-assembly process, the tubes and nacelle components are bolted together and tightened with power-bolting equipment. The bolting equipment may be powered by an electric gear motor or a hydraulic power unit. **Figure 10-2** shows examples of power-bolting tools. The tower tube sections are preassembled with decks at the top, access ladder assemblies, and some cable assemblies. The cables preassembled in each tower tube will be determined by the wind-turbine manufacturer, depending on the wind-turbine model. A turbine manufacturer with the power-conversion system in the down tower assembly (DTA) may have the power cables preassembled in the tower tubes because of their weight. Lifting heavy cables up an erected

FIGURE 10-1 *Positioning a wind turbine tower section*

FIGURE 10-2 *Examples of power-bolting tools*

tube may cause damage to the cables or present a safety hazard to technicians handling them. Manufacturers with a power-conversion system and transformers in the nacelle may not need preassembled power cables because they would not need the larger cable diameter and heavier weight associated with the lower voltage and higher current of the other system. Once the tower components have been assembled, any required power and control cables would be lowered down the tower from the nacelle and attached to the cable attachments mounted to the inside of the tower tubes. **Figure 10-3** shows an example of power cables mounted within the tower tube.

Assembly of a tower design made up of several tube sections reduces the assembly time of the wind turbine at the site. This process also reduces the overall cost of the construction project. Structural assembly being completed in one day saves on the expense of subcontracting the cranes and the rigging crew. The crane cost per day may be $6,000 or more, depending on its size. The daily costs associated with the crane operators and rigging crew will also depend on the numbers and required training. The tube construction method saves not only on construction costs but also in the amount of time required to inspect and maintain the tower after assembly. This chapter will introduce the reader to the types of maintenance activities associated with the typical tube-tower assembly and their requirements. The tower sections to construct a utility-size wind turbine may be anywhere from three to five, depending on the design requirements and overall structure

height. Most of the maintenance activities in each of the upper tower sections will be similar, with the exception of the yaw deck. The variations for each tube section will be pointed out as necessary during the discussion. **Figure 10-4** shows a diagram of a typical wind-turbine tower that points out some of the features and areas to be discussed during this chapter. We will start the chapter with a continuation of the cable-inspection discussions. One important item to keep in mind during the tower preventative-maintenance (PM) process is the use of a checklist to ensure that each activity is thoroughly addressed and documented for later follow-up.

FIGURE 10-3 *Power cables mounted within the tower tube*

INSPECTION OF CABLES AND CONNECTIONS

Preventative maintenance of the tower tube sections should include an inspection of any hanging or suspended power, control, fiber-optic, or grounding cables. Inspection of these cables is very important to ensure they are free of defects that could cause damage to equipment, create a hazard for technicians, or create downtime. Inspection of the tower sections will also include any connections used to electrically bond tube and access ladder sections. The following cable-inspection sections will highlight the required activities for each tower section from above the DTA to the top of the tower assembly.

SAFETY NOTES 10-1—Inspection of Cable Assemblies

Safety hazards

Inspection of cable assemblies in the tower sections will require ascending the ladder assembly to access upper decks. It may also require standing on the ladder support brackets to view cables located between the decks. Always wear a full-body harness, twin-leg lanyard, and ladder safety device along with other personal protective equipment (PPE) required by the site safety policy when climbing and anytime there is a fall hazard.

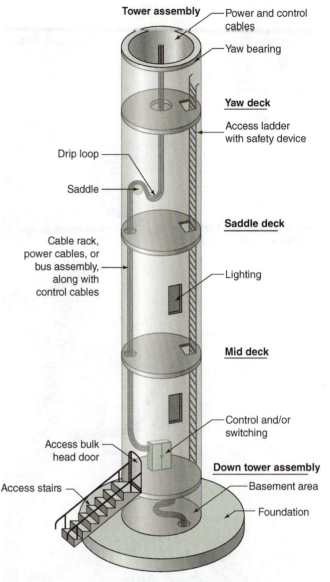

FIGURE 10-4 *Typical wind-turbine tower assembly*

Grounding and Bonding Connections

Grounding and bonding cables along with their respective terminations should be inspected to ensure they are secure. **Figure 10-5** shows examples of grounding and bounding cables that may be seen in the upper tower sections. If any of these connections are found to be loose, then tighten them to the required torque level for the fastener size and grade used for the connection. Bonding connections on the ladder assembly would be another inspection point. These connections may get hit during climbing, potentially breaking them from the crimp termination. Replace the crimp connection and ensure the termination is secure to the ladder sections. A ladder that is not bonded to the assembly may develop a different voltage potential with respect to the tower creating a possible shock hazard.

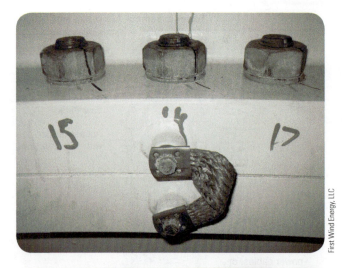

First Wind Energy, LLC

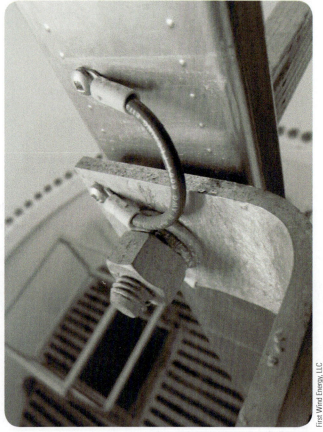

First Wind Energy, LLC

FIGURE 10-5 *Grounding and bounding connections*

Cables Mounted to Tower Tube Wall

The power and control cables are run from the nacelle to the DTA to enable a connection of the power circuits to the farm collector infrastructure and the control circuits to the farm supervisory control and data acquisition (SCADA) system. The power, low voltage control, and fiber-optic communication cables are attached to the inside of the tube through the use of a

SAFETY NOTES 10-2—Power Cables Hung in the Tower

Safety hazards

Inspecting cable assemblies in the tower sections will require ascending the ladder assembly to access upper decks. It may also require standing on the ladder support brackets to view cables located between the decks. Always wear a full-body harness, twin-leg lanyard, and ladder safety device along with other PPE required by the site safety policy when climbing and anytime there is a fall hazard.

- Never remove cable clamps or other cable-supporting structures from long lengths of hanging power cables. Unsupported cable and crimp connections may be damaged, allowing the cable(s) to drop and create a hazard because of falling items.
- Use caution around cables that exhibit damage to the insulation jacket. If the cables are energized, there is a shock or electrocution hazard. Deenergize and isolate cables suspected of insulation damage before continuing the inspection activity.

rack, cable tray, or individual mounting brackets. Each mounting arrangement requires the use of cable clamps or plastic wire ties to evenly distribute the weight of the suspended cable over several locations. Large power cables with higher current-carrying capacity may weigh 3 lbs/ft (4.6 kg/m) or more. Hanging cable of this size 260 to 300 feet (80 to 90 m) may create a load of 900 lbs (414 kg) on the cable termination. Cable-insulation damage from suspending the weight of one of these cables would be a good reason to assemble them to the mounting hardware while the tower tube is laying horizontally on the ground. There may be three of these cables per phase (three phases) along with a ground cable running down the inside wall of the tower tube. This is not a load that could easily be managed even with the aid of a tower hoist.

Each PM cycle should include a visual inspection of the cable clamps or cable ties to ensure they are secure. Every cable-supporting structure should be inspected to ensure they are secure to prevent them from coming loose during operation. If several of the cable clamps came loose during operation, the cables may chafe and create a short between phases or damage a cable splice. Damage to a splice will increase its resistance and in turn create heat that can damage the cable insulation around the splice. Any cables in contact with this splice would also be damaged from the excess heat. This damage may eventually cause the splice to burn through and reduce the current-carrying capability of the cable bundle. This would not be the time to find out there is an issue with a set of phase cables. Inspection with a thermal imaging camera during load operation or right after shutdown may be a good way to determine if there is a suspect splice.

A nondestructive test to find defects in cable insulation created by overheating would be through the use of an insulation-resistance meter. The testing process for insulation-resistance equipment was described in Chapter 7 (Test Equipment). Visual inspection of the cable and mounting hardware may be accomplished during the climb between decks. Areas that are not easily seen from the ladder may be inspected by using a mirror on an extension handle during the climb or by using binoculars or other visual-enhancement tools from an adjacent deck. The use of a crew lift may also help with this inspection process if the tower is so equipped. Another cable section that requires close inspection is the drip-loop area below the yaw deck.

SAFETY NOTES 10-3—Drip-Loop and Saddle Assembly

Inspection of cable assemblies in the tower sections will require ascending the ladder assembly to access upper decks. It may also require standing on the ladder support brackets to view cables located between the decks. Always wear a full-body harness, twin-leg lanyard, and ladder safety device along with other PPE required by the site safety policy when climbing and anytime there is a fall hazard.

- Never remove cable clamps or other cable-supporting structures from long lengths of hanging power cables. Unsupported cable and crimp connections may be damaged, allowing the cable(s) to drop and create a hazard because of falling items.
- Use caution around cables that exhibit damage to the insulation jacket. If the cables are energized, there is a shock or electrocution hazard. Deenergize and isolate cables suspected of insulation damage before continuing the inspection activity.
- Use caution around the drip-loop assembly if the nacelle is rotating. Entanglement within the drip loop may cause a pinch or crush injury to body parts that become trapped.
- Use a predetermined communication method when working around the drip loop and ensure that others know of the inspection activity to eliminate unnecessary risks.

Drip-Loop Cable Assembly

The **drip-loop** cable assembly is the loop of cable that extends from the nacelle through the yaw deck to the saddle deck. The **yaw deck** is the deck located just below the nacelle. It is a stationary platform used to work on the yaw assembly and to access the nacelle. The drip loop provides an extra length of cable so the nacelle can rotate a couple of turns in either direction. **Figure 10-6** shows an example of a drip-loop and

First Wind Energy, LLC

FIGURE 10-6 *Drip-loop and cable-support saddle assembly*

cable-support saddle assembly located on the **saddle deck**. The wind-turbine programmable logic controller uses sensors located by the yaw-drive gear to keep track of the number of turns and rotation direction. When the nacelle reaches two turns in either direction during operation, it will take the turbine offline and rotate the nacelle back to the center position. Typically, the nacelle does not accumulate two turns in any one direction. There is a balance of both clockwise and counterclockwise motion as it follows the wind so that the system rarely has to stop and unwind the drip-loop cable assembly. Original equipment manufacturers (OEMs) may also use a secondary fail-safe switch that disables the yaw-drive system and sends an error message to SCADA if the nacelle exceeds the two turns. This error message informs farm operations that there is an issue with the primary nacelle position sensor. As you may recall, this sensor is one of the SCADA switches activated during the DTA PM safety-switch test. The drip-loop assembly eliminates the need for a slip-ring assembly to transfer power and control signals to and from the nacelle control console. Many residential and mini wind-turbine systems use a slip-ring assembly for the power transfer and radio communications to control the system. The size and cost of a slip ring to do this for a utility-sized system would be prohibitive.

Drip-Loop Inspection Items

Inspections for the drip-loop assembly should include visual inspection for chafes, cuts, oil contamination, and damaged insulation along the length of the bundle. Oil may drip from gearbox seals or from residue left on the bedplate. Cleaning up after each service call will prevent this type of contamination. Ensure that the cable bumper located in the yaw deck opening is in place and secure so that it will not move during operation. This bumper prevents the cable bundle from being chafed as it twists against the yaw-deck frame during operation. A visual inspection of the deck area below the drip loop will indicate whether the bundle has experienced chafing during operation. There will be insulation fragments on the deck below the bundle created by chafing. If this has occurred, inspect the bundle and find the damaged area. Document the extent of the damage and include this in the PM report. Notify farm operations personnel of the damage and let them determine further corrective action if it is necessary.

Oil contamination should be cleaned off the cables to prevent damage to the insulation. Inspect the area below the gearbox and clean the oil residue as part of the PM activity. This will ensure that further contamination of the cable bundle will not occur. If the gearbox is leaking oil around the shaft seals, then document this in the PM report for farm operations to follow up later. Notify farm operations if the leak is substantial enough to compromise the operation of the equipment and let them determine further corrective action.

Inspections of the drip-loop support assembly should include visual inspections of all mounting hardware to ensure that components are secure. Secure loose hardware and tighten fasteners to the appropriate torque level according to size and grade. Replace cable clamps or wire ties as necessary to ensure that the cable bundle remains secure.

INSPECTION OF LADDER, DECKING, AND HATCHES

The access ladder, tower decks, and deck hatches allow safe egress to the nacelle from the DTA. These items should be functional and meet safety requirements regulated by OSHA, CCOHS, or the equivalent federal organization of the country in which the turbine operates. Deck areas also enable inspection of the tower flange bolts. These areas should be inspected for structural integrity and cleaned as necessary. Thorough cleaning of equipment, including deck surfaces, is a sign of good craftsmanship and pride in one's work. Consider this: if you invested several million dollars for a machine, would you not expect it to be appreciated? Cleaning up the dust, oil, and other contaminants from the deck surfaces prevents these contaminants from being carried up the tower to the nacelle. Contaminants in the nacelle can accumulate on electrical components and cause damage.

The ladder and its related safety devices are important to the safety of any technician ascending or descending the tower. Inspection items for these components will be discussed in the next section.

Ladder and Related Safety-Device Inspections

Inspection of the access ladder assembly should include a visual check of all mounting hardware and safety components. All safety hardware and required manufacturer's tags should be present and secure. Any loose structural hardware should be tightened to the appropriate torque level for the size and grade of the fastener. Any missing hardware should be noted on the PM report for farm operations to follow up. If any safety-related hardware is compromised—damaged, cracked, or missing—report it immediately for corrective action.

Two typical ladder safety systems are used for fall protection. One is a wire rope assembly (**vertical lifeline**) that hangs from the top deck of the tower along the ladder and is connected to the bottom two ladder rungs. The other system is a rail that is attached to the center of the ladder. **Figure 10-7**

FIGURE 10-7 *Ladder safety systems*

shows examples of the two ladder safety systems. The vertical lifeline assembly has mounting hardware at the top that also includes a rubber shock absorber to reduce the impact from a slip off a rung. The vertical lifeline system uses a cable-attachment assembly (Lad Saf) to connect to the front D-ring of the full-body harness. It also has several **cable keepers** positioned along the ladder to hold the cable in place during equipment operation. Missing cable keepers will allow the vertical lifeline to chafe against the ladder and damage either the ladder rungs or the lifeline.

A loose lifeline chafing against the ladder will also create an objectionable noise to any residents close to the wind turbine. Over the life of the wind turbine, the rubber clip portion of the cable keepers will wear and allow the cable to slip out during equipment operation. Note any missing or severely worn cable keepers in the PM report for follow-up by farm operations. Replacing damaged cable keepers will prevent damage to the ladder and lifeline. It will also keep the neighbors happy.

The safety-rail system is another style of fall protection. The safety rail is attached to several of the ladder rungs along its length to provide the required anchorage strength. This type of fall-protection system has a slider-attachment device that connects to the front D-ring of the full-body harness. Inspection of the rail system should include secure attachment points along the ladder length. Any loose hardware should be tightened, and any missing hardware should be reported to farm operations for immediate corrective action. If the ladder safety system is compromised, then use the double-lanyard climb method to ascend the ladder. This climb method was described in Chapter 2 (Working at Heights). This climb method will allow access to the next deck for further inspection and PM activities.

Another ladder system that is becoming popular is a climb-assist system. This system can be adjusted to decrease the effort required to climb a tower ladder. This type of system reduces the fatigue associated with climbing multiple towers in one day for service or PM activities. **Figure 10-8** shows an example of a climb-assist system. Consult the climb-assist manufacturer's manual for specific inspection and PM requirements. Document inspection findings and adjustments on the PM report for follow-up later as necessary.

Decking and Hatch Inspections

Decking hardware should be inspected during each PM activity to ensure the hardware is secure and does not present a safety hazard. The decks are typically made of two or three sections bolted together and mounted in the tower tube at the factory or before the tower section is lifted during construction. This makes some of the deck hardware inaccessible after the tower is constructed. Loose hardware may allow the deck to rattle during equipment operation.

FIGURE 10-8 *A climb-assist system*

Power Climber Wind

Inspect any hardware accessible to the top of the deck and use visual-enhancement tools such as mirrors or binoculars to check hardware that is attached to the bottom of the deck. Note any loose hardware that cannot be accessed on the PM report for future follow-up by farm operations.

The decks will also have one or several hatches that should be inspected during the PM cycle. These hatches will include the openings for ladder access along with others used to cover the tower hoist lift area if equipped. Some wind-turbine hoist systems are located in the nacelle on a gantry or swing arm that can be deployed externally to the tower assembly. Each deck hatch should be inspected to ensure that the mounting hardware is secure and that the hatches move freely for easy egress. The tower hoist access hatch or hatches should also have guardrails or a safety chain present when they are open to prevent a fall hazard. Ensure that the guardrails or other safety features are present and functional if the tower was designed for their use. If the tower does not use fall-prevention devices around open deck holes, ensure that fall-prevention or fall-protection PPE is employed. One note of caution around openings in decks: ensure that any tools placed on the deck are away from openings that do not have a safety lip. Most decks have a safety lip around openings to prevent accidental dropping of items to lower decks or down the ladder access area. Dropping a wrench or a bolt down the tower onto another person can cause a serious injury. Hardhats approved by OSHA, CCOHS, or their equivalent are a requirement around towers, but a tool dropping on a shoulder can cause a serious injury that a hardhat will not prevent.

Another large opening in decks may be for crew-lift equipment. This area should have fall-prevention devices around the opening. Ensure that required hardware is present and secured to prevent personnel injury or damage to the lift system. Another PM requirement for each deck area is the inspection and torque check of the bolts holding the tower sections together.

TOWER SECTION BOLT INSPECTION

This chapter previously discussed the construction process for a wind-turbine tower. These towers can be assembled in less than a day because the long tube sections are manufactured with flanges that allow them to be bolted together in the field. This is a vast improvement over the lattice towers used at the turn of the 21st century. **Figure 10-9** shows an example of bolts that capture the tower flange. Lattice towers have many components and bolts that can take days to assemble and days to inspect during a PM cycle. There are still lattice towers supporting operational wind turbines today, so wind technicians may find themselves being called on to do torque inspections on these wind-turbine systems. The inspection and torque check described in this section will focus on the tube-tower construction. The process of checking the torque on a lattice tower is similar, but there are many more bolts to check and accessing them is not as easy.

Typical torque inspection of bolts connecting the tower flange is a *10% inspection*. This means that 10% of the nuts

FIGURE 10-9 *Bolts capturing a tower flange*

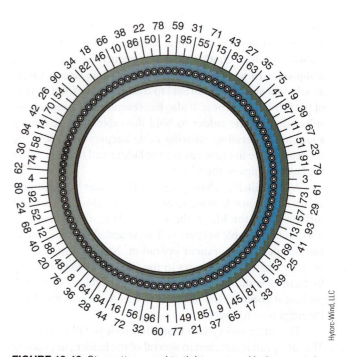

FIGURE 10-10 *Star pattern used to tighten several bolts or nuts in a large flange assembly*

are tightened to a predetermined torque level. It does not mean that the nuts are tightened to 10% of the original torque value. For example, if there are 120 bolts around the inside of the tower flange, then 12 of them will be checked. Check one out of every 10, and the problem is further simplified. The assembly of tower sections requires tensioning of the bolts in a star pattern at a lower level than the finished assembly and then going back over them a second time at the final torque value as discussed in Chapter 6 (Bolting Practices). **Figure 10-10** shows an example of a star pattern used to tighten several bolts or nuts in a large flange assembly. Tightening the nuts in the **star pattern** ensures that the flange seats properly around the circumference of the assembly. This is the same technique used for pipe flanges on industrial fluid systems and with rims on an automobile. Consult the OEM's PM procedure for required torque value for size and grade of the bolts, testing sequence, and acceptance criteria. Appendix F shows suggested torque values for varying SAE bolt grades and sizes.

To complete this process, inspect and set up the power-bolting equipment per the manufacturer's instructions or as described in Chapter 6. Choose the appropriate tool for the size of the fastener and torque-level requirement. Adjust the pressure of the hydraulic power unit or dial on the electric system for the appropriate torque output. One note to remember: the power-bolting manufacturer lists the output torque for each tool and adjustment level, but these may not be accurate enough for each activity. Many OEMs require the use of a torque-calibration instrument to check the output torque of the tool so it is within tolerance. **Figure 10-11** shows an example of a torque instrument that can be used

FIGURE 10-11 *Torque instrument calibrating power-bolting tools*

FIGURE 10-12 *Torque multiplier setup*

to calibrate power-bolting tools. The use of this instrument was described in Chapter 6.

Another method for checking the torque level of the tower bolts is the use of a manual torque wrench and torque multiplier. A torque multiplier can increase the input torque provided by a manual torque wrench by tenfold or higher, depending on the model. A **torque multiplier** is essentially a gearbox assembly that enables the attachment of an impact-grade socket on the output side of the tool and a manual torque wrench on the input side.

SAFETY NOTE: An impact-grade socket should be used with a torque multiplier or power-bolting equipment to prevent the socket from shattering during use at high torque levels. Refer to the power-bolting tool manual for approved socket information.

Impact-grade sockets are machined from carbon steel stock with thick walls and not die cast with thin walls like the ones used with a manual socket set. Impact-grade sockets are a natural steel color or treated with a black-oxide finish to prevent corrosion. An impact-grade socket should be easily distinguished from die-cast sockets by its markings and finish. Die-cast sockets are typically plated with nickel or chrome that leaves them with a shiny or highly polished finish. Power-bolting tool requirements and torque multipliers are discussed in Chapter 6. **Figure 10-12** shows an example of a torque multiplier setup.

Once the tools are chosen and set up appropriately, do not forget the following safety points. These safety points were explained in Chapter 6, along with general tool operation, but they are important enough to mention again. First, never place any body part—fingers, hand, arm, and so on between the reaction arm of the tool and the reaction point on the structure. The tools are designed to achieve the torque level adjustment before they can release pressure. If a finger is caught between these points, it will be crushed. Not a pretty picture. Second, the tool operator should be the one in control of the activation switch. There are only rare occasions when this will not work. For these occasions, there should be distinct words that tell the switch operator when to activate

the tool. *Go* and *No* will not work. Choose words that work for your particular area and dialect. Third, use safety devices such as a clip to prevent the socket from coming off the tool and an appropriate length of cable lanyard to attach the tool to the full-body harness. The cable lanyard should be used whenever there is a possibility of dropping the tool down a deck opening or when used on the outside of the tower.

Place the socket on the nut and activate the bolting tool. Typically, during a PM activity, the nut will not turn because the torque applied by the tool is at the same level as the original construction. If it does turn, then note how much the nut turns before it attains the required torque level and compare this value to the OEM's requirements. One suggestion: if more than two nuts rotate more than two flats, then a 100% inspection of the nuts should be completed on the flange. After the nut is tested and has passed the requirements, a paint pen should be used to mark the nut. Draw a line down one of the flats and as a radial line out across the washer and onto the flange. **Figure 10-13** shows an example of a nut that has been marked after passing a PM cycle torque test. The purpose of the paint line is to indicate the final position of the nut after the PM activity. Each PM cycle should use a different paint color to distinguish which nuts have been tested and when. The initial installation typically is marked with a black paint pen; subsequent PM cycles should be with a different color. This enables a technician to look at the nuts and see which ones were tested during the last cycle. During subsequent PM cycles, a suggestion would be to test the next nut located to the right each time. After the PM activity, a technician should use the same color paint pen to write the date and their initials in small print somewhere that will be easily seen by the next maintenance crew. The results of the inspection should also be documented on the PM report for future follow-up as necessary. This process should be completed for each tower flange and the yaw-bearing bolts located below the nacelle assembly. **Figure 10-14** shows an example of the bolts securing the yaw-bearing assembly.

FIGURE 10-13 *Nut marked after passing a PM cycle torque test*

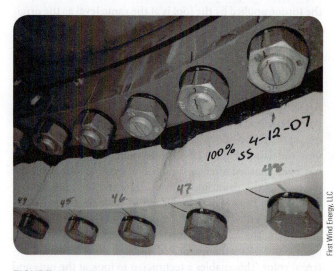

FIGURE 10-14 *Bolts securing yaw-bearing assembly*

INSPECTION OF TOWER LIGHTING AND ELECTRICAL RECEPTACLES

Inspection of each tower deck level includes all of the cables and related electrical components to ensure they are in good condition. Power and control cables were mentioned during a previous section of this chapter; now our discussion turns to the cables and hardware associated with the tower lights and power receptacles used for powering tools and other test equipment. Typically, each tower deck has a light assembly and a **ground fault circuit interrupter (GFCI)** receptacle, which is a specially designed receptacle that has a safety circuit that senses the current flow between line and neutral. If the receptacle circuit senses a current imbalance of as much as 3 to 5 milliamps, it will trip the interrupter within milliseconds and stop current flow from the receptacle. The circuit is designed on the concept that any current imbalance between line and neutral is the result of current flow to ground. Any current flow to ground would be considered a short to ground. The short may be the result of a tool that has been dropped in standing water or an operator who has become part of the circuit and has current flowing through him or her to ground. The National Fire Protection Association's National Electrical Code (NFPA70 NEC) requires GFCI protection for outdoor receptacles or any indoor receptacles placed near a water source. Each GFCI receptacle should be tested before use to ensure proper function. **Figure 10-15** shows an example of a GFCI receptacle used in a tower. The results of the test should be documented in the PM report, and any discrepancies should be passed along to farm operations for immediate correction. Do not use a GFCI receptacle if it does not trip with the manual trip-test button.

Tower decks are equipped with a light assembly to aid in work activities during service or maintenance. The PM cycle should be used to ensure that lights are functioning properly and that diffuser lenses or other safety features are in place. These lights are exposed to temperature and humidity extremes during the operation of the wind turbine, so inspecting the assembly during the PM cycle will ensure they are functional when needed. Follow the site maintenance plan and replace bulbs or other hardware as directed by the procedures. Document findings and any necessary replacement components on the PM report for follow-up later. These

FIGURE 10-15 *GFCI receptacle used in a tower*

activities cover each of the decks of the tube-tower assembly. The yaw-deck section of the tube tower is located at the top just below the nacelle connection. This deck may have equipment such as a hoist, crew-lift system, or other items depending on the manufacturer's model or the customer's chosen accessories. The next sections will describe some of the inspections that are associated with the yaw-deck area.

INSPECTION OF HOIST ASSEMBLY

Some turbine manufacturers include a hoist system on the yaw deck to help bring equipment, tools, and consumables up the tower. The hoist may be designed with a wire rope or a chain that can be deployed down the tower to raise or lower a **lift bag**, which is designed to transport items to and from the nacelle. It also eliminates the need for tying ropes or wrapping items with web straps for the lift. To safely use lift bags, never overfill a lift bag above its maximum rated load capacity and have personnel available at each deck to ensure a smooth transition through each deck opening if the items to be lifted are large. Never stand under a lift bag or any other item that is being lifted. And never use a lift bag that is torn, has damaged straps or damaged threads holding the straps or damaged lift rings. **Figure 10-16** shows an example of a lift bag used to move tools and supplies up a tower.

FIGURE 10-16 *Lift bag used to move tools and supplies up a tower*

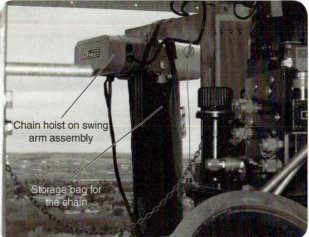

Chain hoist on swing arm assembly

Storage bag for the chain

University of Maine at Presque Isle

FIGURE 10-17 *Chain hoist system used to raise and lower service items*

Inspection items of the tower hoist assembly include integrity of the structural components attached to the tower, cables, cable reel and drive systems, chain assemblies, and chain buckets or bags. The inspection process should also include any electrical connections and safety stop switches that are integral to the system. Ensure that all mounting hardware is secured with safety wire or cotter pins as necessary to prevent fasteners from coming loose. Inspect cables for frays, excessive wear, bird caging, and other defects. Inspect chains for damage to links such as excessive wear, deformation, or cracks. Clean and lubricate cables and chains per the hoist manufacturer's instructions. **Figure 10-17** shows an example of a chain hoist system used to raise and lower service items. Refer to the hoist manufacturer's instructions or local regulations for inspection items and acceptance criteria. Document your findings on the PM report and pass discrepancies along to the farm operations team for immediate correction. Never use a hoist that does not meet documented inspection criteria.

Other items to inspect on the yaw deck include fixed service equipment such as a crew-lift system. Inspection items for a crew lift will vary, depending on model and relevant state or provincial regulations. Inspection of these items is necessary and should be completed by a certified technician.

YAW-BEARING ASSEMBLY

The **yaw-bearing assembly** is the mechanical connection between the tower and the bedplate. This bearing assembly enables the nacelle assembly and rotor assembly to follow the wind. Positioning the rotor assembly into the wind ensures that the wind turbine will optimize the power output of the system. **Figure 10-18** shows a view from the yaw deck looking up at the bearing connection in a Vestas V47. Note the brake-caliper assemblies mounted to the bottom of the bedplate assembly.

Yaw bearing   Brake caliper mounted   Yaw brake ring
inner race   on bedplate

FIGURE 10-18 *View from yaw deck looking up at bearing connection in a vestas V47*

These brake calipers will clamp on the ring attached to the tower to hold the nacelle assembly in place when the yaw-drive system is not activated. There are several different yaw-brake designs, depending on the OEM or model of the wind turbine. Maintenance activities for the yaw-brake assembly will be discussed with the yaw-drive assembly in the next chapter. The next section of this chapter will discuss the remaining inspection and maintenance activities required for the yaw bearing.

Yaw-Bearing Inspection and Maintenance

The yaw bearing is a vital component of the wind-turbine tower to nacelle assembly structural support. If the yaw bearing fails to work properly, the wind turbine will not be able to move the rotor assembly into the wind, which will limit or prevent generator output. A failure of the yaw bearing could also be catastrophic to the structural support system.

The yaw bearing should be inspected for visible corrosion or loose debris that may be present to indicate deterioration of the moving elements and race assembly. Examples of bearing assemblies with different moving elements can be seen in **Figure 10-19**. Moving elements may be either ball or roller, depending on the bearing design and its function. This damage may be because of inadequate or incompatible lubrication or water contamination of the lubrication. If incompatible lubricants were accidentally mixed, the lubricant could break down and not provide adequate lubrication. This issue along with water contamination can create metal-to-metal contact that will cause deterioration of the bearing elements that lead to failure. Consult the wind-turbine OEM or bearing manufacturer for recommended lubrication type. Never mix grease types without consulting the OEM engineers or lubricant manufacturer. The lubrication type should match the environmental conditions that the bearing will be

FIGURE 10-19 *Examples of bearing assemblies with different moving elements*

exposed to during service. The environmental conditions for this bearing include low rotational speed (RPM), temperature extremes, and moisture levels, along with static and dynamic loading. One lubricant formulation for this application may be lithium-soap grease. Grease types and applications were discussed in detail in Chapter 4.

Lubrication of the yaw bearing is through the use of a grease-dispensing system to charge the bearing grease manifold. Always consult the material safety data sheet (MSDS) provided with the grease to ensure its safe handling and disposal. The dispensing system may be a manual grease gun or a powered dispensing unit that can be connected to the grease fitting located on the manifold. The **grease manifold** may have several plastic or metal tubes that are connected around the bearing stationary race to evenly distribute the grease to the moving elements. **Figure 10-20** shows an example of a grease manifold that may be used with a yaw-bearing assembly. Consult the turbine OEM for the recommended quantity of lubricant to replenish the manifold system each PM cycle. Wipe any excess lubricant off the fittings and inspect the manifold and feed tubes for leaks. If a tube leaks, it may not have replenished the lubricant in the area that it feeds.

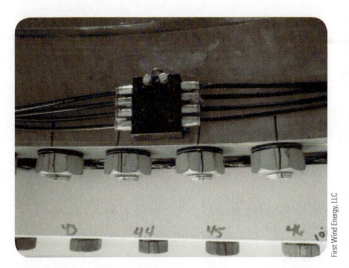

First Wind Energy, LLC

FIGURE 10-20 *Grease manifold used with a yaw-bearing assembly*

Repair or replace the connection and ensure that lubricant was evenly distributed around the entire bearing assembly.

OTHER PM ACTIVITIES

Cleaning is a very important component of any PM cycle. Cleaning of components eliminates the buildup of contamination that can create overheating of electrical components. Removal of dust and other contaminants from coated or bare metal surfaces in the tower reduces the opportunity for corrosion to deteriorate the surfaces. Dust or soil contamination on painted surfaces such as the tower walls will cause blistering of the paint and corrosion (rust) of the underlying steel. Dust and soil can absorb moisture and chemicals from the air that will create and maintain a corrosive reaction. This corrosion process will continue to cycle with changes in humidity levels until the contamination falls off the surface with the deteriorated paint or rust products or is removed during a cleaning activity. Make it a practice to clean all accessible surfaces with a mild cleaning solution to prevent the buildup of contamination that can damage components. Attention to detail in maintenance and service work will ensure that the equipment will operate properly. Cleaning is just one of the activities that show attention to detail.

SUMMARY

Maintenance of the tower assembly includes inspection of cables and mounting hardware, access equipment, structural assemblies, and ancillary equipment such as the hoist and crew lift. Maintenance of the tower assembly also includes inspection of the yaw bearing along with its related lubrication activities. Another equally important activity of maintaining the tower assembly is the documentation of inspection results and related PM activities. Documentation will also enable follow-up of issues by the farm operations team on the next service call to the tower.

Working in the tower portion of the wind turbine also requires special attention to safety guidelines related to working at heights. Maintaining a proper safety attitude will enable the job to be completed efficiently and effectively. Following the OEM or component manufacturer's recommendations will ensure that the wind turbine will operate properly over its intended life.

REVIEW QUESTIONS AND EXERCISES

1. List several PM activities necessary for a wind-turbine tower assembly.
2. What is the purpose of marking the nuts on the tower flange bolts?
3. Why is it important to inspect ladder safety equipment?
4. Is ladder safety equipment used for fall prevention or fall arrest?
5. What are some safety considerations for working around the tower hoist?
6. What is the purpose of the cable assembly draped below the yaw deck?
7. What are some of the methods used to inspect power cables in the tower?
8. What is the purpose of a GFCI receptacle?
9. What are some of the safety considerations associated with power-bolting tools?
10. What are some of the safety considerations associated with working within the tower?
11. Why is it important to document PM activities?

CHAPTER

11

Machine Head

KEY TERMS

bedplate
brake caliper
cable tray
cooling dampers
dye-penetrant test
Hazardous Waste Operation and Emergency
    Response (HAZWOPER)
investment casting
machine head
magnetic-particle test
metal inert gas (MIG) welding
nacelle

nondestructive testing
pillow block
sight glass
solenoid
soundproofing
top box
tungsten inert gas (TIG) welding
wire trays
yaw
yaw brake
yaw ring

OBJECTIVES

After reading this chapter and completing the review questions, you should be able to:

- Describe braking methods used with a yaw system.
- List required maintenance activities for a yaw system.
- Describe methods used to manufacture a bedplate.
- Describe the structural inspection types used on a bedplate.
- Describe regulations that cover handling and disposal of waste petroleum products.
- Describe the reason to treat isolated power systems as live circuits until they are tested.

- Describe the safety requirements necessary when working with electrical systems.
- List some of the maintenance activities necessary with control and power cabinets.
- List maintenance activities necessary with power-conversion cooling systems.
- Describe why it is important to clean the nacelle interior of dust and other contaminants.

INTRODUCTION

The last couple of chapters listed the typical preventative-maintenance (PM) activities that would be encountered in the down tower assembly (DTA) or tower sections of the wind-turbine assembly. This chapter will continue to the next group of PM activities that will typically be encountered in the machine-head assembly of a utility-sized wind turbine. The **machine head** or **nacelle** of a utility-sized wind turbine contains the structural connection for the rotor assembly, drive components, generator, and the control subsystems necessary for operation. Activities in this chapter will be grouped into subassemblies that will require PM. Depending on the model or manufacturer of the wind turbine determines the subassemblies present in the nacelle. For example, recent utility-sized Vestas models have the power-conversion

system and phase step-up transformers in the nacelle. General Electric and other manufacturers typically locate their power-conversion systems in the DTA. Typical PM activities for the power-conversion system are listed in Chapter 9 of this text. This chapter will mention some of the activities necessary for PM of the power-conversion system located in the nacelle but will not list specific activities. Specific activities will refer the reader back to those presented previously.

Whether the transformer is in the nacelle or beside the tower, it provides the same function to match the wind-turbine output voltage to the farm collector system. Other variations in system design that affect equipment present in the nacelle include connection of the rotor assembly to a main shaft or direct connection to the gearbox, direct-drive systems without a gearbox, hydraulic or electric pitch control, direct connection of the generator to the gearbox or through the use of a flexible coupling, and the hoist system used. These are just some of the variations that may be encountered in performing PM in the nacelle. This chapter will cover many of the necessary PM activities for subsystems and structural components such as the yaw system, drive-train mounting, bedplate, control, and ancillary items. One of the first subsystems encountered when entering the nacelle assembly is the yaw system. This subassembly is the means in which the control system can position the rotor assembly into the oncoming wind. Proper function of this system can optimize the power that is collected by the rotor and eventually converted to electrical output by the generator. The yaw system is made up of a large bearing attached to the top of the tower, a ring gear, drive motors, and a means to lock the nacelle from turning when the drive is not operating. Many necessary PM inspections for the items located in the nacelle will be discussed in the following sections.

YAW SYSTEM INSPECTION AND MAINTENANCE

The yaw system of a utility-sized wind turbine has several components that allow the control system to rotate the nacelle and rotor assembly to align with the prevailing wind. **Yaw** refers to motion around a vertical axis. It is typically used in reference to an aircraft's motion, but in the context of a wind turbine it refers to the motion of the nacelle as it rotates around the axis of the tower. The main structural component of the yaw system is a large bearing attached between the bedplate and tower. The **bedplate** is the main frame assembly for the wind turbine. All the drive-train components are attached to the bedplate to ensure structural support and proper alignment. Other components that make up the yaw system include a control system, electric gear motor assemblies, a large ring gear, and a braking system. The ring gear is attached to the tower, and a spur gear is attached to electric motor drives mounted to the bedplate assembly. The spur gear of the motor drive meshes with the ring gear to provide rotational control of the nacelle assembly. **Figure 11-1** shows an example of a yaw-drive gear motor being positioned at the assembly facility.

The braking method for the yaw system will depend on the design from the original equipment manufacturer (OEM). Typical **yaw brake** designs fall into two categories: passive and active. Passive or friction yaw-brake designs rely on drag created by multiple plungers or pistons that are spring-loaded against a yaw ring attached to the top of the tower. The drag produced by several spring-loaded pistons prevents the nacelle from freewheeling when the yaw-drive motors are not activated. Each of the motor drive systems also has a solenoid-activated brake located on the motor assembly to lock the motor when it is not operating. The combination of the two assemblies prevents the nacelle from drifting after it is positioned. The active brake design uses a caliper-brake system similar to that used with an automobile. A ring

FIGURE 11-1 *Yaw-drive gear motor being positioned at assembly plant*

assembly is attached to the inside of the tower with multiple caliper assemblies mounted to the bedplate that lock on the ring to prevent the nacelle from drifting. **Figure 11-2** shows examples of the two yaw-brake variations.

FIGURE 11-2 *Examples of yaw brake variations*

Inspection and Maintenance of the Yaw Bearing

Inspection of the yaw-bearing assembly can be accomplished from the nacelle access ladder or through the use of a short folding ladder used from the yaw deck. Checking the bearing mounting fasteners for proper torque is discussed in Chapter 10, so a visual inspection of the remaining assembly is the focus here. Check the bearing for corrosion, metallic debris, and the presence of any water contamination. If the bearing looks suspect for excessive wear or water contamination of the lubricant, then report the issue to the farm operations group for corrective action. Excessive wear or contaminated lubrication can be a sign of pending bearing failure.

Inspect the grease fittings and manifold block to ensure that all hardware and lines are secure and not damaged. Replace any damaged grease fittings or grease lines feeding the bearing assembly. Large bearing assemblies typically have a central grease manifold with several feed lines to distribute lubrication uniformly around the circumference of the bearing. Consult the OEM specification for the approved grease formulation and required quantity for the yaw bearing. **SAFETY: Consult the appropriate material safety data sheet (MSDS) for any hazards associated with the required grease. The MSDS should include a review of appropriate personal protective equipment (PPE), cleanup requirements, and accidental-exposure protocols.** Appendix I shows an example of an MSDS for an extreme-pressure, low-temperature lithium grease that may be used to lubricate the yaw bearing. Grease application to the yaw bearing can be made using a manual or electric-powered grease gun. Typical maintenance tool kits include a battery-operated grease gun with a rechargeable battery pack similar to that used with a portable drill. **Figure 11-3** shows an example of a battery-powered grease gun. **CAUTION: Never mix grease formulations. It can be disastrous if the formulations are incompatible or the resulting mixture is not stable in the application. The best practice is to clearly identify a grease gun for only one formulation. This will prevent accidentally mixing formulations during the PM activity.**

Application of the grease should be completed while the yaw bearing is in motion. This will ensure that the grease is spread uniformly around the bearing assembly. This activity may require two technicians to complete: one operates the manual yaw controls on the control console (**top box**) panel located in the nacelle and one is positioned on the yaw deck with the grease gun. **CAUTION: Notify all personnel in the nacelle, hub, and tower before activating the yaw-drive system. This will prevent a person from becoming pinched by the yaw-drive gears or causing a person on the top of the nacelle to fall. Maintain clear communications between all members of the maintenance team when performing manual equipment-control functions. Injecting fresh grease into a bearing will also purge existing grease from the assembly, so be prepared to collect and dispose of any waste materials.** Consult the site's **Hazardous Waste Operation and Emergency Response (HAZWOPER)** plan for proper collection and disposal of any waste materials. Disposal and accidental-release procedures should be used for any waste generated during a PM cycle. Waste products and contaminated consumable items include grease, waste oil, lubricant-soaked rags, soiled absorbent granules, soiled absorbent tubes, partial grease cartridges, and used oil-filter assemblies. The site plan should also include information on who should be notified in the event of an accidental release of waste material into the environment. Proper HAZWOPER planning and implementation is a requirement of OSHA 29 CFR 1910.120 and 29 CFR 1926.65 or the equivalent regulatory body of the country in which the wind farm operates.

Inspection and Maintenance of the Yaw-Brake Assembly

Inspect the **brake caliper** mechanism to ensure it is secured to the mounting assembly and functions properly. Consult the OEM procedure for necessary specifications. **SAFETY: This activity may require two technicians to complete. Ensure that there is clear communication between technicians to prevent an accidental pinch of a finger or hand in the caliper assembly during operation.** The gap between the brake pads and ring should be adequate to prevent binding but not large enough to create excessive travel of the caliper. Inspect each brake pad for excessive wear. There may be a visual indicator to show the maximum allowable wear tolerance for the pads. When in doubt, the OEM inspection procedure may have an allowable minimum-thickness specification for the brake pads. Note the findings on the PM checklist and replace the pads as directed by the maintenance procedure.

Inspect the caliper assembly for oil leaks if it is a hydraulically activated assembly. Tighten any fittings that may be leaking and wipe down the assembly of any residual oil

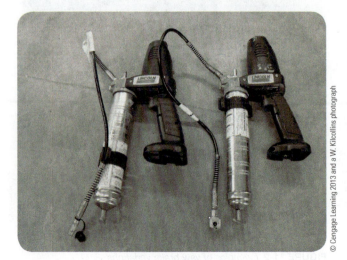

FIGURE 11-3 *Battery-powered grease gun*

© Cengage Learning 2013 and a W. Kilcollins photograph

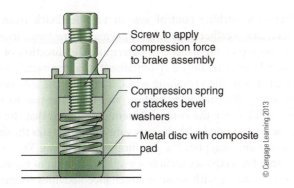

FIGURE 11-4 *Cutaway view of passive yaw-brake assembly*

— Screw to apply compression force to brake assembly

— Compression spring or stackes bevel washers

— Metal disc with composite pad

© Cengage Learning 2013

or other contaminants. Grease the caliper mechanism if required to ensure consistent motion each time it is activated. Consult OEM specifications for an approved grease formulation and the required quantity necessary to lubricate the brake mechanism. The second style of yaw-brake system comprises several spring-loaded plungers that ride against a **yaw ring** mounted to the top of the tower assembly.

Maintenance requirements for a passive yaw brake system include an inspection of the canister and plunger assembly. The plunger assembly may be made up of a steel or bronze piston that may be faced with a composite wear disc, compression spring or set of beveled washers, bolt or threaded rod, plain washers, and a locking nut. **Figure 11-4** shows a cutaway view of a passive yaw-brake assembly. The passive yaw-brake assembly may be made up of 30 or more plunger assemblies, depending on the size of the turbine. The plunger assemblies will be located on the bottom portion of the bedplate, and many can be accessed as you enter the nacelle from the yaw deck or located around the outside of the bedplate. Some of the assemblies may be located below the back of the gearbox toward the generator. These back assemblies may have limited access and require a bit of body contortion to gain access. Follow the OEM procedure for visual inspection requirements of the friction system. **SAFETY: Some composite disc materials may produce dust as part of the wear process. Always refer to the MSDS for the disc material and use appropriate PPE such as a dusk mask. Visual inspection requires disassembly, cleaning, and lubrication of all or some of the plunger assemblies, depending on OEM requirements. Removal of the plungers for inspection should be done with only a few at a time to ensure there is always braking action maintained between the nacelle assembly and the tower.** Once the assembly is cleaned, inspect the components for excessive corrosion, pitting, excessive wear, and damage that may cause the plunger to not slide within the canister. It is important that the plunger move freely to maintain a contact force with the yaw ring. Damaged or severely corroded plungers should be replaced. If the plunger has a composite wear disc located on its lower surface, inspect for excessive wear. Consult the OEM specification for minimum usable thickness and replace the disc as necessary. If the disc is within specification, compare the remaining thickness to the last PM activity.

How much has the disc worn since the last inspection? For example, if the disc has 3/32" (2.4 mm) remaining before it reaches the minimum allowable thickness and it has worn 1/8" (3.2 mm) since the last inspection, it should be replaced before the next PM cycle. Note the disc thickness on the PM checklist and inform farm operations for follow-up. If the plunger wear disc is worn below the minimum thickness, then replace the disc and make note on the PM checklist. Reassemble the plunger assembly and apply lubrication to the spring and outside surface of the plunger cylinder before placing it into the canister cap and inserting it back into the canister body. Consult the OEM procedure for the appropriate grease formulation. **CAUTION: General purpose grease may not work in this application because of environmental conditions and incompatibility with the assembly materials. Secure the cap to the canister and turn the bolt or threaded rod to create force between the plunger and the yaw ring.** Follow the OEM procedure for initial torque of the plunger threaded assembly. The torque value applied to the threaded assembly will correlate to a compressive force between the plunger and the yaw ring. This compressive force is the means to create drag and prevent the nacelle from turning when the yaw-drive motors are not operating. Complete these steps for each canister that requires inspection. Once all of the required canisters are inspected, torque each assembly to the specified final value using power-bolting equipment. The torque process should be completed using one of the star patterns discussed in Chapter 6. **Figure 11-5** shows an example of a pattern used to torque multiple plunger assemblies around the bedplate. **CAUTION: Tightening the plunger assemblies starting at one location and proceeding around the bolt circle may cause an uneven loading of the bearing assembly that could lead to bearing damage.**

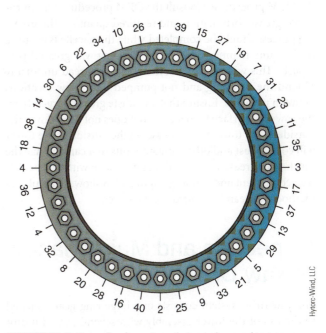

FIGURE 11-5 *Example of a pattern that may be used to torque multiple plunger assemblies around a bedplate*

Hytorc-Wind, LLC

FIGURE 11-6 *Methods to clean outside of tower*

Once the plungers are tightened to the required torque value, the assembly will need to be lubricated according to the OEM procedure. Consult the OEM procedure for appropriate grease formulation and required quantity. The greasing process should be completed while the nacelle is rotating in the same manner that the yaw bearing was greased previously. This will ensure that the grease is spread uniformly around the yaw ring and not pumped into a few locations between plungers. Ensure that any waste grease purged from the yaw ring surface is collected so it does not drop down the outside of the tower tube. Grease on the outside of the tower will collect dust and other contaminants that can damage the finish. This grease would be difficult to clean without specialized equipment and training. **Figure 11-6** shows examples of methods to clean the outside of the tower.

Inspection and Maintenance of Yaw Drives

The yaw-drive system consists of a large ring gear attached to the top of the tower assembly with several geared motor assemblies, each with a spur gear meshed with this gear.

The wind-turbine control system uses feedback from an electronic weather vane (also known as a *yaw vane*) located on the top of the nacelle to determine the direction of the wind. When the control system detects that the nacelle is not aligned with the prevailing wind, it releases the yaw brake, energizes the yaw-drive motors, and turns the nacelle into the wind. Once the control system determines that the nacelle is positioned with the yaw vane, it deenergizes the drive motors and reapplies the yaw-brake mechanism. Yaw-drive system PM activities include a visual inspection of the components along with some maintenance activities to ensure proper operation of the system between maintenance cycles.

Visual inspection of the yaw-drive gear motors includes determining the oil level in each of the gearboxes. The oil level can be determined by viewing the **sight glass** on a supplemental oil reservoir if so equipped or the sight tube located on the side of the gearbox assembly. The oil level should be located above the minimum level marker of the sight window or tube and may have a light amber color. **Figure 11-7** shows an example of an oil-reservoir sight glass. Compare the original color of the OEM-specified oil used in the motors as the guide for the inspection. The gear oil color may vary with different manufacturers and grade properties. If the level is below the minimum, add the appropriate oil to bring the level back within tolerance. If the level is substantially low, determine the leak location and repair the components before refilling the

FIGURE 11-7 *Oil-reservoir sight glass*

reservoir. **SAFETY: Review the MSDS for the oil to be used before continuing the procedure. Tighten any loose oil-line connections or replace seals as required to ensure proper function of the gear motor between maintenance cycles. If the oil is dark brown or black, then determine whether the drive assembly has been running at an elevated temperature or has damaged gears or bearings.** Make note of the findings on the PM checklist and alert the farm operations group for corrective action. A damaged gear motor assembly should be removed from service so that it does not fail, lock up, and damage the remaining drive motors. A damaged gear drive motor will typically cause its circuit protection device to trip giving the supervisory control and data acquisition (SCADA) system feedback of the drive problem that in turn will shut down the wind turbine. The SCADA system fault code may show up as a generic motor protection issue or specific problem, depending on the manufacturer's software code written for troubleshooting diagnostics.

The goal of the PM cycle is to prevent such issues, but occasionally failures can happen despite the best intentions and planning. If the gear drive motors have a built-in solenoid-activated brake assembly, inspect them for excessive dust or wear products that may settle on or below the motor assembly. A **solenoid** assembly uses an electrical current passing through a coil of wire to produce a magnetic field to move an iron core within the coil. Motion of this iron core can activate a switch or engage a brake disc or contact assembly. Activate the yaw-drive gear motors and determine if they are operating properly. Previous activations of the yaw-drive system during the greasing process of the yaw bearing may have given an indication of a drive problem. Situational awareness is always a great trait to have when working around equipment. It will help a good technician notice problems. **SAFETY: Should the brake assembly need to be disassembled, follow the approved lockout–tagout (LOTO) procedure before attempting an internal inspection of electrical connections or moving components. Remove the cover, clean the assembly, inspect for excessive wear, and inspect electrical components for issues.** Measure the gap between the discs along with the thickness of the discs to determine if they are within the OEM specification. Make adjustments to the assembly or replace components as directed by the PM procedure. Complete the inspection for each gear drive motor assembly and wire harnesses and replenish gear oil as necessary. Reassemble the components after completing the inspection and adjustment process. Make note in the PM checklist of any action taken to correct issues. If the issue cannot be corrected, notify the farm operations group for follow-up.

The next inspection process for the yaw-drive assembly is the large ring gear, gear motor spur gears, automatic lubrication equipment (if equipped), and feedback sensors for the control system. This process may require climbing below the bedplate to view the assembly. **SAFETY: Ensure that other technicians are aware of the inspection process being performed. If the activity is under the bedplate, some assistance may be necessary for extra lighting, tools,**

FIGURE 11-8 *Automatic grease system for ring gear*

or safety checks. If the inspection process requires working with hands around any of the spur gears or removing covers from an automatic grease dispenser, then implement an approved LOTO procedure to eliminate pinch hazards. **CAUTION: Climbing under the bedplate can be a dirty job, so wear coveralls or a disposable Tyvek painter suit to reduce or prevent contaminated work clothes with any encountered grease or oil. When the inspection is complete, the disposable suit can be discarded and work clothes will not be contaminated.**

Prolonged exposure to lubricants can cause health issues. Always review the MSDS for the grease or oil to be used. Inspection of the yaw-gear system includes removing excessive grease from the gears, sensors, and cleaning of any excess grease from the bottom of the nacelle. **Figure 11-8** shows an example of an automatic grease system for the ring gear. Dispose of lubricant-soiled rags or waste material appropriately. If the system is not equipped with an automatic grease system, use a paint brush to apply a thin layer of approved grease to the ring gear. If the system has an automatic lubrication system, then inspect the reservoir and replenish the grease as necessary. When the inspection is completed, notify others that the task is complete and remove the LOTO items. Document your findings on the PM checklist for follow-up later as necessary.

INSPECTION OF STRUCTURAL COMPONENTS

Structural components located in the wind-turbine nacelle include the bedplate that serves as the backbone for the drive-train components and a frame used to mount the generator. The bedplate assembly is used to transfer the weight of the machine head to the tower. The bedplate may be a welded assembly of many steel components or it may be a large steel

First Wind Energy, LLC

casting. Whether the bedplate is cast or welded together, it should be inspected during the PM activity to ensure it has not developed cracks because of fatigue or stress produced by dynamic loading. Typical areas that may develop cracks during operation include any area with a change in shape or geometry. These areas include webs, gussets, holes, and fillets. Each of these areas creates what is known as a mechanical stress riser. A stress riser is created when the stress of a load has to be distributed through a reduced cross section or through a sharp transition of a component. The ideal situation is to carry a load through a uniform cross-sectional area.

Anytime there is a transition or change in cross-sectional area, the stress through that section increases. Remember: stress is the load applied divided by the cross-sectional area. For example,

$$\text{Stress (psi)} = \frac{\text{load (pounds)}}{\text{cross-sectional area (in.}^2)}$$

International Organization for Standardization (ISO) units may be given in newton/mm^2, which is equal to a megapascal (MPa). Refer to Chapter 6 on bolting for more details on stress and loading.

Typical engineering design practices are to reduce the number of shape changes or provide large radii in these areas to reduce the stress-riser effect. These design practices are useful when the loading scenario is understood and the components are made to specifications. Occasionally, imperfections occur during the casting process or in the welding process. These imperfections may not be seen until the components are placed into service. Any areas with holes, gussets, fillets, or webs should be inspected on a regular basis to catch cracks before they become failures. When an imperfection or crack is suspected, it may be difficult to determine the extent of the problem by visual means alone, so other means of nondestructive testing may be needed to detect imperfections without damaging the component to determine the extent of the problem.

Three typical types of **nondestructive testing** available for field use are visual, magnetic particle, and dye penetrant. First-line nondestructive structural inspections are done visually with the aid of magnification equipment if necessary. If a crack is suspected, it is important to remove any paints or coatings from the suspected defect to ensure it is not a crack that has formed in the paint layer. It would be embarrassing to have a specialized team come on site to determine that there is only a crack in the paint over a welded joint. Once a crack is visually verified in a joint, the other two inspection processes may be necessary to determine the extent of the defect. **SAFETY: If grinding is necessary to remove the paint or coating, follow the site's hot-work policy to ensure that the inspection can be completed without injury or equipment damage because of a fire.**

Magnetic-particle and dye-penetrant inspection processes require specialized equipment along with trained technicians to correctly interpret the results. A **magnetic-particle test** works through the interaction of magnetism with iron. Magnetic flux lines travel through a magnetic material, such as iron or an iron alloy, in a uniform pattern from the magnet's north pole to its south pole. If there is a break in the magnet material, the two edges formed by a crack will become another set of poles. The discontinuity in the magnetic material will cause iron fillings or magnetic particles to spread over the defective area and cluster between the regions of these new poles. The size of the defect will determine the extent of the disruption in the localized magnet field and the amount of magnetic particles that will cluster into the crack area.

The next nondestructive test is a **dye-penetrant test** process. It uses a colored or ultraviolet (UV) luminescent dye to make cracks highly visible during the inspection process. **Table 11-1** shows the basic steps of the magnetic-particle and dye-penetrant tests. The next couple of sections highlight some of the areas of the bedplate and generator subframe assemblies that should be considered for inspection.

TABLE 11-1

Basic Steps of Magnetic Particle and Dye Penetrant Nondestructive Tests	
Magnetic Particle	**Dye Penetrant**
Remove paint or coating from the area of the defect.	Remove paint or coatings from the area of the defect.
Clean the area of contaminants with a solvent and dry thoroughly.	Clean the area of contaminants with a solvent and dry thoroughly.
Magnetize the area of defect using permanent magnets or electromagnets.	Spray the dye over the area and allow capillary action to draw the dye into the defect.
Spread a thin layer of iron fillings (magnetic particles) over the region.	Remove excess dye from the area with an appropriate solvent.
Photograph area to record the pattern of particles.	Spray developer on the area.
Interpret results: • Particles will cluster over an area with an imperfection.	Illuminate with appropriate light source and inspect: • Use high-powered light source for standard dye. • Use UV light source for luminescent dye.
	Interpret results: • Dye will wipe off an area that is not cracked.

Bedplate-Construction Concepts

Wind-turbine bedplates may be constructed through several different manufacturing processes. These processes may include bolting, welding, or casting, or some combination of these processes. The manufacturing process determined by the OEM will depend on available equipment and cost-effectiveness of the process. Welding or bolting of structural channels, I-beams, or flat stock would be one method of construction.

Another process may be an investment casting of the bedplate assembly. The process of **investment casting** is a means of manufacturing a metal component by use of a mold to hold molten metal during the solidification process. The mold for the process has an internal cavity similar to the shape of the finished part. An investment-casting process can reduce overall cost of the bedplate if multiple assemblies will be made over the life of the production model. Investment casting has a substantial upfront cost to set up the process, but this cost can be spread over the number of assemblies made. Casting is typically not profitable for short production runs. These typically would use bolted or welded component assemblies. The construction process of bolting or welding several components together will always have an associated labor cost for the unit.

In either case, there are inherent quality items to track with the processes. Proper control of the casting process produces uniform material properties along with good dimensional control. After the casting is made, it will go through a final cleaning and machining process to produce a finished product ready for shipment to the wind-turbine assembly facility. A couple of items that can affect the final product may result from raw-material issues or casting process control. The casting process issue may be the result of poor temperature control. Improper casting temperature can create poor material fill within sections, voids because of trapped gases, or cracks because of improper cooling of the sections. These are typically caught in the production process, but minor defects may occur that could get by process control and show up later during the service life of the component.

Construction of bolted and welded assemblies can also have good dimensional controls through the use of large machining centers to produce the component parts and fixtures to position components during the welding process. Bolted joints can be visually inspected and checked for the torque value of bolt and nut assembly. Inspection of welded joints during the manufacturing process can be monitored for proper current levels, penetration, fill appearance, and bead consistency. These factors, along with statistical data from the required destructive tests performed during the production run, are good indicators of weld integrity. Unlike bolted joints, welded joints can still be difficult to inspect once they are painted and ready for final assembly. Cracks in a welded joint can be masked by paint or other coatings applied for cosmetic or corrosion-resistance purposes. Understanding the nature of the joint can aid in inspecting for defects in the field.

Welding Process

A welded joint is formed by heating two mating parts to their melting temperatures. When the joint is hot enough to melt and flow, the material will puddle into the space between the components. The addition of a filler material during the melting process will ensure a uniform mix (alloy) of the materials and the correct filling of the joint. Three forms of welding—**metal inert gas (MIG)**, **tungsten inert gas (TIG)**, and stick—use an electrical-arc heating method, filler material, and some means to prevent the molten metal from oxidizing. MIG and stick welding use an electrode made of the filler material to supplement the fill of the two components during the welding process. TIG welding uses a tungsten electrode that is not consumed as a filler material during the electric arc process, so a filler material is supplied through a secondary process. MIG and TIG use an inert gas blown over the molten metal to prevent oxidation during the melting and solidification process, whereas stick welding uses a flux coating on the electrode that will vaporize and form a protective atmosphere over the molten metal. **Figures 11-9** and **11-10** show examples of the welding equipment and typical joint configurations. Typical defects that may be created

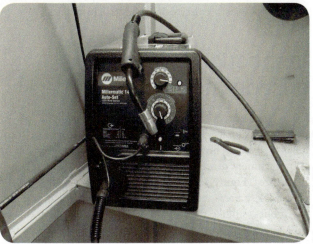

FIGURE 11-9 *Examples of welding equipment*

© Cengage Learning 2013 and a W. Kilcollins photograph

© Cengage Learning 2013 and a W. Kilcollins photograph

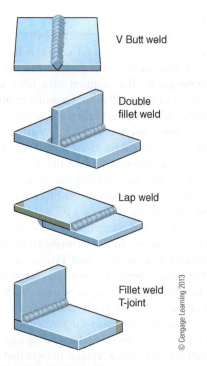

V Butt weld

Double
fillet weld

Lap weld

Fillet weld
T-joint

© Cengage Learning 2013

FIGURE 11-10 *Weld joint configurations*

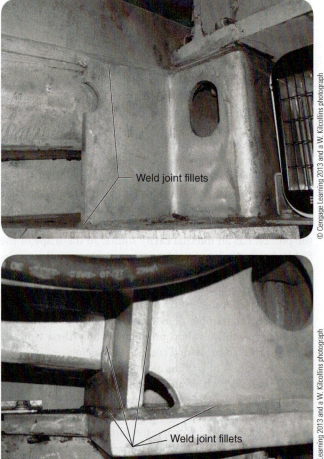

Weld joint fillets

© Cengage Learning 2013 and a W. Kitcollins photograph

Weld joint fillets

© Cengage Learning 2013 and a W. Kitcollins photograph

FIGURE 11-11 *Examples of weld transition areas*

during the welding process are minimal penetration of melt zone into components; blow-through, where holes or voids form from excessive heating; contamination of melt materials, which causes sections that do not melt or alloy properly; incorrect filler material or mismatch of joint materials, which creates a weaker-strength alloy in the joint; lack of an inert atmosphere, which causes oxidation of the molten metal; uneven cooling of the molten metal if metal sections are not similar in cross section; and cracks in the joint if the metal is heated up or cooled too quickly during the process. These are just some of the issues that may be produced during the welding process. Manufacturing process control for welding operations minimizes the chance of these issues reaching the field, but it is still wise to take time to inspect and catch any minor cracks before they can create structural issues.

Inspection Points for a Bedplate Assembly

The type of inspection process for a bedplate assembly will depend mainly on its manufacturing process, but there are some areas that should be noted for changes in geometry, including the attachment of webs, gussets, and reductions in cross-sectional area. **Figure 11-11** shows a few examples of transition areas. These areas can be present on both welded and cast assemblies, so take the time to visually inspect them during a PM cycle to ensure that minor issues do not become failures before they are detected. Use a high-powered inspection

light to see whether cracks are present around welded joints or thinner cast sections. Remember: there may be a crack in the painting or coating over the joint or thin section because paint in this area may puddle and crack during the curing process. If a crack is found in the paint, take time to carefully remove the paint and inspect the underlying metal joint. Cracks may form and propagate through the paint from the metal below, so determine the extent of the crack and document the findings on the PM checklist. Documentation will allow for follow-up later by the farm operations group or specialty team that is trained in structural inspection techniques.

Generator Subframe

The generator subframe assembly may be bolted or welded together, depending on the requirements of the system. The frame design requirements for an in-line drive system of a generator connected to a gearbox with a coupling are going to be different than a system that does not have a gearbox. A system that uses direct coupling of a generator to the gearbox will not need a separate frame for the generator. If the system to be inspected during the PM cycle has a generator subframe, then the same inspection criteria mentioned previously is still valid. Inspect areas that have been welded together for possible cracks and use caution if a crack is seen in the paint finish. Remove the paint and determine if the crack is in the paint or caused by a crack in the welded joint below. Inspection of the bolted joints should be completed to the OEM procedure to ensure that the fasteners are secure. If the information is not documented, then request the lead technician or site manager call the OEM engineer responsible for the system structural design for recommendations.

INSPECTION OF CABLE, GROUNDING, AND BONDING CONNECTIONS

Cable connections used for grounding and bonding should be inspected for secure connections. It is important to ensure that structural components be maintained at the same potential and that there is a ground connection in case of an electrical fault or a lightning strike. Bonding connections in the nacelle interior include connections between the nacelle components, connections between walk-platform supports, and any items that may be mounted within vibration-isolation mounts. **Figure 11-12** shows an example of a vibration-isolation **pillow block**. Vibration-isolation mounts such as those used to support the gearbox will prevent vibration from being transmitted down the tower and causing undesirable noise and damage to sensitive equipment.

Grounding connections that should be inspected during PM of the nacelle include grounds located on motor assemblies, the interiors of control consoles, and any junction (J) boxes mounted on major components such as the gearbox, bedplate assembly, or generator. **Figure 11-13** shows an example of a J-box located on the back of a gearbox. Inspection activities for the gearbox and generator will be covered in later chapters, so the focus in this section will be on J-boxes attached to the bedplate and grounding connections in control consoles. Grounding connections located on the outside of motor assemblies, bedplates, and external surfaces of control consoles should be checked for a good wire-crimp connection and to ensure that the fastener holding them is secure. This inspection may be completed without the need of a LOTO procedure. Inspection of grounding connections within J-boxes and consoles should follow an approved LOTO procedure to ensure that the activity can be performed safely. Document findings on the PM checklist and perform corrective actions as necessary. If items are missing or cannot be corrected during the PM activity, then notify farm operations for follow-up on the next service request.

FIGURE 11-12 *Vibration-isolation pillow block*

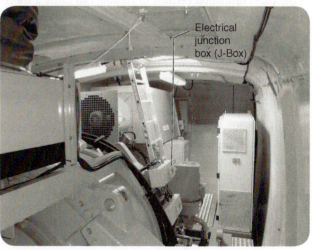

Electrical junction box (J-Box)

FIGURE 11-13 *J-box on back of gearbox*

INSPECTION OF CABLE TRAYS, DECKING, AND CABINETS

Cable trays, decking, and cabinets should be inspected to ensure that the components are fastened securely to the bedplate or other supporting assembly. Bolting used to support the components may loosen with repeated temperature cycling during the day and from seasonal variations along with the added help of vibration. Visually inspect all fasteners and check the torque level as instructed by the OEM procedure. If the number of fasteners to be checked is not defined, then do a random sampling of the fasteners and ensure that the minimum torque-level requirement for the size and grade is used. A random sampling does not mean every fastener has to be checked for torque level. If the component has six or eight fasteners, then choose two or three to check. Check the torque level of all fasteners if the majority of the sample is found to be loose. Refer to Chapter 6 and Appendix F for suggested torque values for fasteners. Document the findings on the PM checklist and note all corrective actions taken.

Inspection of cable trays should also include a visual inspection of the cables in the trays. A **cable tray** is an assembly used to organize, protect, and support power and control cables. The cable tray may be run above or below the control consoles and equipment in a wind-turbine nacelle. Ensure that cables are positioned within the trays as necessary to prevent damage to the insulation and attached with cable clamps and wire ties as documented in the OEM specification. Improperly supported cables, especially on vertical runs, may cause damage to the cable insulation or the termination assembly if the load is not supported evenly over the run. Inspect for damage to the cable insulation because of chafing or excessive heating. Discoloration of insulation around a crimp joint may indicate a high-resistance connection that needs to be corrected. Excessive abrasion of the cable insulation because of chafing can cause a shorting problem with time. Reposition the cable and repair the insulation if possible. If the issue cannot be corrected, then inform the farm operation team for follow-up as necessary. A badly damaged section of insulation may need to be corrected before the system can be returned to service. Ensure that the farm operation team is aware of any condition that will keep the system out of service.

Clean cable insulation of any contamination such as oil that may have dripped or sprayed from a hydraulic system or the gearbox. Wipe the cables down with an approved cleaning solution to prevent damage to the rubber insulation from the oil. **SAFETY: Never use a cleaning solution that is not approved for the type of insulation. Incompatible solvents can damage the insulation and create a shorting issue. Never apply a water-based cleaning solution to a cable if there is a suspicion of damage to the insulation.** This can cause an electrocution or fire hazard. When in doubt, perform an approved LOTO procedure on the cable assembly before doing an inspection and cleaning. Make corrections to the hydraulic system or gearbox to reduce or stop the leak and prevent oil contamination as necessary. Document all findings on the PM checklist for follow-up at the next service call.

INSPECTION OF PLC, CONTROL, AND SENSOR COMPONENTS

Inspecting the programmable logic controller (PLC) and electrical control components will require working in the control console or top box. This activity should be completed using an approved LOTO procedure to isolate the entire console from power or by sections as the technicians' training and authorization levels dictate. The control console will have several levels of voltage, depending on the control systems that are present. PLC control modules, transformers, terminal strips, contactors, and circuit-overload devices may all have different voltages, so proper training of what voltages are present, knowledge of how to isolate the devices, understanding the use of electrical schematics, and authorization to work on these systems is a requirement before any work can be performed in a control or power cabinet. Inspection of the control cabinet should include cleaning of fan assemblies, replacing any air-filtration elements, and ensuring that required insulation is present and secured to the inner walls of the console. Another inspection should be made for the presence of rodents or other wildlife that may have made their way up the tower. It is amazing what can decide to make a turbine nacelle home. If the wind is low for a period when birds are nesting, they will make nests in any area they can access. This may include cooling vents, under the bedplate, and in control cabinets. This can have disastrous results when the wind picks up and the nest is located where it can be damaged or ingested into a fan assembly.

Inspection of wire connections and component support assemblies should also be completed. **SAFETY: Always apply an approved LOTO procedure as necessary to ensure wire connections are secure for the control components. Visually inspect and ensure that grounding connections are secure in the cabinet. Make sure wire tray covers are in place and that wires have not slid out of the trays if covers were not present. Wire trays** are box-shaped trays with covers that are used to route control and power cables within a console. Their purpose is to protect the wires and organize their distribution path. Ensure that wire components and control devices are secure to the DIN rail or any other support assembly. Wipe excess dust or other contaminants from electrical components especially those that rely on air cooling during operation. When work is complete in the control console, notify the team that the LOTO will be removed so it may be returned to service for other PM activities.

Inspection of sensor devices present in the nacelle should be verified before completing the PM activity of the control console. When the console door is open, the PLC input modules can be visually verified for sensor status along with the status of other input devices such as manual controls and emergency stop switches. Chapter 9 mentions a check of critical safety devices located in the tower, including emergency stop, hydraulic pressure, and nacelle rotation, among others. This is a great time to compare the PLC input data to analog gauges or through visual inspections. Compare the SCADA input values for activated emergency stop switches, pressure readings seen on hydraulic gauges, temperature measurements of components with a handheld meter, proximity switches that indicate the rotation of drive-train components, and proximity switches that indicate the position of devices such as air-vent dampers. **SAFETY: Ensure that everyone in or on the nacelle is aware of the system checks—and that actuating an emergency stop switch will lock the parking brake and feather the blades to 90°. Anyone not aware of the activity could be pinched by the activating brake caliper or stumble because of the sudden stop of a pin wheeling rotor assembly.** Document the findings of each test and make notes on any discrepancies that are encountered on the PM checklist. If a discrepancy is related to the system safety device, then it should be corrected before the system is returned to service.

POWER-CONVERSION SYSTEM

Some OEM wind-turbine models design the power-conversion system in the nacelle assembly instead of the DTA. Inspection and maintenance activities for the conversion system should be completed according to the manufacturer's procedure. **SAFETY: Before attempting any work in the power-conversion cabinet, the system must be isolated and LOTO should be performed to an approved procedure. If the system includes a capacitor bank as part of the inverter switching, then ensure that it has had adequate time to discharge before verifying energy state.** Consult the site safety policy or NFPA 70E tables for the appropriate hazard risk category (HRC) arc-flash PPE. Arc-flash gear should be rated for a fully energized cabinet when opening it for the verification process. Remember: the system should be deenergized, but what if something went wrong in the circuit breaker used to isolate the conversion system? Better to be safe than a statistic: verify that the system is at zero potential using an approved meter rated for the voltage level and any transient present before proceeding with any work.

Maintenance Activities

Maintenance activities in the nacelle power-conversion system will be similar to those encountered in the DTA. Follow the manufacturer's procedures for verifying cable connections bolted to bus bars and terminal strips. Electrical connections need to be secure with minimal resistance in the interface to prevent overheating and damage to components. This can be verified visually for presence of corrosion on connections and through a torque test of the fasteners. Torque the fasteners to the recommended level provided by the OEM procedure. These fasteners are typically copper or an alloy of copper, so they cannot achieve a torque value the same as an equivalently sized steel fastener.

Cooling System

The cooling system for the power-conversion cabinet will be either liquid or forced air as described in Chapter 9. Air-cooled systems use convective airflow for cooling components, so ensure that everything is clean and that heat-sink assemblies are free of any obstructions that will prevent airflow. Liquid-cooled systems will use a 50:50 glycol–distilled water solution as described previously. Ensure that the expansion reservoir is filled to the appropriate level with cooling solution and that any leaking hoses or couplings are tightened as necessary to prevent loss of solution. Wipe up any liquid or dried solution residue that may be on or by control boards to prevent corrosion of electrical traces or wire connections. **Figure 11-14** shows an example of a coolant leak that should be corrected during the PM activity. Check the cooling solution with a hydrometer to determine if the specific gravity of the solution is correct. This will ensure that the cooling solution will not boil under maximum temperature conditions and will not freeze on cold days when the system is down with low winds. Before removing LOTO and reenergizing the power-conversion system, the transformer assemblies should be inspected and any required maintenance should be performed.

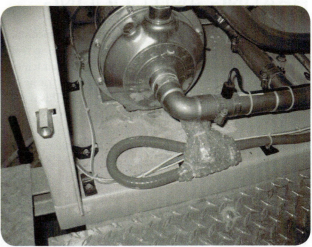

FIGURE 11-14 *Example of a coolant leak*

First Wind Energy, LLC

Inspection of Transformer Connections and Hardware

The power-conversion system and generator feed low-voltage power to the can-style transformers located in the back of the nacelle for some wind-turbine systems. The Vestas power-conversion system and phase transformers are located in the nacelle assembly. When the power-conversion system is isolated from the main power contactor during the LOTO procedure, take the opportunity to inspect and perform maintenance on the phase transformers. **SAFETY: Consult OEM documentation for the recommended isolation and approved LOTO procedure. Follow the approved verification procedure to determine if there is zero energy at the transformers. Use the recommended arc-flash PPE as directed by the OEM procedure or NFPA 70E HRC requirements.** Inspection of the phase transformers should include verification that the power-cable connections are secure by a torque test of the fasteners. There should also be a visual inspection of the cans to ensure that there are no visible oil leaks, areas of discolored paint that show excessive heating, or visible corrosion that may create problems.

Document the findings on the PM checklist and make corrections as necessary. The generator and related power cables are other items that should be inspected and PM activities completed before removing the LOTO devices. These areas will be covered in a separate chapter later in the text. When all of the power-system components have had their PM activities completed, notify all affected personnel in the nacelle that the LOTO devices are going to be removed and the system will be reenergized. Energize the power-conversion system and ensure that it is ready for operation before exiting the nacelle at the end of the maintenance activity.

INSPECTION OF NACELLE HEATING AND COOLING

Heating and cooling components in the nacelle encompass equipment used to vent hot air and bring in cool air to maintain the temperature of the nacelle. When outside air temperature is below the required temperature range, electric heaters are used to supplement the nacelle temperature. This is especially important when the wind is low for a period of time and the system has to be brought up to the minimum operating temperature for electronic systems to function properly. When the wind-turbine system is operating and producing power for the utility grid, the generator and other components such as the gearbox are producing heat that will warm the nacelle. This is the time when the nacelle cooling system is blowing warm air out of the nacelle and drawing in cool air. Vents on the nacelle exterior typically have **cooling dampers** or louvers attached to them so that the control system can

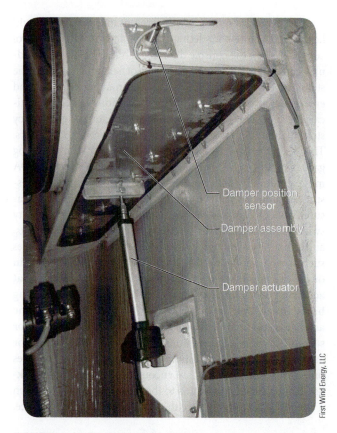

Damper position sensor

Damper assembly

Damper actuator

First Wind Energy, LLC

FIGURE 11-15 *Cooling-vent assembly*

automatically open or close the vents with motor drives as necessary to maintain temperature. **Figure 11-15** shows an example of a cooling-vent assembly that may be used to control the temperature in the nacelle.

The control cabinets also have supplemental heaters to ensure that the electronics are maintained at a minimum temperature during times when the outside air temperature is below the operation range. Many of the control cabinets have adjustable thermostats to control heaters when the cabinet temperature drops and thermocouples that are used for feedback to the PLC to activate exhaust fans to cool the cabinets as necessary. Previous discussions in this chapter highlighted the cleaning activities for the control-console fans and air filters so these activities may have been completed. The items that were not highlighted were the electric heaters located in the nacelle and the exhaust fans used to move air in and out of the nacelle. **Figure 11-16** shows an example of a heater that may be used to maintain the temperature in the nacelle. These components should be inspected to ensure that they are securely attached to their mounting structures.

SAFETY: Fans and heaters are typically on automatic start systems, so they should be isolated and LOTO should be performed to an approved procedure. Remove covers and panels to gain access to the components for cleaning and removing debris such as dust or the remnants of wildlife activities. This will ensure adequate airflow through the

FIGURE 11-16 *Heater for maintaining nacelle temperature*

fan and heaters when they are needed to control the temperature of the nacelle. Replace covers and panels after the cleaning process has been completed. Document findings on the PM checklist and complete corrective actions before removing LOTO devices.

INSPECTION OF NACELLE INTERIOR AND STRUCTURAL HARDWARE

Inspection of the nacelle interior includes verification of any hardware required to hold the nacelle sections together, mounting hardware for lighting fixtures, wire conduits, bulkhead seals for cables, bonding straps as mentioned previously, skylight hatches for access to the top of the nacelle, and large panels used to cover lifting points for the nacelle assembly. Ensure the necessary hardware is in place and secure; this includes soundproofing foam that is located on the interior walls of the nacelle. **Soundproofing** foam is used to reduce equipment noise from passing to the surrounding area. Replace any lights and lighting diffuser covers as necessary. Consult OEM documentation for details and specification. Inspect seals and skirts on hatches, panels, and blower assemblies to make sure they are in place and thus minimize water and snow from entering the nacelle during storm conditions. Adjust the seals and skirts as necessary and tighten clamping rings that secure them to the nacelle.

Inspect the nacelle for required safety signage and tie-off hardware. Primary exit hatches or skylights used to gain access to the top of the nacelle should be marked with a statement such as "Fall Hazard Beyond This Point," and fall-arrest PPE should be worn. Inspection of mounting hardware on the ceiling of the nacelle that connects the safety rail or tie-off

eyes to the nacelle top should also be checked for integrity. Tighten any loose fasteners and replace ones that have become corroded. Document the findings on the PM checklist and make corrective actions as necessary. Any missing hardware should be documented and passed on to the farm operations team for follow-up.

RECOMMENDED CLEANING ACTIVITIES

Cleaning of the nacelle assembly is important to minimize dust and debris from entering generator air-cooling vents, liquid heat-exchanger assemblies, and control-console vents. Wipe down external surfaces of the cabinets and equipment with an approved cleaning solution and rags. **SAFETY: Consult the MSDS for any cleaners or other chemicals that may be used during the cleaning process and use the recommended PPE such as gloves and safety glasses or goggles. Clean up any spilled oil or grease to prevent slips on walking surfaces and prevent the buildup of dust on equipment and other interior surfaces. Use a vacuum cleaner to remove dust and plant matter from vents and areas below the equipment. Dry plant matter will invite unwanted guests and create a fire hazard if it builds up around heaters or other devices that operate at elevated temperatures. Clean up any oil or grease that may have spilled under the bedplate assembly.** This area can collect rainwater during a storm and may wash the lubricant out of the nacelle and down the tower. Not a pretty sight. If the lubricant residue makes it to the ground, it is considered an environmental release, according to the Environmental Protection Agency. Preventative measures now can save big headaches later. Document findings and corrective action on the PM checklist for follow-up as necessary.

SUMMARY

There is a considerable amount of work necessary to maintain the machine-head assembly of a utility-size wind turbine. Items such as the yaw system, structural components, control systems, and cleaning can take several hours to complete. Each activity is critical to the overall operation of the wind turbine; if not taken seriously, it can translate to a costly machine failure or equipment downtime. Either situation may create downtime waiting for parts or adjustments that could have been completed during a routine PM activity. Lost time translates to lost revenue for the farm operator and the operations group if farm availability is tied to their group earnings. Take the time to do the job right. If something cannot be completed during the PM activity, then document the issue so it can be scheduled for the next service call by the operation group.

REVIEW QUESTIONS AND EXERCISES

1. List several of the nacelle-inspection activities necessary during a PM cycle.
2. What are two methods of braking used with a yaw system?
3. Describe three nondestructive test methods used to inspect for cracks in structural components.
4. List several defects that may be created during a welding process.
5. List five requirements that are necessary for a technician to work in an energized cabinet.

6. Why is it necessary to have communication between technicians when performing manual override of the yaw control system?
7. What two OSHA standards cover HAZWOPER requirements?
8. Why is it important to ensure that cooling systems have proper airflow?
9. Why is it important to inspect mechanical and electrical connections during a PM cycle?
10. Why is it necessary to clean the interior of the nacelle, especially the area below the bedplate assembly?

12 Drive Train

KEY TERMS

accumulator
compression-coupling
desiccant
electric solenoid
filter element
gearbox
hub adapter
hydraulic cylinder
main bearing

main shaft
oil analysis
oil-cooler
oil sampling
parking brake
rotor lock
slip ring
wear elements

OBJECTIVES

After reading this chapter and completing the review questions, you should be able to:

- Describe the different inspection activities necessary for the wind-turbine drive train.
- Describe some of the variations with wind-turbine drive-train systems.
- Explain why it is necessary to add sufficient grease to a large bearing assembly during a PM cycle to purge the existing grease.
- Explain the need for the elastomeric material in the gearbox mounting pillow blocks.

- Explain the need for periodic lubricating-oil analysis.
- Explain why it is necessary to obtain an oil sample that has not passed through the filtration system.
- Explain the need for an accumulator in a wind-turbine hydraulic-power system.
- Describe the need for a parking brake.
- Explain some of the necessary safety considerations when working around a drive train.

INTRODUCTION

Many of the utility-sized wind turbines in operation today have drive trains that include gearboxes to bring rotor speeds up to the synchronous speeds of typical utility generators. Wind-turbine system designs for many 3,000-kW (3-MW) or larger systems in the past few years have been moving toward a drive system that eliminates the need for a gearbox. The goal with this change is to eliminate a component that requires considerable maintenance and can also be one of the typical drive-train failure sources. Despite the design move of larger systems to a drive train without a gearbox, the current drive-train design will continue to be serviced and maintained produced because it remains the industry workhorse. There are many variations to drive trains which include a gearbox in service today. These variations depend on the design philosophy of the wind-turbine manufacturer. Variations encompass the technical goal of developing an optimum mix between structural integrity, improved system reliability, serviceability, and optimized power production.

Drive-Train Components

The typical drive-train components in utility-sized wind turbines include the following: hub adapter, main shaft, main bearing, gearbox, and parking or secondary brake. A brief description of each of these would be useful to help someone not familiar with wind-turbine systems to get an understanding of the components. The **hub adapter** (**Figure 12-1**) is used as a structural attachment point between the rotor assembly and the drive train. The hub adapter may be used to connect the rotor assembly directly to the gearbox input shaft or to a main shaft and bearing that would be used to help support the weight of the rotor. This variation would reduce the size of the gearbox assembly because it

FIGURE 12-1 *Hub adapter*

FIGURE 12-3 *A main bearing and shaft assembly*

would not need to support the entire weight of the rotor. The **main shaft** is a large steel-shaft assembly used to connect the hub adapter to the gearbox input shaft. This shaft assembly is typically hollow to enable electrical cables or hydraulic connections or both to feed between the rotor assembly and equipment in the nacelle. The use of a **slip-ring** assembly (**Figure 12-2**) mounted on the opposite end of the gearbox main shaft allows for the connection of communication and power cables while still enabling the rotor to move freely during operation. A slip ring is made up of several conductor rings that have spring-loaded electrical contact brushes riding on these rings. The **main bearing** assembly is typically a large spherical roller bearing. This bearing assembly can be used to support a major portion of the rotor assembly weight. The use of a spherical roller assembly enables the bearing roller elements to self-align during assembly of the wind turbine and allows for limited

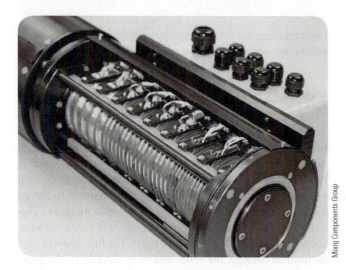

FIGURE 12-2 *Slip-ring assembly used for communication and power circuits between hub and nacelle (Note the copper rings and brushes used to complete the electrical connections.)*

flexing of the support structure during changes in wind loading. A spherical roller bearing assembly can also transfer the axial load from the rotor assembly to the bedplate to reduce the wind load on the gearbox. Axial loading is created by a force applied directly down the axis of the main shaft from the oncoming wind. **Figure 12-3** shows an example of a main bearing and shaft assembly.

A gearbox designed for direct connection to a rotor assembly would also include the same bearing style integral to the gearbox chassis for structure support. The **gearbox** is used to transfer low rotational speed (RPM) and high torque from the rotor assembly to a higher rotational speed and lower torque output needed for proper generator operation. Remember the relationship between rotational speed (RPM) and torque is expressed as

$$P = \frac{(T \times S)}{9,549}$$

where P = power (kW), T = torque (N-m), S = shaft speed (RPM), and 9,549 is a constant for these units. This means that power into the gearbox main shaft will equal the power output to the generator minus any losses because of inefficiency. Another point to make on gearbox design is the relationship between the number of stages in the assembly and the output rotation. A two-stage assembly will have a rotation opposite the input direction. For example, a clockwise (CW) input would have a counterclockwise (CCW) output. The typical wind-turbine gearbox assembly has three stages, so the rotation of the output shaft would be the same as the input shaft. Some wind-turbine models are designed with a two-stage gearbox assembly, so do not assume that the input shaft will always rotate in the same direction as the output shaft.

Another important component on the wind-turbine drive train is a parking brake or secondary brake. The **parking brake** is used to keep the drive shaft from rotating during operational shutdown, service work, or an emergency stop situation. This brake system is considered secondary because pitching or feathering of the blades to 90° is the primary

FIGURE 12-4 *Typical parking-brake systems*

shaft with a compression coupling. Gearbox output shaft configurations may include a couple of design variations. The gearbox output shaft may be coupled to a single generator, or there may be two or four generators mounted directly to the gearbox chassis such as used by Clipper (UTC) for its Liberty system design. **Figure 12-5** shows examples of drive-train configurations.

These designs may have different steps in their maintenance procedures, but the typical requirements will be similar. Bearings will need to be inspected and lubricated. Structural connections will need to be inspected and checked for integrity by use of a fastener torque check. The gearbox will need to be inspected and the lubricating system serviced. Consult the manufacturer's procedure for specific steps and requirements for the system being serviced. Use discussions in this text as a reference of typical preventative-maintenance (PM) activities. The activities discussed in this text will be generic in nature and are not specific to any manufacturer's equipment. The goal is to provide a range of activities that may be encountered during a PM cycle of drive-train components. Discussion topics will be presented as the activities may be encountered starting at the hub adapter and working toward the generator assembly.

braking method for a wind turbine. Pitching of the blades to 90° can typically stop the drive assembly within a few revolutions at which time the parking brake could activate and hold the assembly from further rotation. The parking brake is a larger design version of the disc brake used with automotive applications. **Figure 12-4** shows examples of typical parking-brake systems.

Drive-Train Design Variations

The design of wind turbines with a gearbox drive-train assembly come in a several variations, depending on the OEM system requirements. These variations include the attachment method for the rotor assembly to the gearbox and the method by which the generator may be connected to the gearbox. Variations for the rotor attachment to the gearbox may be through a direct coupling of the rotor assembly to the input shaft of the gearbox or through the use of a main shaft and bearing assembly to help support the rotor assembly. The main shaft for this design is attached to the gearbox input

FIGURE 12-5 *Drive-train configurations (Continued)*

FIGURE 12-5 *(Continued)*

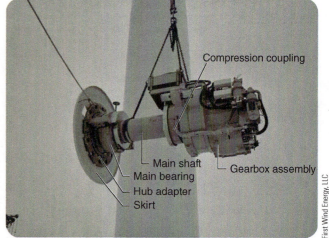

FIGURE 12-6 *Gearbox and main shaft assembly being lifted to nacelle*

HUB ADAPTER-BOLT TORQUE INSPECTION

Preventative-maintenance activities for the hub adapter include a visual inspection of the adapter for corrosion, missing hardware, and structural integrity of the fiberglass skirt assembly used to block rain and snow from entering the nacelle. **Figure 12-6** shows a view of a gearbox and main shaft assembly being lifted to the nacelle. This image allows a nice view of the skirt, hub adapter, main bearing, shaft, and gearbox assembly. Another maintenance activity for the hub adapter includes a 10% torque check of the nuts that secure the hub to the adapter during each PM cycle. Follow the manufacturer's specifications for the torque value required for these nuts. These studs may be M55 or greater in diameter and manufactured to ISO grade class 10.9 material specifications, so they need to be tested with power-bolting equipment to achieve the necessary torque requirements.

Ensure that the power-bolting equipment is set up with the proper tool, socket, and torque level for the size and material grade of the nut securing the hub to the adapter. Ensuring the equipment is set to the necessary torque level may require testing on a torque calibration station. Use of a torque test station is discussed in Chapter 6. Follow the manufacturer's procedure for the verification process. Choose a starting location on the adapter and torque the necessary nuts to the required value. If the system has had a previous PM cycle performed, then choose the next nut on the adapter from the previous test. Remember: a 10% check of the nuts is 1 out of every 10 on the adapter. Mark the stud end, nut, and washer as mentioned previously. **Figure 12-7** shows an example of the hub adapter nuts marked after a torque-level verification test. Move to the next nut and continue the process until completed. Further details are listed next in the discussion.

Figure 12-8 shows an example of the power-bolting tool setup for the torque testing procedure. There will be nuts that cannot be accessed without rotating the rotor assembly, so work as a team to manually release the parking brake to rotate for access to these nuts. Ensure that the parking brake is re-engaged before continuing to the next nuts. **SAFETY: Ensure clear communication between technicians performing this process as with any manual override procedure. Remember to keep tools, safety gear, and body parts away from the adapter assembly while it is in motion.** The momentum of a spinning rotor assembly will not stop instantly, so pinching or entanglement can cause a severe injury. Always follow the manufacturer's safety precautions when working with power-bolting equipment. The tool operator should be the person who operates the control pendant. Use appropriate sockets

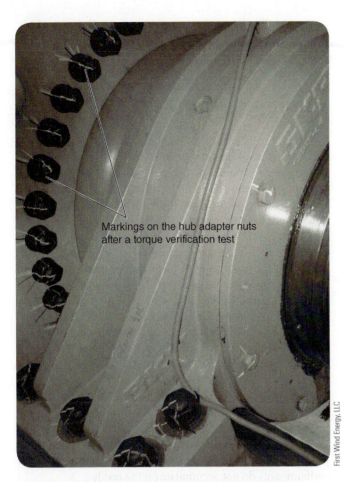

FIGURE 12-7 *Hub adapter nuts marked after torque-level verification test*

FIGURE 12-8 *Power-bolting tool setup for torque testing procedure*

and never place body parts between the reaction arm and the reaction point. Complete the PM activity for the required number of fasteners on the adapter and mark each fastener with the appropriate color paint pen for the PM cycle. Follow

the manufacturer's requirements for any fastener movement. Document the hub adapter findings on the PM checklist before moving to the main bearing assembly.

MAIN BEARING INSPECTION AND MAINTENANCE

Typically, the next assembly on the drive train will be the main bearing assembly unless the wind turbine has the hub assembly bolted directly to the gearbox input shaft. If that is the case, then the procedure would move to an inspection of the gearbox. The main bearing is mounted in a pillow-block assembly that secures the bearing to the bedplate assembly. **Figure 12-9** shows an example of a main bearing mounted in a pillow-block assembly. The main bearing requires a visual inspection, torque of the mounting fasteners, and lubrication. Visual inspection includes a check for corrosion, wear products, water contamination of the lubricant, and hardware or sensors that may be mounted to the assembly.

The main bearing assembly is exposed to rain and snow that can blow in around the hub adapter skirt. Surface corrosion may not look great, but it will not create structural problems. Inspect the assembly for signs of damage because of corrosion of the bearing rolling elements. There should be no particles—metal or rust—around the base of the bearing assembly or any exposed area of the moving elements. The moving elements may not be visible for an inspection, depending on the cage or seal design of the bearing and covers that may be present on the pillow-block assembly. Refer to the manufacturer's inspection procedure for the specific system. Check for signs of water contamination of the bearing assembly. Rust-stained water mixed with lubricant running from the bearing assembly is not a good sign. If present, inspect for possible damage. Notify farm operations if there is a suspicion of damage to the rolling elements and

FIGURE 12-9 *Mounting main bearing in pillow-block assembly*

race of the bearing assembly. Inspect the temperature probe mounted to the bearing to ensure it is secure. If the temperature probe comes out during operation, the supervisory control and data acquisition (SCADA) system will not detect a bearing issue associated with an elevated temperature. Visually inspect hardware around the bearing assembly for mechanical integrity and function. Refer to the manufacturer's procedure for a list of specific items located in this area of the bedplate and nacelle.

Main Bearing Bolt Torque Inspection

The main bearing pillow-block assembly is secured to the bedplate through the use of several threaded studs and nuts. The nuts should be inspected to ensure they are secure during each PM cycle. **Figure 12-10** shows main bearing fasteners being verified for torque level. Choose the appropriate power-bolting tool, socket, and torque level for the size and material grade of the fastener used to secure the pillow block to the bedplate. Use the bolting tool manufacturer's recommended pressure level or setting on the tool for the torque output. As mentioned previously, refer to the wind-turbine

FIGURE 12-10 *Main bearing fasteners being verified for torque level*

manufacturer's procedure or specification for torque specification and acceptance criteria. Note inspection findings on the PM checklist for follow-up later.

Lubricating the Main Bearing

Lubrication of the main bearing assembly during each PM cycle will ensure that the bearing is lubricated sufficiently and that any wear products or water contamination present have been purged from the assembly. Refer to the wind-turbine manufacturer's specifications for required grease formulation and quantity necessary. **SAFETY: Consult the grease manufacturer's MSDS for PPE requirements along with any cleanup and handling precautions. Adding sufficient lubricant to the main bearing to purge contaminants means that there will be an equal quantity of waste lubricant discharged from the bearing assembly. Depending on the bearing size, this may be a couple of pounds (1,200 grams) or more of waste grease.** The bearing should be greased while the main shaft is rotating slowly to ensure that the lubricant is uniformly distributed around the entire bearing. **SAFETY: This activity requires a team effort to complete, one person operating the bypass control for the park brake and another person operating the grease gun. Keep body parts, safety gear, and tools away from the rotating assembly to prevent accidental entanglement.** Wipe excess grease from the bearing and fitting to ensure that dust and other contaminants do not accumulate on the residue.

Lubrication Waste Disposal

Refer to the site hazardous waste policy for cleanup and disposal of waste materials generated during lubricating activities. Typical requirements call for the disposal of oil-soaked rags, soiled absorbent pads, filters, and lubricants to prevent accidental contamination of groundwater. This may require storage in a sealed compatible container that will be scheduled for pickup by a reputable waste hauler for proper disposal. Refer to federal and state regulations along with industry standards for proper marking and storage details on waste-lubricant containers such as NFPA 704 and DOT 22 CCR 66262.31 and 32 or U.N. requirements. These requirements may include labeling items such as "Hazardous Waste" and "Flammable—Keep Fire Away" and include information on lubricant composition and accumulation start date, along with name, address, and phone number of the organization using the waste-storage container.

COMPRESSION-COUPLING INSPECTION

The compression coupling is used to secure the main shaft to the input shaft of the gearbox. A **compression-coupling** design uses two tapered sleeves that are forced together

FIGURE 12-11 *Compression coupling used to attach main shaft to gearbox*

Screws used to compress the coupling onto the main shaft

Compression coupling on input shaft of the gearbox

through the use of several fasteners to tighten or compress the inner sleeve onto the main shaft, which locks it into position. **Figure 12-11** shows an example of a compression coupling used for this purpose. The coupling should be inspected for corrosion, missing hardware, or damage. This assembly is typically covered with a guard assembly to prevent accidental injury. **SAFETY: Inspection of the compression-coupling assembly may require removal of safety guards, so the drive train should be locked to prevent rotation. Follow the site safety policy for proper lockout–tagout (LOTO) of the drive train.** Choose the appropriate power-bolting tool, socket, or bit and the torque setting for the fasteners clamping the main shaft. Check each fastener for the proper torque level and mark with the appropriate paint pen color. Note findings on the PM checklist and notify farm operations of issues that need to be followed up on during the next service activity.

GEARBOX INSPECTION AND MAINTENANCE

Inspection of the gearbox assembly will depend on the manufacturer's PM activity requirements. A typical gearbox inspection and maintenance cycle may include checking for lubricant leaks, an oil sample, filter replacement, checking structural connections, adjusting the park brake, and cleaning oil cooling-system components. The inspection activity may also require opening the access panels on the sides of the gearbox to view the gears, shafts, and bearings within the assembly. Refer to the wind-turbine manufacturers requirements for

the type of PM cycle. Requirements for the semiannual and annual PM cycles cover many of the same activities, but the annual PM may add other items such as inspecting the gearbox internal components.

Gearbox Housing Inspection

Inspection of the gearbox housing includes checking for loose hardware, verifying gauges, pump mounting assembly, cooling components, junction boxes, cables, and the parking-brake assembly. Inspection of the gearbox should also include a check of seals around bearings, shafts, sensors, valves, access panels, and sections of the gearbox housing. Loose hardware components on the gearbox assembly will shift and chatter against the housing and possibly damage external components or the housing. Structural components that secure the pump, oil-cooling components, and the parking brake can cause serious damage to the wind turbine if they come loose during operation. Failure of the oil-cooling system or pump assembly would allow oil to leak out of the system and prevent lubricant from flowing into the gear assembly. This would cause overheating of the system and failure of the bearings. Temperature, flow, and pressure sensors will typically catch problems before serious damage occurs, but there would still be a loss of lubricating oil that requires cleanup in the nacelle and on the tower. Failure of the parking-brake assembly during maximum output could lock up the drive train, causing serious damage to the gearbox or generator. **Figure 12-12** shows some of the components typically attached to the gearbox housing that should be inspected for structural integrity during a PM cycle.

Inspection for leaks from the pump, filter canisters, hoses, or seals on the gearbox should be corrected immediately to prevent oil loses from the system. Even minor leaks can cause oil to accumulate in the bottom of the nacelle and eventually leak out with rainwater during a storm.

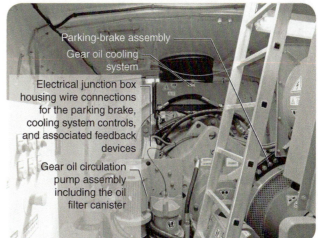

Parking-brake assembly
Gear oil cooling system
Electrical junction box housing wire connections for the parking brake, cooling system controls, and associated feedback devices
Gear oil circulation pump assembly including the oil filter canister

FIGURE 12-12 *Components typically attached to gearbox housing*

You may recall the inspection for oil under the bedplate was mentioned in the nacelle-inspection process. Oil leaks can create system issues along with environmental issues if not corrected promptly.

Pillow-Block Bolt-Torque Inspection

Visual inspection of the pillow-block assembly mounting the gearbox to the bedplate should include a check for missing hardware along with any shift in the pillow block that may have occurred during operation. The gap between the gearbox assembly and each pillow block should be equal. **Figure 12-13** shows a view of the gearbox pillow-block assembly. A shift of the gearbox in either or both of the pillow-block assemblies that secure the gearbox could cause serious damage to the drive-train assembly. The gearbox is mounted in an elastomeric isolator to absorb vibration and shock created in the drive train during operation. Failure of this material would allow the gearbox to shift, so a thorough inspection of these assemblies will prevent minor issues from becoming serious problems during operation. The pillow-block assemblies will also need to have the mounting fasteners checked during the PM cycle. Choose the required power-bolting tool as mentioned with the main bearing torque inspection. Refer to the manufacturer's maintenance procedure for specific torque-level and acceptance requirements. Each fastener should be checked during every PM cycle to ensure they are secure. Note findings in the PM checklist and mark each fastener assembly with the appropriate paint pen color. **SAFETY: Always be aware of hazards associated with power tools such as power-bolting equipment. Accidents can happen quickly, and they can leave lasting scars, especially if a finger or part of a hand is amputated by a pinch between bolting tool and a reaction point on an adjacent nut.**

Inspection of Lubrication System

The lubricating oil is the lifeblood of a gearbox assembly, and the pump is the heart of the system. Without proper lubrication, the gears and bearings may overheat from the friction of moving components and eventually fail. The gearbox of a utility-sized wind turbine has a lubricating pump to move oil from the sump, through a filter assembly, and through a cooling assembly before directing it onto gears and bearings within the assembly. Proper function of the lubrication system will keep the gearbox operating for many years without issue.

Inspection of the lubrication system included the steps mentioned previously to determine if there are any leaks that would cause the loss of oil. Maintenance of the system includes determining if there is an adequate supply of oil in the sump assembly, replacing filter elements, and collecting a sample of the oil for analysis. A visual inspection of the sight glass located on the gearbox chassis will indicate whether the oil level is within specification. **Figure 12-14** shows a typical oil-level sight glass used for a lubrication application. If the level is low, add an approved oil to bring the quantity within the operation range of the sight-glass markings. Never add oil to the full mark if the system is cold. Oil added to this level while cold will cause a sump to run above specification when the oil is brought up to the operating temperature because oil expands when it heats up. Raising the oil level above the full line also may create other issues during operation. Refer to the wind-turbine manufacturer's lubrication specification for proper oil formulation and grade to ensure proper lubrication properties. **SAFETY: Refer to the oil manufacturer's MSDS for information on appropriate PPE along with handling and storage requirements.** Other items to take from an inspection of the oil are color and the presence of contaminants such as water. Compare the color of the

First Wind Energy, LLC

FIGURE 12-13 *Gearbox pillow-block assembly*

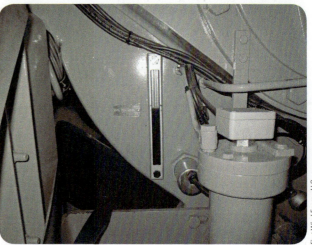

First Wind Energy, LLC

FIGURE 12-14 *Oil-level sight glass used for lubrication application*

oil to a fresh sample of the same oil. A significantly darker color may indicate deterioration of the oil from excessive heating or oxidation because of time in service. Oil does not last forever, so refer to the manufacturer's requirements for operation time before replacement. This time may be anywhere from a few years to five years, depending on the service conditions of the equipment. Another item that may be seen with an inspection of the sight glass or oil in the gearbox assembly would be excessive foaming of the oil. Foaming may be caused by entrapped air or moisture. Either condition will reduce oil's lubrication properties. Report findings such as a significantly dark color or foaming to the farm operations group for assessment and correction.

Oil Sampling

Oil sampling of lubricant used in a gearbox is vital to determining its condition. A previous chapter on lubricants mentioned **oil sampling** as a method for determining the overall condition of a lubricant, including wear elements, pH, moisture, viscosity, and levels of additives. **Wear elements** such as iron, copper, tin, lead, and zinc can indicate the extent of wear for bearings, gears, and bushings within the lubrication system. Tracking these elements along with viscosity and pH from one PM cycle to the next will give a clear indication of the condition of the gearbox components and circulation system. Use the oil sample results and test laboratory interpretation along with guidance from the lubricant manufacturer to assess the condition of the equipment. This information is useful in determining a schedule for oil change, major equipment inspection, and prediction of component failures.

To ensure proper **oil analysis** results, the oil sample should be pulled from the system before the filter assembly. This will ensure that the filter element has not eliminated or significantly reduced the presence of wear products from the sample. The oil sample container should be marked with the site name, machine number, date, and hours of service. Refer to the wind-turbine manufacturer's PM procedure for other necessary labeling items and the approved collection process. **Figure 12-15** shows a pump and filter assembly. Some equipment manufacturers have relied on an analysis of the oil-filter element for wear products along with the oil sample to get an overall picture of the equipment condition. Large particles may be trapped in the oil-filter element during operation time between oil samples. These particles would normally not be seen without an evaluation of the filter element. These particles can yield a picture of damage to gears and the pump assembly that will need to be addressed before further damage is created.

Another sampling method that can be performed is passing a magnet around the oil sump area of the gearbox. **SAFETY: Follow the site-approved LOTO process to ensure that the gearbox is locked to prevent entanglement by moving components. Never wear loose clothing or safety gear that could be caught in the gearbox during the inspection process.** The magnet will attract large particles or pieces of steel that may have broken from the gears during

First Wind Energy, LLC

FIGURE 12-15 *Pump and filter assembly*

operation. Some wind-turbine manufacturers have installed magnetic sensors within the gearbox sump to indicate the presence of excessive numbers of steel particles or pieces. These sensors will provide feedback about a pending problem to SCADA through the programmable logic controller (PLC) so service teams can investigate the issue before there is excessive damage to the equipment. Note the findings of the oil sump inspection on the PM checklist and forward issues to the farm operations team for follow-up as necessary.

Replacement of the Oil-Filter Element

Replacement of the oil-filter element during each PM cycle will ensure that contaminants are removed from the lubricating system before they can block the cooling system heat-exchanger element or distribution lines and damage moving elements in the gearbox. Blockage of the cooling system element will prevent adequate cooling of the oil during operation and reduce the flow of oil to the distribution lines. Both of these components are vital to normal operation of the gearbox. Changing an oil-filter element is completed by draining the canister located on the side of the gearbox, removing the canister cap, pulling the filter elements, placing an approved new element in the canister, replacing the canister cap, and placing the drained oil back into the gearbox sump. The **filter element** is the cylindrical component located in the canister that may be constructed of several layers of pleated paper or fabric, synthetic fiber, wire mesh, or a combination of these materials. Refer to the wind-turbine manufacturer's PM procedure for filter replacement for details on the required component part numbers and specifics on completing the activity. **Figure 12-16** shows examples of oil-filter elements that may be used in a fluid-power system.

FIGURE 12-16 *Oil-filter elements used in fluid-power systems*

FIGURE 12-17 *Gear-oil cooling assembly*

SAFETY: Several items should be considered during this process, including approved LOTO of the pump assembly, proper PPE as directed by the oil MSDS, clean collection containers for the oil, and proper disposal procedure for saturated oil-filter elements. Refer to the previous section on disposal of lubricant waste or the site hazardous waste plan for details on proper container labeling and other requirements.

If the drained oil is going to be returned to the gearbox sump, ensure that precautions are used to prevent contamination of the oil. LOTO of the pump-assembly circuit breaker will prevent the PLC from energizing the pump when the canisters are drained. The control schematic for the system being serviced will indicate the sensor locations used to detect oil level and pump activation. Always LOTO and play it safe! Energizing the pumps while the canisters are open may spill oil around the nacelle, creating a slipping hazard not to mention the need for extensive cleaning after the pump is shut down. When the activity is completed, remove the LOTO devices as directed by the procedure and note materials consumed and any other items as necessary on the PM checklist.

Inspection of Oil-Cooler Assembly

Inspection and maintenance of the oil-cooler assembly require checks for oil leaks, security of attached hardware, and cleaning of the element to remove dust and other organic materials. The oil-cooling system typically has a flexible skirt connected to the assembly that directs the airflow through the element. This ensures the heat from the oil is exhausted out of the nacelle to reduce the temperature in the nacelle. The **oil-cooler** assembly is made up of a heat exchanger (radiator), exhaust fan, and a hood or louvers mounted on the nacelle assembly. **Figure 12-17** shows an example of a gear-oil–cooling assembly. **SAFETY: If the procedure requires removal of the skirt or any other guards around the fan assembly, ensure that a site-approved LOTO procedure is followed to reduce the hazard of being exposed to moving** fan blades. **The exhaust fan located on the cooling system is controlled by the PLC system, so its function will be automatic.** The fan will energize when the temperature sensor on the cooling system reaches a preset value. Turn off the circuit breaker powering the fan and LOTO as required for safety. Wipe down the fan assembly of dust and any other contaminants. Carefully spin the fan assembly by hand and listen for any issues with the fan bearings. Lubricate the fan assembly as necessary or as directed by the PM procedure. Vacuum the cooler element of dust and other organic matter that may have accumulated in the assembly since the last PM cycle. This assembly may be a few hundred feet (80+ meters) up in the air, but pollen, leaves, and plant seeds have been known to accumulate in this assembly. If they are not removed, the cooling system will not function properly. Replace the fan guards and skirt, remove the LOTO devices, and turn on the circuit breaker as directed by the procedure. Note findings on the PM checklist before continuing.

Replacement of Gearbox Air Filter

The filter located on the gearbox air vent is used to prevent contaminants from entering the assembly during operation. As the oil in the gearbox heats up and expands during operation, it forces air out of the assembly. When the system is not operating, the oil cools, contracts, and pulls air back into the assembly through the filter. The air-filter element prevents the air entering the gearbox from bringing in contaminants during the contraction phase. **Figure 12-18** shows examples of air-vent filter assemblies. Typically, these air filters may contain a pleated fabric or paper element to trap particulates, but some manufacturers may add another line of defense to protect the oil. The addition of a **desiccant** cartridge to the air-filter assembly will reduce any moisture

that may be drawn in during the contraction phase. The filter element should be changed during each PM cycle to ensure proper venting of the gearbox during operation. The desiccant cartridge assembly has another benefit: it will change color to indicate it requires replacement, so there is no guesswork. Some new desiccant cartridges are yellow or pink and turn dark green, blue, or black when consumed. Refer to the notes on the side of the filter assembly for details on the specific filter being used. Note materials used during this activity on the PM checklist for inventory control.

PARK BRAKE SYSTEM INSPECTION AND MAINTENANCE

The **parking brake** is used to prevent the drive train from moving during an emergency shutdown or for service work. The brake assembly may be mounted on the back or front of the gearbox, depending on the design of the drive train. For purposes of this discussion, the back of the gearbox is the end closest to the generator. Either position will require similar inspection and maintenance activities. The park brake is made up of a frame, actuator assembly, pads, and a disc. The frame assembly holding the pads and actuator will be mounted to the gearbox chassis, and the disc will be mounted to an output shaft. The actuator may be an electric solenoid or a hydraulic cylinder that moves the brake pads against the disc. An **electric solenoid** is a device that uses a magnetic field to produce force or motion within a mechanical system. A **hydraulic cylinder** uses fluid pressure acting on a piston to produce motion or force within mechanical system. A hydraulically actuated park brake assembly may have the power unit mounted to the brake frame or located in another location within the nacelle assembly. **Figure 12-19** shows examples of park brakes and hydraulic-power unit variations.

Inspection of Electrical Connections

Inspection of the electrical connections to the park brake will include any valves and sensors located on the assembly that enable control of the brake and feedback to the PLC that the unit is functioning properly. Control of a solenoid-actuated park brake would be through a relay or contactor controlled by the system PLC. Inspection items for the brake system would include any electrical devices, control cables to the actuators, and any feedback devices. Systems that use a hydraulic actuator would have a solenoid valve to actuate the caliper cylinder along with sensors for feedback to the PLC that the brake is activated and when there is excessive wear of the pads. Another sensor that would be present on a hydraulic actuator setup is a pressure transducer. This would provide feedback

FIGURE 12-18 *Air-vent filter assemblies*

First Wind Energy, LLC

© Cengage Learning 2013 and a W. Kitcollins photograph

Ciantro Coporation

University of Maine at Presque Isle

to the PLC that there may be insufficient pressure in the hydraulic circuit to maintain or activate braking force. Any one of these feedback signals would indicate to SCADA through the PLC that the system is working properly or needs attention. Visually inspect all electrical and hydraulic connections for integrity and proper function.

Inspection of Power Unit

The power unit for the park brake or the blade-pitch system should be visually inspected for leaks and the security of valves, accumulators, sensors, gauges, and wire devices. The reservoir should be inspected to ensure that there is sufficient fluid. The fluid-level inspection can be completed by viewing the sight tube located on the side of the reservoir. Hydraulic fluid is typically a light amber color when new that will darken with oxidation and contamination. Refer to the manufacturer's guidelines for specific details on fluid quantity, grade, and any other requirements. The power unit will have an accumulator in the circuit for the park brake system as well as the blade-pitch system (if equipped) to ensure that they will activate even after loss of utility grid power. An **accumulator** is designed to supply a specific amount of fluid to the hydraulic actuator at a predetermined pressure. Think of an accumulator in a hydraulic circuit functioning like a capacitor in an electrical circuit. They both can store energy and are able to supply that energy back to the circuit as necessary. Both devices have a finite amount of storage capacity, so the circuit designer has to select a device of appropriate size to provide the necessary energy to complete a desired

University of Maine at Presque Isle

FIGURE 12-19 *Park brake and hydraulic-power unit variations*

task. There were three types of accumulators discussed in the fluid-power chapter of this text: spring-loaded piston, gas-charged piston, and gas-charged bladder. **Figure 12-20** shows examples of two accumulator types. The typical accumulator used in a wind turbine is a gas-charged type. Storage pressure within the gas-charged accumulator may be 70 bar (1,015 psi) for a park brake circuit. **SAFETY: Working around these types of pressures requires caution. Never loosen hydraulic fittings without draining pressure from the circuit and never place a body part over a leak that is spraying hydraulic fluid. The leak can inject hydraulic fluid into the body that will create a toxic reaction and possible infection.**

Replacement of Filters

The oil filter located on the hydraulic-power unit is used to remove particulate contaminants from the hydraulic oil. During each PM cycle, this oil filter should be replaced to ensure that contaminants do not block small solenoid valves or damage the pump and actuators in the system. Refer to the manufacturer's guide for the appropriate replacement filter. Replacing the filter assembly may require draining the system of pressurized fluid, so follow the approved site LOTO procedure for this activity. Also follow the site's hazardous waste procedure for proper disposal of the oil-saturated filter. Once the oil filter is replaced, ensure that any trapped air is removed from the system safely. Hydraulic fluid is technically incompressible, whereas air can be compressed and cause issues when it becomes trapped in an actuator, accumulator, or pump assembly. Follow the manufacturer's procedure to ensure that the system is ready for operation after the oil filter has been replaced.

Bleeding Air from the System

Air can be bled from the hydraulic circuit by pressurizing the circuit to a lower level of pressure and slightly loosening a fitting upstream of the oil-filter assembly. This will allow the air to escape. Bleed the circuit until hydraulic oil starts to flow out of the fitting and then retighten the fitting. This will ensure that trapped air will not create a problem with the system once it is placed back into service. Note materials and components used for this activity on the PM checklist for inventory control later.

Inspection and Adjustment of Brake Pads

Inspection items for the park brake include a visual for hardware and wear on the brake pads. It may be two or three years between brake-pad replacements, depending on the wind turbine's operations profile. Continuous running of the wind turbine will not wear the brake pads as much as several stop and start cycles with large variations in wind conditions. Part of

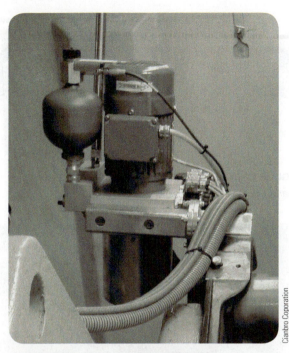

Cianbro Corporation

University of Maine at Presque Isle

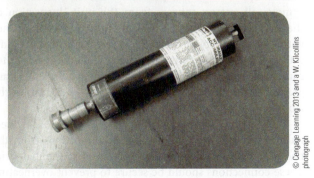

© Cengage Learning 2013 and a W. Kilcollins photograph

FIGURE 12-20 *Gas-charged piston, and gas-charged bladder accumulators*

each PM cycle should include an adjustment of the fixed brake pad to minimize the gap between it and the disc. This gap is created by wear on the pad during operation. The air gap between the pad and the disc should be wide enough to prevent rubbing during normal operation but not so wide as to cause the disc to deflect when the brake is operated. This gap may be 1.5 to 2 mm (0.062–0.093") wide, depending on the condition of the disc assembly. A warped disc may have enough waviness that this gap will need to be widened to prevent rubbing of the pads at the high point of the disc. Refer to the manufacturer's specifications for acceptance criteria on disc flatness and pad wear limits. Note findings on the PM checklist for follow-up later by the farm operations team as necessary.

Inspection of Rotor Lock Assembly

The **rotor lock** is a mechanism used to prevent the rotor assembly from rotating during service work on or in the hub. The site safety plan should include activating the rotor lock as part of the hub-entry procedure. This lock mechanism may be part of the parking-brake assembly or connected to the input shaft of the gearbox. Inspection of the lock mechanism includes a visual for missing hardware, a torque check of fasteners securing the hardware, and a check for proper motion of the hardware as indicated by the manufacturer's procedure.

Inspection of Flex Coupling Guard

The flex coupling guard is used to prevent accidental contact of technicians with the coupling assembly and the parking-brake components during service work in the nacelle. A visual inspection of the assembly will ensure there are no missing components or fasteners that may allow it to drop onto the flex coupler during operation. If the cover dropped onto the flex coupler during operation, it may jam the coupler and possibly damage the coupler and parking-brake assembly.

INSPECTION OF THE PITCH SLIP RING

Some wind-turbine systems use a set of electric gear motor drives in the hub for blade pitch, whereas other manufacturers may use a common hydraulic cylinder or multiple cylinders for blade pitch. The use of electric gear motors for the pitch system requires communication and power between the nacelle assembly and the motor control assemblies within the moving hub. The pitch slip-ring assembly is used to provide power and communication circuits to the hub through movable rings and stationary brushes. **Figures 12-21 and 12-22**

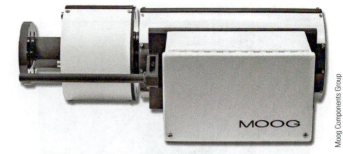

FIGURE 12-21 *Blade pitch slip-ring assembly located on the gearbox in the nacelle*

FIGURE 12-22 *One of three pitch-gear motor assemblies in the hub*

show examples of a slip-ring assembly located in the nacelle and one of the three pitch-gear motor assemblies located in the hub.

Inspection of the pitch slip ring includes a visual of any electrical connections, mounting hardware, and movable components within the assembly. **SAFETY: The pitch slip ring enables the connection of voltage levels from 5 V$_{DC}$ to 240 V$_{AC}$ or higher, depending on the circuits necessary to control hub functions during operation. Before removing the slip-ring cover or junction box cover, ensure that the site-approved LOTO procedure is used to isolate and remove all energy sources from the unit.** Inspection of the pitch slip ring includes a visual of the rings and brushes within the assembly for wear. Replace worn brushes as indicated by the manufacturer's procedure and acceptance criteria. Clean rings and brushes and apply the required lubricant grade and quantity as necessary. **Figure 12-23** shows the rings and brushes present in a typical slip-ring assembly.

Inspect the electrical terminal connections to the internal brushes and the umbilical cable connections from the nacelle control console to the junction box terminal strip. These connections should be secure to prevent overheating and loss of communication to the hub during operation. The electrical connections in the assembly may also supply power

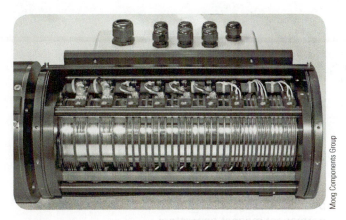

FIGURE 12-23 *Rings and brushes present in typical slip-ring assembly*

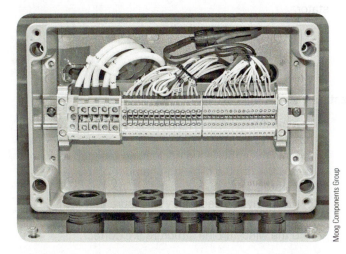

FIGURE 12-24 *Example of terminal connections within junction box of pitch slip-ring assembly*

to a battery charge circuit located in the hub that is used for emergency pitch control in the event of a loss of utility grid power. **Figure 12-24** shows a view of some terminal connections within the junction box located on the pitch slip-ring assembly. Once the inspection and maintenance activity is complete, replace the covers on the slip ring and remove the LOTO devices as covered in the site's safety procedure. Note any issues and replacement parts needed for the maintenance activity on the PM checklist. This information is necessary for inventory control and follow-up later if necessary.

OTHER RECOMMENDED CLEANING

Cleaning activities for the drive-train assembly include wiping down exterior surfaces of all components to remove dust and other contaminants. Use an approved cleaning solution

as specified by the wind-turbine manufacturer or site safety plan. This will ensure that the finish on the equipment will not be damaged and that technicians will not be exposed to toxic chemical vapors. Ensure adequate ventilation of the nacelle during the cleaning process by opening hatches and vents as necessary. Clean up any oil or waste lubricants from the PM activity and dispose of them according to the site's waste-handling procedure. This ensures that technicians will have minimum exposure to the waste and avoid an accidental release to the environment. Waste-grease cartridges, used filters, and oil-soaked rags will need to be packaged for removal from the tower. Remember: these will have to be taken down the tower in the lift bag, so package them carefully so they cannot spill and contaminate tools, safety gear, and documentation that will be brought down later.

SUMMARY

Several variations of a wind-turbine drive train are in use with utility-sized systems. The variations will depend on the manufacturer and the model being serviced by the maintenance team. Even with these variations, each drive-train assembly will have similar components such as bearings, pumps, brake components, filters, electrical connections, and cooling components that require maintenance. Always consult available manufacturers' information, site policies, industry standards, and government regulations to ensure that the activities are performed safely.

REVIEW QUESTIONS AND EXERCISES

1. List some the necessary inspection activities for a wind-turbine drive-train assembly.
2. Why is used grease purged from large bearings?
3. Describe why it is necessary to obtain an oil sample from the system before it passes through a filter assembly.
4. List some contaminants that may be found in an oil sample.
5. Why is it necessary to track changes in oil samples over several maintenance cycles?
6. List some of the safety precautions necessary when working on a drive-train assembly.
7. Why is it necessary to remove air from high-pressure fluid lines after replacing an oil filter or other component?
8. What is the purpose of an accumulator in the park brake system?
9. Why is it necessary to adjust the fixed parking-brake pad during a PM cycle?
10. What is the purpose of the rotor lock on the wind-turbine drive train?

CHAPTER

13 Generator

KEY TERMS

conduction
convection
electromagnet
flex coupler
job-safety analysis (JSA)
National Fire Protection Association (NFPA)

permanent magnet
permanent magnet generators
rotor
slip ring
stator

OBJECTIVES

After reading this chapter and completing the review questions, you should be able to:

- Describe the different inspection activities for a generator.
- Describe the importance of proper generator alignment.
- Explain the need for an approved LOTO procedure when working on power connections.
- Explain the need for using an appropriate generator bearing-grease formulation.

- Explain the need to properly seat new generator brushes to the slip ring.
- Explain the importance of cleaning the carbon deposits from the slip-ring chamber of the generator.
- Explain why it is necessary to ensure that the grounding connections between the generator and the wind-turbine structure are correct.

INTRODUCTION

The generator of a utility-sized wind turbine is the means in which the system can transform the mechanical energy of moving air into an electrical output that can be transferred to the utility grid. Recall that three things are necessary to produce an electrical output from a generator: a coil of wire connected to a load, a magnetic field, and the motion of the coil within the magnetic field. The magnetic field may be located around the moving coil, or it may be rotating within a fixed coil of wire to produce an output. **Figure 13-1** shows two variations of magnetic field with respect to the generator assembly.

Typical components for a generator include the **stator**, or stationary component of the generator; **rotor**, or rotating component of the generator; and a method to connect the output of the generator to a load. A stationary magnetic field with a moving internal coil requires a slip ring to connect the moving coil to the load. A **slip ring** uses spring-loaded stationary brushes to transfer power from rings mounted on the rotor shaft. **Figure 13-2** shows a generator slip-ring assembly. Reversing this setup and rotating the magnetic field within the

stationary coil (stator) would allow for direct connection to a load and eliminate the need for a slip ring. Typical wind-turbine generators use the rotor assembly to supply a moving magnetic field to the stator. The generator rotor assembly may be constructed in one of two ways to produce the moving magnetic field: using permanent magnets or using electromagnets.

Electromagnets or Permanent Magnets

Passing an electric current through a conductor will produce a magnetic field around that conductor. Winding the conductor into a coil will focus the magnetic field through the center of the coil and create distinct north and south poles similar to a permanent magnet. **Figure 13-3** shows a diagram of the magnetic field produced around a coil of wire with current flow. This method of creating a magnetic field is known as an **electromagnet**. A traditional **permanent magnet** is composed of an iron alloy that exhibits a magnetic field. Some magnets available today have very strong

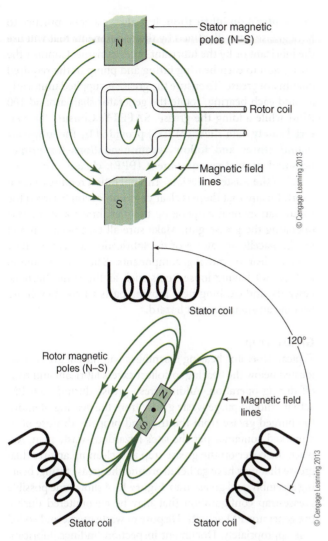

FIGURE 13-1 *Two variations of magnetic field in a generator assembly*

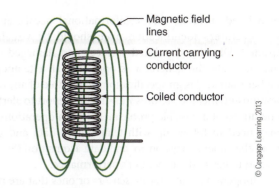

FIGURE 13-3 *Magnetic field produced around coil of wire with current flow*

produce an electrical output. Windings of the rotor are connected to a slip ring to provide an electrical circuit to an external power supply. This electrical connection enables current flow through the rotor windings to produce a controlled magnetic field. Attaching permanent magnets to the rotor is the other method used to produce a rotating magnetic field. Both of these generator designs use the stator assembly to supply electrical current to an external load or utility grid. The generator type used for any particular wind turbine depends on a manufacturer's design requirements of size, operating speed (RPM), efficiency, and equipment availability. Generators with electromagnets have been used commercially for years with electrical utility projects, so they are readily available. Utility-sized **permanent magnet generators** are gaining popularity because they can be made smaller and lighter than an equivalent electromagnetic unit with the same output rating.

Another benefit of permanent magnet types is that they can produce power in remote locations that do not have access to existing power sources. Generator size and weight are very important factors for direct-drive wind turbines under development for larger megawatt-rated land-based systems and offshore projects. Preventative-maintenance (PM) for either system is very important to ensure that a generator will have trouble-free operation over the life of the wind turbine. Many of the maintenance requirements for these generators are similar, so this chapter will identify many of the typical activities that may be encountered during a PM cycle.

magnetic fields compared to the traditional iron alloy types that are known as *rare earth magnets*.

Typical utility-sized generators use an electromagnet rotor assembly to spin the magnetic field within the stator to

FIGURE 13-2 *Generator slip-ring assembly*

GENERATOR VISUAL INSPECTION AND MECHANICAL MAINTENANCE

Visual inspection for a typical generator includes verification of hardware attached to the generator. All fasteners that attach cooling components, guards, junction boxes (J-boxes), sensors, cables, and structural mounts should be present and secure. Loose components, guards, J-boxes, or cable clamps can create problems with the equipment over time, so they should be corrected when they are discovered. The condition of painted surfaces, cables, informational decals, and the manufacturer's identification plate should be visually inspected. Discolored painted surfaces and damaged decals may indicate the generator is operating at a temperature higher than its design specification. Make note of any discoloration and include pictures with the PM report to alert farm operations of a possible problem. The farm operations team may need to follow up with temperature data and trends from the supervisory control and data acquisition (SCADA) system to verify the generator's performance.

Inoperative temperature sensors or ones that are not located properly will not accurately sense operating temperature. The input module for a temperature sensor with loose or intermittent wire connections will read an out-of-range high temperature (i.e., 99.99 °C) that the programmable logic controller (PLC) will detect. Elevated or out-of-range temperature readings will cause the PLC to shut down the wind turbine, and SCADA will send an error signal for a technician to evaluate. Temperature sensors that are not seated within their holders, however, will give a lower temperature than if they were seated properly. This may allow temperatures to become too high without the PLC sensing a problem. Temperature sensors on the generator may be located on each shaft bearing housing and embedded within the stator assembly. Ensure that each accessible temperature sensor is seated correctly within its holder. **SAFETY: If the inspection of a sensor connection requires opening an access panel to the generator housing or J-box, then ensure that the site lockout–tagout (LOTO) policy is followed.** More details on inspection of loose sensor connections will be covered later when J-box and other generator guarded access areas are discussed.

Bearing Lubrication

The generator assembly has bearings located on either end of the frame to support the shaft and rotor assembly. These bearings should be inspected for corrosion or discoloration, which may indicate damage to the bearing. Damaged moving elements such as rollers or balls will cause the bearing to exhibit excess vibration, overheating, and debris because of abrasion between moving elements and the race assembly. Typical bearing failures are the result of improper lubricant formulation, lack of lubrication, improper bearing selection, or corrosion. Unless the bearing was replaced with the wrong bearing during a service call, it should be correct for the application. Most issues typically can be caught during an inspection before bearing failure occurs. Refer to the wind-turbine's original equipment manufacturer (OEM) or the generator manufacturer's specifications for the proper lubricant formulation and quantity to be added during each PM cycle. Generators operate at higher temperatures and speeds (RPM) than other wind-turbine bearings such as the main, yaw, and pitch bearings, so they use a different grease formulation. **WARNING: Mixing grease formulations may cause a breakdown of the lubricating properties. Follow the OEM recommendations or those referenced**

in Chapter 4 on lubrication. It is always a good practice to have grease guns identified by the components that will use the lubricant or by the lubricant product name. Connect the grease gun to each bearing fitting and pump in the required quantity of grease. To ensure the grease is supplied uniformly around each bearing, rotate the generator shaft around 100 RPM while adding the grease. **SAFETY: Consult the material safety data sheet (MSDS) provided by the lubricant manufacturer and follow recommendations for proper personal protective equipment (PPE), cleanup, and disposal of the waste materials.** This operation requires a team effort. Ensure that there is clear communication between the technician manually operating the park brake and the one operating the grease gun. Make sure all technicians in and on the nacelle are aware of the lubrication activity so they remain clear of moving components. Safety steps should include not having loose items such as the climb harness, lanyards, and clothing near moving parts where they could become an entanglement hazard.

Grease Trap

Typical generator bearing assemblies will have a grease trap located below the bearing to collect excess lubricant and runoff during operation. Adding fresh grease during each PM activity should purge waste grease from the bearing. Remove the purged grease from the trap and properly dispose of in a marked container per the site's hazardous-waste plan. Replace the cover on the grease trap and clean up any residual grease that may have gathered around the grease fitting, bearing assembly, and grease trap. **Figure 13-4** shows one possible grease-trap configuration that may be encountered during the generator PM activity. Dispose of waste grease and soiled rags appropriately. Document inspection findings, lubricant part number, and quantity consumed on the PM report.

GENERATOR ALIGNMENT

Alignment of the generator to the gearbox is an important PM activity that ensures the generator will operate properly throughout the life of the wind turbine. Improper generator alignment can create excessive stress on the shafts and coupling used to transfer power from the gearbox to the generator. Alignment issues can also create excessive equipment vibration, heat, and damage to bearings of the gearbox and generator. Chapter 8 discusses several issues associated with improper generator alignment and describes a generic alignment process to ensure that the generator will operate properly. This section will discuss some of the activities and safety hazards associated with the generator-alignment process for a PM cycle.

Always follow the wind-turbine manufacturer's requirements for generator-alignment practices and PM schedule. The generator-alignment procedure should be on an annual schedule at a minimum unless a semiannual visual inspection indicates that an alignment is required. Items that may indicate a generator-alignment issue include loose mounting

FIGURE 13-4 *One possible grease-trap configuration*

FIGURE 13-5 *Flexible coupling and other components located inside a guard assembly*

fasteners between generator and subframe; visual compression or distortion of flex coupler; visible damage to the coupler, shafts, and bearings; and increased vibration trends in SCADA data. Generator-alignment information discussed in this section shows a generator coupled to a gearbox through the use of a flexible coupler. The direct connection of generators to the gearbox such as that used by Clipper or other manufacturers would not need this process. A direct-drive system such as those incorporated with many of the larger megawatt systems would also not require a generator-alignment activity. These systems, however, require visual inspections along with mechanical and electrical maintenance activities listed in this chapter.

Removal of Flex Coupling Guards

Access to the flex coupler for inspection and generator alignment requires removal of the guards from the drive train between the gearbox and the generator. This is accomplished by removing fasteners that secure the guard halves to the mounting hardware located on the gearbox and generator frame assembly. Removing these guards will expose technicians to moving components of the drive train and park brake assembly. **Figure 13-5** shows an example of a flexible coupling and the other components located inside the guard assembly.

Place the guard-assembly components in a secure location away from the area that will be used for any inspection and alignment activities. **SAFETY: Follow site safety procedures for necessary LOTO requirements during the guard-removal process. Placing the guard components in a secure location will ensure that they do not become a tripping hazard during alignment.** Once the removal process is completed, the coupling assembly will be required to move for inspection and alignment activities. This may be accomplished by having one of the technicians operate the manual brake override as necessary or other approved procedure defined by the OEM. Ensure that anytime the drive train is allowed to rotate, everyone involved in the activity is alert and focused. This will prevent accidental pinching or entanglement.

Inspection of the Flex Coupler

Inspection points for the **flex coupler** include the manufacturer-specified distance between shaft connection collars, the condition of flexible links mounting fasteners, and the condition of the dropout isolation component. The coupler manufacturer will typically have a specification for the distance between mounting shaft collars. This dimension shows the nominal distance of the coupler without excess compression or extension of the flexible links. Distance greater than or less than the specified length indicates the coupler is subjected to forces that may damage the components during operation. An out-of-specification measurement indicates the generator or one of the coupling collars has shifted and should be corrected. This would be one of the inspection items that would initiate an unscheduled generator-alignment procedure during a semiannual PM activity.

Flexible links should be inspected for corrosion, cracks, or distortion that may create issues during wind-turbine operation. Suspect link components should be replaced.

Consult the flexible coupling manufacturer's documentation for recommended inspection criteria and replacement part numbers.

All mounting hardware should be present and secure. Loose hardware will create issues during system operation. The coupling manufacturer should reference the proper torque specification for mounting and component assembly fasteners. If these values are not listed in the documentation, then consult the OEM's engineering staff or appropriate torque table for size and material of the fastener used. Never overtighten fasteners because this will degrade the strength and may cause a material failure with time.

Inspect the dropout isolation component of the coupling for damage such as distortion or delaminating of the material. Typical wind-turbine flexible couplings have a nonconductive fiber and resin section between the flexible links. This section is used for electrical isolation of the generator and may be designed to break away when there is a gearbox or generator mechanical failure. Cracks, distortion, or delaminating of the material may indicate a material defect or excessive stress on the component. Premature failure of this section or the links holding it during operation can cause serious damage to any electrical or fluid-power components nearby. Note any discrepancies on the PM report and notify farm operations if a coupling replacement may be necessary. Include pictures of any defects that may be present during the inspection process to aid in assessing the component before the next climb.

Alignment Requirements

Alignment requirements for the generator include specifications for axial, parallel, and angular misalignment. Consult the wind-turbine OEM recommendations for proper generator-alignment criteria. **Table 13-1** shows typical recommendations for generator alignment to drive components. Following documented alignment specifications will ensure that stress levels on shafts are normal, that vibration levels for the drive system are minimal, that excess heat from abnormal loading on shaft bearing assemblies is not created, and the possibility of premature component failure is reduced.

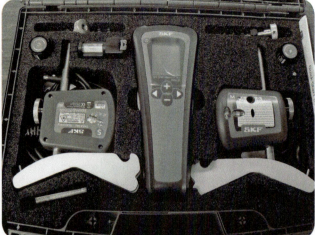

FIGURE 13-6 *Typical laser-alignment kit*

Generic Alignment Steps
Refer to Chapter 8 of this text for details on the generator-alignment process. **Figure 13-6** shows an example of a typical laser-alignment kit. Basic alignment steps include the following.

- Follow the site's safety procedure for the specific generator-alignment process. This will ensure that the procedure can be completed safely and efficiently.
- Mount the manual drive equipment for positioning the shafts during the alignment process. Refer to the OEM procedure for recommended and necessary drive unit setup requirements.
- Mount the head brackets to the gearbox and generator shafts at each end of the flexible coupler and ensure that they are visually level and perpendicular to the shafts. Use a small spirit level to ensure that they are located at the 12 o'clock position during this step. Secure the brackets with the magnetic clamp or chain assembly as required. *Note:* This step is a rough setup process. The fine adjustment of the head brackets will be completed as necessary once the laser heads are installed and energized.
- Mount the laser heads to their appropriate head bracket. Note that the head located on the gearbox is considered

TABLE 13-1

Suggested Maximum Shaft Misalignment Values				
	Angular	Parallel	Angular	Parallel
RPM	mm/100 mm*	Mm	0.001"/1"**	0.001"
0-1000	0.10	0.13	1.0	5.1
1000-2000	0.08	0.10	0.8	3.9
2000-3000	0.07	0.07	0.7	2.8
3000-4000	0.06	0.05	0.6	2.0
4000-6000	0.05	0.03	0.5	1.2

*Gap at coupling edge per 100 mm coupling diameter

**Gap at coupling edge per 1" coupling diameter

the stationary (S) unit and the head located on the generator is considered the movable (M) unit.

- Connect each head cable to the control unit. Ensure the cable connector for the S and M units are attached to their proper location and marked with the same letter.
- Energize the laser control unit and do a final alignment of the heads. Use the thumbwheels located on each head for vertical placement and the head-bracket assembly mounting hardware for the angular alignment. The heads should be located at the 12 o'clock position for the initial setup.
- Measure the distance between (A) laser heads, (B) movable head and the front-mounting foot bolt center line of the generator, and (C) distance between bolt center lines located in the generator mounting feet. Make note of these dimensions on the generator-alignment sheet. Appendix H shows an example of an alignment report supplied by an alignment equipment manufacturer, and **Figure 13-7** shows the setup of the equipment. These dimensions will be used later when setting up the laser control unit.
- Scroll down through the control unit menu for input of setup dimensions. Input the values for A, B, C, and press Enter as directed by the setup menu. **Figure 13-8** shows the display menu for one laser-alignment control unit.
- Follow the procedure for positioning the laser heads at the required clock positions such as 9 o'clock, 3 o'clock, and 12 o'clock as directed by the control unit menu.
- Once this step is completed, it will indicate the angular and parallel offset for the vertical generator misalignment when the heads are in the 12 o'clock position and the horizontal misalignment for angular and parallel offset when the heads are in the 9 o'clock position. Note these values on the alignment report and compare them to the OEM specifications. If all four dimensions are within tolerance, the job is complete. If not, make the necessary vertical adjustments by loosening the mounting bolts and adding shims under each foot (F1 and F2) as directed by the control unit or sliding the generator side to side for the horizontal adjustment as necessary. Refer to the OEM's alignment procedure or

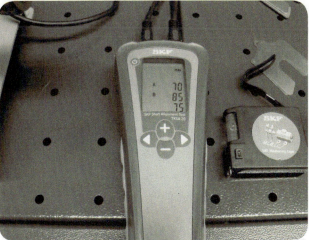

FIGURE 13-8 *Example of a display menu for one manufacturer's laser-alignment control unit*

refer as needed to the steps in Chapter 8. Remember: this process takes time to master, so always work with a qualified technician until you are qualified to do the process. **SAFETY: This process requires a minimum of two technicians to complete. Refer to the site's safety procedure for details.**

- The process is complete when the generator alignment is within specification. Ensure that the generator mounting bolts are secure to the recommended torque value and then recheck the alignment values. The values should not change during this process. Note the final values for horizontal and vertical misalignment along with shim thickness values under each generator foot on the report under the appropriate heading and attach the report to the PM checklist.

The procedure listed here outlines the basic steps to complete a laser-alignment process. Refer to the wind-turbine OEM, laser-alignment manufacturer, or Chapter 8 of this text for further details. Once the generator-alignment process is complete, remove the alignment kit and manual drive equipment from the drive train and assemble the coupler guard before moving to further activities. Ensure that all technicians working in or on the nacelle are notified that the alignment activity is complete before removing any locking mechanism. Always follow the approved site safety plan for LOTO procedures. The next PM activity for the generator is an inspection and maintenance process of cooling-system components.

GENERATOR COOLING ASSEMBLY

Like most electrical or mechanical systems, a generator is not 100% efficient. Inefficiency in a system can be the result of mechanical friction or electrical losses in the power-generation

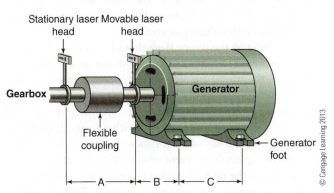

FIGURE 13-7 *Setup dimensions of laser equipment assembly*

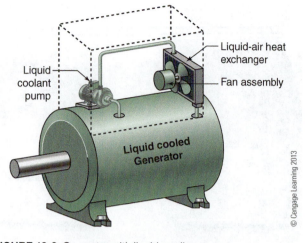

FIGURE 13-9 *Generator with liquid-cooling system*

process. Electrical inefficiency or power loss in this type of system is observed as heat or as an increase in equipment temperature. Manufacturers may use forced-air cooling, liquid cooling, or a combination of both methods to maintain the optimal generator operating temperature. Inspection and maintenance activities for a generator cooling system are similar to those used for power-converter equipment described in previous chapters. **Figure 13-9** shows a diagram of a generator with a liquid-cooling system. Other generator cooling systems include forced air cooling.

Air-Cooled System Activities

Forced-air cooling systems use blower assemblies to direct cool air into the generator frame assembly. This allows air to circulate around the stator and rotor assembly to draw off excess heat (thermal energy) through processes known as *convection* and *conduction*. **Convection** is created by air becoming warm, decreasing in density, and being lifted as cool denser air moves in to replace it. You may recall this process described in conversations about convection cells and weather processes used to balance the temperature in the atmosphere. Convection is a natural mechanism used to create thermal equilibrium. **Conduction** is the process of air molecules taking on energy from a warm surface during contact with the surface. Forced airflow over a warm surface accelerates the processes of convection and conduction by maintaining cooler air around the surface and increasing the number of molecules contacting the surface. As air passes around heated components, it increases in temperature; this warmed air is then discharged out of the nacelle through a set of vents to maintain the operating temperature of the generator. Controlled airflow out of the nacelle or recirculation back into the nacelle can maintain the nacelle interior temperature without extra heating equipment. This is an inexpensive process for cooling electrical systems and controlling the temperature of the nacelle interior.

Maintenance activities for a generator air-cooling system include cleaning of all intake and exhaust assemblies. This includes motor assemblies, fan blades, and associated ductwork.

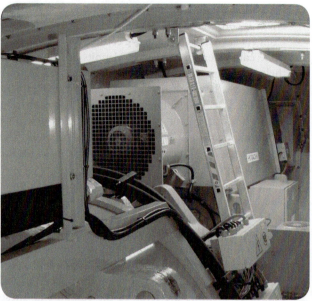

FIGURE 13-10 *Blower assembly used to cool utility-scale generator*

Figure 13-10 shows an example of a blower assembly used to cool a utility-scale generator. **SAFETY: Cooling systems are automatic, so use an approved LOTO procedure to notify affected personnel and isolate energy sources to the fan motor circuit. Follow the OEM's electrical control schematic to open the circuit breaker that supplies power to the fan motor and apply an approved LOTO device.** Use an approved cleaning solution to wipe down all equipment surfaces. Refer to the cleaning solution MSDS for necessary and appropriate precautions such as PPE, exposure limits, ventilation requirements, and cleanup in case of an accidental spill.

Inspect each component of the blower, mounting assembly, electrical connections, and ductwork for loose fasteners and damaged hardware. Tighten loose fasteners to ensure safe and reliable operation of the cooling system. Inspect the motor assembly for blocked cooling vents, discolored paint, damaged bearings, and chafed electrical wires. Blocked cooling vents will prevent the fan motor from operating within the required temperature range. An excess operating temperature will cause damage to the windings, bearings, and electrical connections. Damaged bearings will create excess heat and vibration during operation.

Pushing on the motor shaft should only allow for minimal movement side to side and in and out on the shaft axis. Refer to the motor manufacturer's specification for maximum allowable shaft travel. Excess travel in any of these directions indicates a bearing issue. Spinning of the shaft assembly should be smooth and without any dragging or grinding noise. If any of these issues are observed, the motor should be replaced to prevent unscheduled downtime should it fail before the next scheduled maintenance. Lubricate the bearing assembly with light oil if required. Take care not to overoil the assembly because excess oil spilled into the motor can degrade the wire insulation of the windings and cause a premature failure. Wipe off excess oil from the bearing housing to prevent dust buildup during operation. Some motor assemblies are designed with sealed bearings. These assemblies will still require a visual inspection for damage, discoloration, and excess shaft travel.

Inspect the motor mounting hardware and fan assembly for damage such as cracks, distortion, and missing hardware. Damage to mounting hardware may be an indication of excess vibration during operation. If there were no issues discovered with the bearing inspections, examine the fan assembly and ensure it is balanced. The fan should spin free of the enclosure and without any of the blades tending to stop at the bottom of the rotation. Should the blades contact the enclosure during rotation, then make necessary adjustments to prevent this from occurring. If one of the fan blades has a tendency to stop at the bottom, then check for missing hardware or damage to the assembly. An unbalanced fan blade will create vibration and uneven loading on the motor bearings and shaft that will cause premature failure. Make note of any issues discovered during the inspection process. **SAFETY: Notify personnel working in the nacelle before removing LOTO from the generator air-cooling system.**

Inspect the vent assembly used to direct exhaust air from the generator cooling system. Ensure that the vent mounting hardware is secure and panels move freely during operation. Ensure the actuator assembly is functioning properly. Follow the OEM procedure to manually operate the vent assembly for the inspection process. **Figure 13-11** shows an example of a vent assembly used to control the flow of exhaust air from an air-cooled generator. **SAFETY: Notify personnel around the vent before manual operation to ensure no one is pinched. If the actuator or vent assembly does not operate properly, then adjust the assembly as necessary for smooth operation. If it cannot be adjusted, then notify the farm operations team for follow-up and note the findings on the PM checklist. Failure of this assembly to operate properly can cause damage to the generator assembly because of inadequate cooling.**

Liquid-Cooled System Activities

Liquid-cooling systems are another method of controlling the operation temperature of a wind-turbine generator. Liquid-cooling systems typically use a solution of water and glycol like that used for the power-conversion systems mentioned in Chapters 9 and 11. The advantage of a liquid-cooling system over an air-cooled system is reduced size. Air-cooled systems require considerable surface area to cool an assembly compared to the surface area used for a liquid system. Making electronic packages smaller has been a goal for many years; this philosophy works with power electronics as well. Liquid-cooled systems can reduce the overall size of the generator housing, increase cooling efficiency, and free up nacelle space that can be used for other control systems. Maintenance activities for a generator liquid-cooled system include checking the specific gravity of the solution, checking for leaks in the system, flushing the cooling solution and replenishing as required, and cleaning the heat-exchanger assemblies.

Checking the specific gravity of the cooling solution can be accomplished with a hydrometer and a sample of the cooling solution in the system expansion reservoir. Access the cooling-system expansion reservoir and draw out a sample of the solution into the hydrometer. The typical solution mixture for the cooling system is 50:50 distilled water and glycol. Refer to the OEM's specifications for the type of glycol solution used with the equipment. **SAFETY: The cooling-system expansion reservoir may be located in a cabinet with power electronics and an automatic fan assembly. Follow the approved site safety plan for the LOTO procedure and training qualification necessary to work in this control cabinet. Always review the MSDS information before using any chemicals.** Make note of any required PPE, first-aid requirements, exposure limits, and cleanup requirements on the safety checklist or **job-safety analysis (JSA)** before starting an activity. A JSA is a systematic method of determining the possible hazards for an activity and determining required action to reduce or eliminate the risk before starting the activity. Use of a JSA is a proactive approach to workplace safety. For further details on safety related topics refer to earlier discussions in Chapter 3.

Make note of the hydrometer reading on the PM checklist. If the reservoir is low, add the recommended solution mixture to bring the level back up to appropriate full level. *Note:* The full levels for a cold system and a hot system are different.

FIGURE 13-11 *Vent assembly used to control flow of exhaust air from an air-cooled generator*

First Wind Energy, LLC

Filling a cold system to the hot level will cause the solution to run over when the system heats up because the cooling solution expands with the temperature increase. If the hydrometer indicates that the solution's specific gravity is out of specification, then the OEM requirement may be to drain and flush the system and replenish it with a fresh mixture of coolant. Follow the OEM-approved process to complete this activity.

Check the system for leaks around heat exchangers, the pump assembly, and plumbing connections made around the generator assembly. Leaking coolant can build up on components as the water evaporates, causing damage to sensitive electrical components through shorting or overheating. Buildup of coolant residue also allows buildup of dust around components and connections. Leaking coolant will also cause corrosion issues with metal surfaces around the control cabinet and generator assembly. Loss of coolant during operation can also cause overheating of the generator, so make sure leaks are corrected to avoid unscheduled downtime.

Cleaning the heat exchanger, fan assembly, and vents is also necessary to ensure proper cooling of the liquid in the system. Heat exchangers can be cleaned using a vacuum cleaner and an approved cleaning solution to wipe down surfaces. Use the cleaning solution to wipe down the fan assembly and vent components. This will remove buildup of dust on the assemblies that can reduce the cooling efficiency of the system. Make note of the finding from the inspection activity and cleaning process on the PM checklist. Ensure that consumables such as coolant solution added to the system are noted for inventory control adjustments later. Follow the site's LOTO procedure for removal of devices and notification of personnel after the cooling-system maintenance activity is completed.

GENERATOR ELECTRICAL INSPECTION AND MAINTENANCE

The generator electrical inspection and maintenance activities include inspection of power-cable connections, terminal strip control and feedback connections, winding insulation resistance, slip-ring assembly inspection, and maintenance activities. Access to power cables and the generator slip-ring assembly will require opening panels to gain access to the interior of the generator assembly, so follow the site-approved safety plan and training requirements before proceeding with these PM activities.

Lockout–Tagout Considerations

Remember that the generator has connections with cables at elevated voltages. Depending on the system design, the generator, conversion system, and phase transformers may

operate at voltages in excess of 1,000 V_{AC}. In each case, use appropriate precautions as needed to perform a safe generator PM activity. Work on the generator will need to be completed while the generator is isolated and LOTO devices are in place to prevent accidental energizing. This will require opening any contactors, circuit breakers, and disabling control circuits that connect the generator or conversion system to the utility grid. Follow the site-approved LOTO procedure to identify all sources of power and required isolation steps. Safely verifying that the generator circuits are at zero potential requires verification of the circuits with appropriate voltage-measuring equipment, proper PPE, and management approval for working with these systems. Refer to OEM procedures, the **National Fire Protection Association (NFPA)** electrical workplace-safety standard (NFPA 70E), or other country-specific industry standards for guidelines and PPE necessary for elevated voltage activities. The verification process should always be approached as if the circuits are live until proven otherwise. Many technicians have been injured or killed working around live circuits that were thought to be deenergized. Estimates from workplace safety organizations indicate that more than 2,000 electrical accidents occur each year in the United States. Do not become a statistic! Once the generator and associated power circuits have been isolated and verified, remove the panels from the generator assembly to begin work.

Inspection of Control and Feedback Wire Connections

Control and feedback wire connections can be inspected after removal of access panels on the generator cabinet or through junction boxes located on the generator. Access to these connections should be used to ensure that the connections are secure. A loose connection for a temperature probe will cause the PLC system to sense an elevated temperature condition within the generator and shut down the turbine. Do not wait for SCADA to flag a down turbine for generator overheating when the problem could be a loose connection. **Figure 13-12** shows an example of a junction box with control and feedback connections for a utility-scale generator. Replace the cover when the PM activity is completed and make notes on any findings on the checklist.

Inspection of Power-Cable Connections

Inspection of the power cables may include cable lug-crimp connection, mounting hardware, spacing between cable insulation and crimp lug, cable-insulation resistance, and torque level of the fasteners holding the connections. The cable lug-crimp connection should be secure with no signs of cracks or conductor damage. A loose, cracked, or damaged

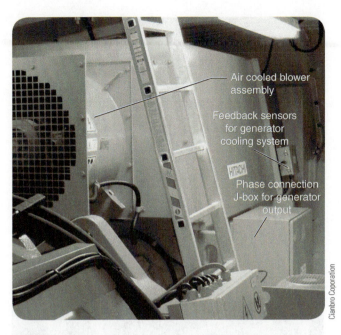

Air cooled blower assembly

Feedback sensors for generator cooling system

Phase connection J-box for generator output

Ciantro Coporation

FIGURE 13-12 *Junction box with control and feedback connections for utility-scale generator*

crimp joint can create excessive heating that can break down the connection during operation of the generator. The cable insulation should be clean cut, with minimal exposed conductor between it and the crimp lug. Many power connections such as this include a shrink-tube jacket applied over the cable insulation and barrel portion of the crimp lug to minimize exposure between electrical conductors and other components within the cabinet. Hardware assemblies such as bus bars that hold the phase conductors should be secure with required electrical insulation components in acceptable condition. Mounting insulators should not be cracked or stripped by the mounting fasteners. These conditions will allow the phase connections to become loose during operation. Phase cable insulation should be verified for integrity as recommended by the OEM. This procedure was described in Chapter 7, and portions of that discussion will be summarized in this section.

Verification of the phase cable insulation requires removal and isolation of the cables from the generator terminal before proceeding with the test procedure. Follow the meter manufacturer's test procedure for appropriate voltage level and test duration and record the results on the PM checklist. **SAFETY: Ensure that ends of the phase cables are isolated and that personnel in the area of the testing are cautioned about the hazard associated with insulation-resistance test. Use appropriate signage and barricades to ensure that personnel are restricted from the test area and cable ends are not in proximity to sensitive control circuits that may be damaged if the energized cables accidentally came in contact with them.** After the phase cables are tested for insulation resistance, the generator stator winding insulation resistance should be performed before the phase cables are reconnected.

Stator Winding Insulation Resistance Testing

Testing the stator windings is a good practice to track the condition of the generator. Over time, the stator windings insulation will decrease because of oxidation and exposure to elevated temperatures, dust, oil, and atmospheric contaminants such as ozone. Insulation resistance for phase windings should be performed between phase windings and each phase winding to ground. This gives a good representation of each phase winding's condition during the service life of the generator. Data gathered from insulation testing should be compared as a trend based on time and not as an isolated test. If a sudden decrease is noted, then consult the wind-turbine OEM's or generator manufacturer's engineering staff for recommendations and corrective action. Follow the trend of the insulation resistance until readings are presented that indicate it is time for replacement of the unit. This process is known as *predictive maintenance*. The trend data can indicate that there is a possibility of problems to come, so corrective action is scheduled before a failure occurs. Note test findings on the PM checklist for later input into a database or spreadsheet software. Notify the farm operations team of any unusually low readings for immediate follow-up and corrective action before returning the system to service.

Once the phase cable insulation and stator winding insulation tests are completed attach the cable lugs to the appropriate terminal connection and tighten each. Refer to the OEM specifications for the torque level required for the size and grade of fastener used to secure the connectors to the bus bar or terminal. **CAUTION: Verify the fastener and bus-bar material especially if the bus bar has threaded holes. Tightening a copper alloy fastener to the torque requirement of a steel alloy fastener will have disastrous results. Replacing fasteners may require extra time, but replacing the bus bar may cause the wind turbine to be down until a replacement arrives. A few moments to verify what you are doing now can save time and money by eliminating a mistake.** Note findings on the PM checklist before starting the next inspection process.

INSPECTION OF GROUNDING CONNECTIONS

Inspecting generator grounding connections is important to ensure that ground currents are discharged through the cables and not through bearings. This is especially true for the ground connection between the slip-ring assembly and the generator chassis. Ground currents passing from the rotor assembly to the chassis through the bearings will cause arcing between the moving elements of the bearing and the race assembly. Bearing damage caused by ground currents can appear pitted or as a white haze over the surface of the bearing race and moving elements. The roughened surface of the elements in turn will cause increased friction, overheating, material erosion, and premature failure of the bearing.

Check each ground cable assembly to ensure the ends are crimped securely and the fasteners holding them in place are at the required torque level. Refer to the OEM procedure or published torque chart for the size and grade of the fastener used with the application. Tighten the fasteners as necessary to ensure a suitable connection. Note findings on the PM checklist for follow-up as necessary.

INSPECTION OF GENERATOR SLIP-RING ASSEMBLY

The generator slip-ring assembly is used to provide power to the electromagnet poles mounted on the rotor assembly. Depending on the equipment manufacturer or model of the generator, there may be one or more rings per phase on the slip-ring assembly. These rings are made from a copper alloy to improve the conductivity of the connection between the brush assembly and the ring. Inspection of the rings should include observation of the surface finish, color, and buildup of carbon dust from the brush material. Surface finish of the rings should be smooth and without score marks created by contaminants or excessively worn brushes. A rough surface of the ring assembly will cause the brush material to wear excessively or cause sparking between the components. Excessive sparking between the brush and ring will create a pitted surface that will further deteriorate the brush material and ring surface. **CAUTION: When inspecting or cleaning the ring surface, wear nitrile or lint-free cotton gloves to prevent oils from your fingers from tarnishing the surface. Always use an approved cleaning solution to prevent damaging the ring surface from mild acids or lubricants that may be present in some cleaning solutions. SAFETY: Refer to the MSDS supplied with the brush assemblies for proper PPE, first-aid, and disposal requirements. Use of a dust mask during the cleaning process will prevent inhalation of carbon dust. Excess exposure to carbon dust can cause respiratory issues with time.** Cleaning of the slip-ring assembly area should be accomplished with oil-free cotton rags and a vacuum cleaner. Remove any carbon buildup between rings brush holders springs, wires, bearing assemblies, interior surfaces, and ventilation openings. Carbon dust is conductive, so a buildup between phase assemblies may cause a short to occur. A buildup of carbon dust around bearings may cause damage to the bearing if the dust contaminates the lubricant. Buildup around ventilation openings of the slip-ring cabinet will reduce necessary airflow for proper cooling of the assembly.

FIGURE 13-13 *Slip-ring brush-holder assembly*

Inspection of Brush Holders

Ring-assembly inspection should include a visual for discoloration of the rings and brush holders. Contaminants on the ring surface or damage to the brush or holder assembly can create excess heating of the assembly. Excess heating may be created by a low contact force between the brush and the ring surface or one of the brush assemblies not contacting the ring because of wear. This will cause excess current flow through the remaining brush assemblies. Excess heating may appear as a darkening or bluish color change to the brush holder materials. Check each of the phase and ground brush assemblies for freedom of movement within the holder. Ensure that the spring assembly is contacting the brush within the holder and the brush is positioned so that the curvature of the brush matches the radius of the slip-ring surface. **Figure 13-13** shows an example of a slip-ring brush-holder assembly. The brush assembly may have a maximum wear line located on the side of the assembly. Refer to the OEM procedure to ensure that this line is sufficiently above the reference marker on the brush holder. Some OEM procedures indicate a minimum measurement specification from the slip-ring surface to the top of the brush. If this is the case, then use a steel machinist scale to measure and document this value on the PM checklist. Follow the wind-turbine OEM's or generator manufacturer's recommendations for the change-out interval or minimum brush length required before replacement.

Replacement of Brushes

There are several different brush and brush-holder assemblies on the market. Follow the generator manufacturer's procedure for the system being maintained or serviced. The following is a generic procedure that may help in preparation for the brush change. As mentioned previously, the brush is held against the slip-ring face with a spring assembly. The spring may be either a coiled torsion spring or a compression style, so there will be different precautions necessary when lifting it from the brush. Lift the spring from the brush and park it in the slot located on the top of the holder or remove it from the assembly and place it in a secure location so it does not get lost. Loosen and remove, if necessary, the fasteners holding the braided conductor and place them in a secure location. Some brush assemblies are supplied with braided conductors that have fork terminals crimped to the ends instead of ring terminals. This type will only require loosening the fasteners and not removal. Remove the brush assembly from the holder and make note of the orientation of the brush within the holder. Use the same orientation for the new brush when inserting it into the holder.

CAUTION: **Some brush manufacturers supply the brush assembly with a radius on the end face that approximates the curvature of the slip ring, whereas others supply the brush with a tapered angle that aligns with the slip-ring face. No matter which brush is supplied, it may have to be machined to provide a continuous contact surface with the slip ring. Gaps between the brush and the slip ring will create sparks and excess heating that will cause damage to the brush, holder, and slip-ring face.** Refer to the manufacturer's recommended machining procedure to achieve the required brush mating surface. Once the brushes are installed into the holder with the curved end matching the slip-ring face, install the spring against the brush and secure the braided conductor to the bus-bar assembly. Ensure the brush slides freely within the brush holder. If it does not, make adjustments to the holder or spring to allow proper movement. If the brushes are not supplied with a curved face that matches the slip-ring face, the following suggestions may help. Always follow the manufacturer's recommendations and use these steps as a guide only.

Seating of Brushes

Machining procedures that may improve the mating of the brush to the slip ring include machining each brush in the operations shop before going to the tower or machining it in the field. Machining the brushes in the operations shop requires the use of a custom setup that includes a movable drum assembly of the same diameter as the slip ring and a holder for positioning the brush against the drum. **Figure 13-14** shows an example of the brush-machining process that may be used to prepare the brushes before use in the slip ring. The use of 100 or finer grit cloth secured to the outside edge of the drum may be used to rotate against the brush to remove

FIGURE 13-14 *Brush-machining process*

excess carbon until the curvature of the brush face matches the drum radius. Use a vacuum cleaner and the precautions listed previously to reduce exposure to carbon dust produced by this operation. The field method is very similar to the procedure in the operations shop. Securing 100 or finer grit cloth to the slip ring and rotating the generator with the brushes in place will remove excess material from the face of the brush. After this process, clean the carbon dust from the slip-ring cabinet. Cleaning the carbon dust can be accomplished using a vacuum cleaner and lint-free cotton rags to wipe down the assembly. Once the slip ring is inspected and the brushes are changed as necessary, do a final inspection for tools and close up the cabinet.

CAUTION: **Tools or other materials left in the slip-ring cabinet may cause serious damage to the generator if not removed. Always make a final visual sweep of the equipment for missed articles or components that have not been secured before closing up and moving to the next operation.** Make note of inspection findings, parts used, and quantities on the PM checklist for follow-up later and adjustments to inventory levels. It cannot be stressed enough that listing consumed parts on the PM checklist or an inventory sheet can make the site operation run more efficiently. Trying to remember the next day or the next week what was done in a particular tower does not work, especially if the team is completing a maintenance cycle on a tower every day.

PERMANENT MAGNET ROTOR ASSEMBLY

The preceding sections list the inspection activities for a slip-ring assembly as part of the PM cycle. For generators that use a permanent magnet rotor assembly, these steps are not necessary. Refer to the generator manufacturer's maintenance procedure for the requirements of these systems.

RECOMMENDED CLEANING ACTIVITIES

Recommended cleaning activities for the generator assembly include wiping down all exterior surfaces, cables, and mounting hardware. Use an approved cleaning solution and observe the safety recommendations listed on the cleaner MSDS supplied by the manufacturer.

SUMMARY

The particular maintenance activity required for wind-turbine generator assembly will be determined by the design of the equipment. Generators connected directly to the gearbox or direct-drive systems will not need the alignment activities listed in this chapter. The visual inspections, lubrication, and cooling-system activities may be similar in nature. Always refer to the OEM's recommendations for maintenance and inspection activities necessary to keep the equipment operating efficiently.

REVIEW QUESTIONS AND EXERCISES

1. List some PM inspection activities for a generator.
2. List a few generator configurations that may be seen in wind turbines.

3. List two methods for producing a magnetic field for a generator.
4. Why is it important for proper generator alignment to the gearbox?
5. Why is it important to maintain a good ground connection between the slip ring and the generator chassis?
6. What is the purpose of using an approved LOTO procedure and safe work practices when working with and around generator power connections?
7. List a national organization that recommends safety procedures for working with electrical equipment.
8. Why should the stator insulation resistance be checked during multiple PM cycles throughout the service life of the generator?
9. Why is it important for proper brush seating to the generator rotor slip ring?
10. Why is it important to document inspection results and materials used during each step of the PM activity and not waiting until the end of the day?

Rotor Assembly

KEY TERMS

axis cabinet
blades
communication link
confined space
emergency medical service (EMS)
hub
job-safety analysis (JSA)
National Electrical Code (NEC)
Occupational Safety and Health Administration (OSHA)
permit-required confined space
pinwheeling

pitch
pitch bearing
pitch control
positive contact
rabbit eared
rotor assembly
rotary encoder
rotor lock
spinner
standard operating procedure (SOP)
tachometer-generator (tacho-generator)

OBJECTIVES

After reading this chapter and completing the review questions, you should be able to:

- List different maintenance activities necessary within the hub.
- Describe maintenance activities related to a hydraulic pitch system.
- Describe maintenance activities related to an electric pitch system.
- Describe lubrication activities associated with the rotor assembly.

- Describe visual inspections associated with interior systems of the hub.
- Describe bolting torque inspections associated with the rotor assembly.
- List safety considerations for rotor-assembly maintenance activities.

INTRODUCTION

A wind-turbine **rotor assembly** is made of several components used to capture energy from the wind and translate it to usable mechanical power to spin the generator. Components that make up a horizontal axis wind-turbine (HAWT) rotor assembly include the hub and blades. The **hub** is a large steel casting machined to accept pitch bearings, blades, connection to the drive train, pitch-control system, and a spinner. **Pitch bearings** are large bearings that enable the angle of the blades to be adjusted with respect to the hub. The **blades** are airfoils designed to convert linear motion of the air mass—wind—into rotary motion for the turbine. A variable-pitch blade design enables the wind-turbine control system to vary the blade angle—that is, the **pitch**—to adjust for changes in the system's need for torque and rotational speed (RPM).

Remember that power is a function of torque and RPM as discussed previously with the drive-train components:

$$P = \frac{(T \times S)}{9{,}549}$$

where P = power (kW), T = torque (N-m), and S = rotational speed (RPM); 9,549 is the constant for these units.

The **pitch-control** system is used to translate control signals from the programmable logic controller (PLC) to required changes in blade position. The basic pitch-control system is made up of components for communication, control, drive, and feedback. These components are similar for each wind-turbine system with one notable difference—the drive component. Some manufacturers use electric gear motors to pitch the blades, whereas other systems may use hydraulics actuators. The function of these systems will be discussed during the chapter as well as the maintenance requirements

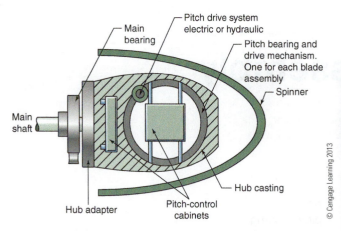

FIGURE 14-1 *Typical components within the hub assembly*

FIGURE 14-2 *View from the top of a nacelle*

for each. Another component that typically makes up a rotor assembly is a **spinner**, also known as a *nose cone*, which is a shell used to improve aerodynamic flow around the hub and nacelle. The spinner is made of a polyester resin that is typically reinforced with polyester or glass fibers. **Figure 14-1** shows typical components within the hub assembly.

Maintenance activities for the rotor assembly will vary, depending on the design of the assembly. Activities required for an electric pitch drive will be different than those of a hydraulic pitch. This chapter will discuss typical equipment configurations that a technician may be exposed to during maintenance activities. The discussions supplied with this text will be generic in nature. They will broadly explain several of the common activities but will not go into detail for any one manufacturer's system. Always refer to the wind-turbine manufacturer's maintenance procedures for details on settings, material specifications, and specific safety procedures. This chapter will begin with safety hazards a technician should recognize before considering a hub entry and then move on to maintenance activities by subsystem and variations of subsystems.

SAFETY REQUIREMENTS

Several work activities are associated with the rotor assembly such as visual inspections, verification of electrical parameters, fastener torque verification, and lubrication, along with cleaning activities. These activities may vary between each manufacturer's equipment design and model. Common among turbine designs is the safety issues that should be considered when working on the rotor assembly. The rotor assembly, as its name implies, rotates or spins during operation of the wind turbine, so lockout–tagout (LOTO) is a consideration.

Hub height is around 300 feet (92 meters) in elevation from the surrounding terrain, so issues from working at height are a consideration. These considerations are not necessarily the same as the climb to the nacelle. **Figure 14-2** shows a view from the top of the nacelle looking over the hub. The interior of the hub may also meet the requirements

of a confined space, as defined by OSHA, so those issues should be considered as well.

These safety considerations are in no way the only ones to consider when working in or on the rotor assembly. Always consult recommendations from the original equipment manufacturer (OEM) for electrical, mechanical, environmental, and other hazards when preparing for work activities with the rotor assembly. Technicians should always be aware of safety hazards associated with a work assignment to prevent injuries. A great time to reduce exposure to safety hazards is *before* an activity. Always consider, document, and analyze hazards during the **job-safety analysis (JSA)** phase of a tailgate meeting. This ensures that appropriate precautions are addressed before a work activity and not after an injury occurs. Refer to Appendix J for items to consider while reading the following sections related to service work with a hub-entry activity.

Confined Space Considerations

The **Occupational Safety and Health Administration (OSHA)** 29 CFR 1910.146 designation of **confined space** includes the following considerations. The space is not designed for continuous human occupancy and has limited means of ingress and egress, but it does allow a person to enter for an assigned work activity. A wind-turbine manufacturer's safety requirements or the site's safety policy may indicate the hub interior classification as confined space or as **permit-required confined space**, which OSHA describes as having the addition of a safety hazard that can cause serious health issues or death. This hazard may be created by an atmospheric condition, a physical configuration, or another health issue. Wind-turbine manufacturers typically place signs or placards near the entrance of the hub to indicate whether the interior is classified as a confined space. They may also include this information in bold print within work procedures to alert technicians of possible safety hazards. Covering this information during the tailgate meeting will reinforce the need for safety precautions. Each of these safety issues should be considered before preparing for hub activities along with personnel responsibilities, training, types of activities, and rescue requirements.

Personnel responsibilities for OSHA confined space entry include the minimum number of personnel required for a safe entry along with the responsibility of each person participating. The minimum personnel requirement for a confined space entry includes authorized entrant, attendant, entry supervisor, and emergency rescue. The authorized entrant must be trained to recognize hazards associated with the entry activity as well as be authorized by management to enter a specified hub.

Hub assemblies for multiple turbine models are typically different. Being authorized for one manufacturer's hub assembly does not make a technician authorized for all models. The attendant for the entry activity must be trained to recognize hazards associated with a specific hub and rescue requirements, as well as be stationed to assist the entrant during the entry activity. The attendant must not enter the hub to assist the entrant unless another authorized attendant is available. The third person required for a hub entry is the entry supervisor. The entry supervisor is responsible for ensuring that the activity follows company policies and to provide backup should an emergency situation arise. Hub-entry rescue protocol should be specified in the company's safety policy and each participant should be trained to respond accordingly. Typically, hub rescue is performed by the same technicians authorized to perform service and maintenance activities on a site's wind-turbine models. Should something happen to the authorized entrant, the attendant would provide rescue after assessing the hazards and alerting the entry supervisor. Technicians should never attempt a rescue without first making positive contact with someone to provide backup support to the rescue. **Positive contact** means requesting assistance, getting a direct reply (the emergency situation is understood, I am calling for help, etc.), and taking action. Asking for assistance through a radio call does not mean someone has heard and will respond. Maybe the radio is not working and quite possibly there is no cell-phone reception in the area. One never knows when the rescuer may become another victim.

First-line rescue support provided by wind-service technicians is the reason for each technician to be qualified and proficient with first aid, cardiopulmonary resuscitation (CPR), an automatic external defibrillator (AED), and tower rescue. Many wind farms are located in remote locations with limited access to the turbines. An **emergency medical service (EMS)** is not typically located close enough to a wind-turbine farm to meet the minimum time requirement for initiating a rescue operation. Time is of the essence for any rescue to be successful in saving a life in an emergency. Even if the EMS is close by, typically the personnel are not

trained in tower-rescue operations so they will only be available for support once the victim is on the ground. Refer to the site's safety policy and OSHA regulations for further details on confined space entry and rescue requirements. Along with confined space issues associated with hub entry, a technician should also consider the method used to move from the nacelle into the hub, necessary LOTO procedures, and weather-condition requirements.

Fall-Arrest Considerations

Hub entry for utility-scale wind turbines is either by entry from the top of the hub or from within the nacelle through a crawl space to the hub interior. Each access method requires considerations for a safe entry and exit. Access from the top of the hub requires that a technician climb from the top of the nacelle, onto the hub, and enter through an external hatch. Remember in Chapter 2 the top three safety hazards associated with the wind industry:

1. falling from the tower or nacelle,
2. being struck by a falling object, and
3. shocks and electrocution.

To keep technicians from suffering the first hazard, the risks associated with working on top of the nacelle, hub, blades, and outside the tower deserve another reminder. OSHA 29 CFR 1910.66 and 1926.502 regulations require the use of fall-prevention or fall-arrest systems for activities more than 4 feet and 6 feet above the ground, respectively. Work activities on top of a utility-scale wind turbine are well above either of these limits, and wind turbines typically do not have fall-prevention equipment such as handrails mounted on top of the nacelle or hub. Not having fall-prevention equipment available requires technicians to incorporate fall-arrest equipment into their required personal protective equipment (PPE). Access to the hub from the top of the nacelle requires that a technician don an appropriate full-body fall-arrest harness, twin-leg lanyard, and any other required PPE and also use approved nacelle and hub tie-off anchors.

CAUTION: Twin-leg lanyards are required for 100% tie-off. This means that one of the large snap hooks needs to be connected to an anchorage at all times. The other "free" snap hook may need to be stowed during part of the activity. Whenever the free snap hook is stowed, it should never be stowed in a harness metal D-ring such as a side D-ring. It should be stowed in a breakaway lanyard keeper. See **Figure 14-3** for an example of a breakaway lanyard keeper. A lanyard keeper is attached to the body harness with Velcro or snaps so that the free snap hook will separate from the body harness should a fall occur. Securing the free snap hook in a side metal D-ring creates a rigid connection that may prevent the lanyard shock absorber from activating properly or it may become entangled during the fall. Not a good thing! **Figure 14-4** shows a preferred lanyard snap hook stowing position with a breakaway keeper.

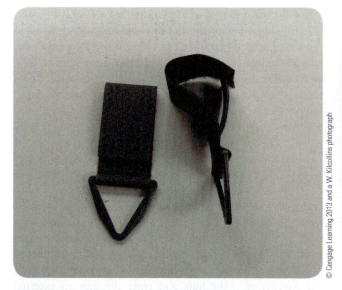

FIGURE 14-3 *Example of a breakaway lanyard keeper*

FIGURE 14-5 *Small cable lanyard*

FIGURE 14-4 *Preferred lanyard snap hook stowing position with breakaway keeper*

CAUTION: Communication devices such as radios and cell phones should be attached to the body harness with a small cable lanyard or stowed in a pouch with a hands-free communication setup. This precaution will prevent technicians on the ground or lower levels from being hit by a falling communication device. Figure 14-5 shows an example of a small cable lanyard that may be used for this function.

The second method to enter the hub assembly is through a crawl space or walkway from within the nacelle. This access may be located by the gearbox or main bearing, depending on the configuration of the drive system. This type of hub access would not require the fall-arrest considerations mentioned previously. A common factor between these two access methods is how to prevent the hub from rotating freely (**pinwheeling**) during the entry activity. Preventing the rotor

assembly from pinwheeling requires LOTO of the park brake system or the use of a mechanical locking component to keep the drive train from rotating.

Lockout–Tagout Considerations

Preventing the rotor assembly from moving during access to the hub for maintenance and service is a critical requirement. Pinwheeling of the rotor assembly would allow any loose items to tumble and injure the technician. A technician tumbling within the hub assembly may also be exposed to energized open control cabinets, gears drives, and other equipment that may cause serious injuries if they were operational. This is not the ride at the amusement park you may have enjoyed when you were younger. This author has heard stories from new technicians who made improper hub entries shortly after being hired. Several of them discovered that they had forgotten to lock the rotor assembly when returning from the hub. Fortunately for them, they did not end up as victims of their mistake. Being attached to a spinning rotor assembly by a lanyard also would not be a ride to envy. Proper LOTO procedures should be stressed during daily tailgate and weekly safety meetings.

Rotor Lock

The **rotor lock** assembly may vary in design between equipment models, but the main function is to mechanically lock the rotor assembly from moving during service activities. Always refer to the manufacturer's safety procedures for the specific lock style installed in the system being serviced. Typical rotor locks use a pin or a clamp assembly to lock the parking brake disc or engage the hub adapter for a positive lock. The rotor lock assembly should be designed to accept LOTO device such as a hasp, lock, cable lock, or chain. **Figure 14-6** shows examples of these devices.

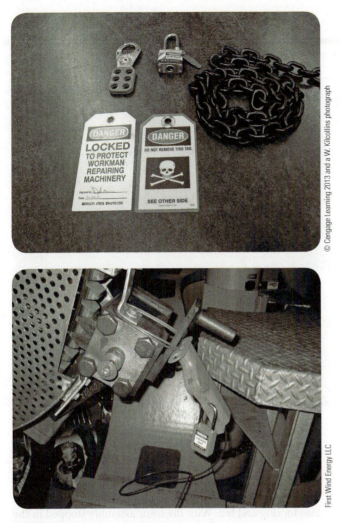

FIGURE 14-6 *Rotor lock LOTO devices*

Other LOTO considerations for the hub include any energy source that may be present that could accidentally be released during service or maintenance. This may include electrical or hydraulic systems present in the hub. If the energy source can be isolated, deenergized, and remain in the qualified technician's control, then it may not require LOTO. Always refer to the site-approved LOTO procedure for guidance. The National Fire Protection Association's **National Electrical Code (NEC)** requires the energy control to be within the qualified technician's control for this to be considered a safe practice. By "technician's control," NEC means that the energy control switch or disconnect must be within the line of sight and within 50 feet (15.4 meters) of the technician (NEC Article 430). These two requirements cannot be met if the power control is located in the nacelle and the technician is within the hub. If this is the case, then LOTO is required. These are some of the considerations for LOTO with a hub entry. There may be other considerations, depending on the wind-turbine system or activity that is planned. Another hub-entry safety consideration is the weather conditions anticipated during the service or maintenance activity.

Weather Considerations

Weather conditions during any activity on the wind farm or during wind-turbine access are important to monitor. Chapter 3 discusses many of the items that should be considered during farm operations. A few of these items are worth repeating for hub entries, including high winds, freezing rain, and thunderstorms. As many of us know, wind farms are positioned strategically at windy locations, so working with moderate winds during maintenance or service activities is **standard operating procedure (SOP)**. Any wind load on the rotor assembly creates stress, but a wind 15 m/s (33.6 mph) or higher may create excess stress on the rotor and drive-train assembly when it is not allowed to pinwheel. Refer to the OEM's procedures for details on acceptable wind levels for locking the rotor or performing activities on the nacelle or in the hub.

Monitoring for changing weather conditions is another important activity for the entry supervisor. The morning tailgate meeting should include a check of the weather forecast for the work activity time period. Notable hazard conditions include high winds, thunder showers, flash floods, freezing rain, snowstorms, and low visibility. These may have not been listed in the morning forecast, but they can be showstoppers if they move in during a maintenance activity or during the commute back to the service center. The National Weather Service (NWS) and other weather alert service providers have phone, text, and e-mail alert notifications if severe weather conditions are moving toward a predefined geographic area. To receive weather alerts, a wind farm may subscribe to an online service provider that has coverage for the area. Typical services include text messaging, automated voice messaging, and follow-up e-mail alerts.

As the reader can see, there are many issues to consider for a safe hub entry or any work activity on the outside of the wind-turbine nacelle. These items are discussed in Chapters 2 and 3 of this text, but they are reviewed here as part of the preparation for rotor-assembly maintenance. A good way to ensure that each consideration has been evaluated is by using a checklist. **Figure 14-7** shows a sample checklist that may be used for hub-entry activity.

PREPARING THE HUB

Entering the hub requires more preparation than a climb up the ladder to the nacelle. The safety considerations are listed in previous sections, so here we look at the equipment preparation. When the turbine is offline, the normal mode for the rotor assembly is pinwheeling to reduce stress on the drive train. This means the rotor assembly will be spinning or stopped in a random blade orientation. External access to the hub may be through one of three hatches located between the blades or a single access hatch on the nose of the spinner. Refer to the operations manual for the wind turbine located on the farm for specific hub-access requirements. For

Date: _____

Wind Farm Name: _____

Wind-Turbine Location: _____

Technician Names: _____

Service Activity Anticipated: _____

Activity	Y/N	Corrective Action
Required team training up to date		
Wind turbine set to "Service" status		
Remote access communication disabled		
Weather briefing - Acceptable		
Rotor assembly positioned for entry		
Rotor assembly lock engaged		
LOTO performed per approved entry procedure		
Approved PPE available		
PPE donned correctly		
Entrant trained for anticipated activity		
Attendant available in nacelle		
Rescue back up available		
Defined communication method functional		
Required service tools available and functional		

Any response listed as "No" shall have a corrective action in place to proceed with hub-entry activity.

© Cengage Learning 2013

FIGURE 14-7 *Sample hub-entry checklist*

these configurations, a technician must walk out between the blades and open a hatch to access the hub interior. This requires positioning the rotor assembly so that two of the blades are directed up and one straight down. This orientation is referred to as **rabbit eared** because the blades resemble the ears of a rabbit. **Figure 14-8** shows an example of the rabbit-eared orientation. Positioning the blades in this orientation requires manually operating the park brake to stop the rotor assembly. Before doing this procedure, all technicians working around the nacelle need to be notified that the rotor assembly will be stopped. A sudden stop of the rotor

FIGURE 14-8 *Rabbit-eared orientation*

REpower—Dennis Schwartz Photograph

assembly will cause the tower assembly to sway because of its change in momentum. Anyone not ready for the sway may lose footing and fall. With the rotor assembly stopped in the rabbit-eared position, the rotor will need to be locked to prevent accidental motion during the hub entry. **SAFETY: The LOTO process will be required whether access is made internally to the nacelle or externally through the hub. Final steps to enter the hub safely require:**

- review of the hub-entry checklist,
- proper use of fall-arrest gear (external entry),
- use of approved anchorage rails or tie-off points (external entry),
- required LOTO performed and verified, and
- exit the nacelle and enter the hub.

ENTERING THE HUB

No matter which method is used to enter the hub, the view within the hub will be similar for many of the utility-scaled wind turbines. The view will include large bearings, access to the blades, control cabinets, pitch-drive components, cables, conduits, lighting, and structures to mount the components. The bearings may be 5 feet (1.5 meters) or larger in diameter, with large fasteners securing them to the blades. There may be one or several control cabinets to house the drive-system

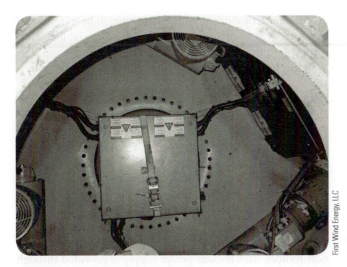

FIGURE 14-9 *Example of a hub interior layout*

electronics. The interior layout of some hubs includes a control cabinet mounted toward the nacelle along with cabinets mounted over each blade opening. The cabinet mounted toward the nacelle encloses the PLC and control components that interface the pitch motor controllers and the PLC located within the nacelle. Pitch-control systems that use hydraulics may only have one control cabinet. Refer to **Figure 14-9** for an example of a hub interior layout. The cabinets mounted over each blade opening are known as **axis cabinets**, and they enclose the DC drive controllers, contactors, circuit breakers, fuse assemblies, backup batteries, and other components used with an electric pitch-drive system. Blade-pitch actuators will also be located at the blade openings. These actuators may have a spur gear meshed with the large ring gear at the root of each blade. Another configuration may use actuators connected to a linkage used to adjust the position of the blades. Other items mounted around the blade-pitch drive components will be sensors that provide feedback to the drive systems for positioning and rotation speed. Inspection and maintenance activities for these items will be covered in the following sections.

INSPECTION AND MAINTENANCE OF BLADE-PITCH DRIVE SYSTEMS

A blade-pitch drive subsystem may include a PLC controller with a rack of associated modules, multiaxis drive control unit, DC motor drives, battery backup, electric gear motors, and feedback sensors for an electric pitch system such as that shown in **Figure 14-10**. Or the system may have a PLC control system operating a power unit, valves, fluid-power actuators, and feedback sensors for a hydraulic blade pitch. **Figure 14-11** shows a couple of examples of electrohydraulic blade-pitch systems. Each system works effectively to control

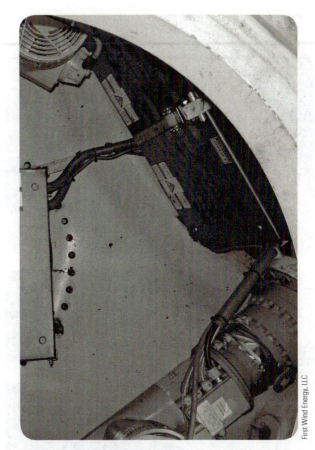

FIGURE 14-10 *Electric pitch system*

wind-turbine power output through blade-pitch control. The operating differences between the systems will determine the required maintenance necessary to ensure reliable operation. Each system will be made up of wire harnesses and cables used to connect electrical components for power, control, or communication. Maintenance for wire connections includes visual inspections and verification of connection integrity as it did with systems discussed in previous chapters.

Inspection of Wire Connections within the Axis Cabinets

Inspection of wire connections should be made for components located within the axis cabinets as well as any external components associated with the pitch system. **SAFETY: Before working within axis cabinets, the cabinets should be powered down using an approved LOTO procedure. Powering down the cabinets will prevent a possible shock or electrocution hazard for the technician as well as protecting the components from an accidental short circuit.** Inspection of wire connections and components within the axis cabinets should include a visual search for discoloration, chafing, and loose or missing cable tray covers. Take time to

University of Maine at Presque Isle

University of Maine at Presque Isle

First Wind Energy, LLC

FIGURE 14-11 *Electrohydraulic blade-pitch systems*

ensure that the wire connections to terminal strips, relays, contactors, fuse blocks, and drive control units are secure. Include an inspection of mounting hardware securing all wiring devices and components within the cabinets. Remember: the hub assembly spins during operation, so a component that comes loose within an axis cabinet will create considerable damage before the system senses the fault and shuts

down. This author cannot emphasize enough how important this activity is to ensuring maximum farm availability. Note inspection findings on the PM checklist for follow-up later as necessary.

Inspection of Battery Backup Components

During normal operation, the electric blade-pitch drive-system controls run on utility grid power. The system motor drive units use AC voltage supplied from the utility grid to develop DC voltage to operate the DC gear motors. Loss of utility grid power to the wind turbine causes the system to run an emergency shutdown procedure that will pitch the blades and set the park brake. Running the shutdown procedure without the utility grid requires that the pitch-drive system have a backup supply of DC power to operate the DC gear motors. This backup power supply is provided through a battery pack or packs designed to supply the DC voltage level necessary to pitch the blades back to the 90° or feathered position. Aside from the normal operation of blade pitch, the drive system charges the battery packs, monitors battery voltage, and runs scheduled battery-load tests. If the control system determines a failure of any of the battery functions, then the wind turbine will shut down and send an error message through SCADA for follow-up by a service technician. PM of the battery backup system should include verifying that all components are secure within the battery cabinets. Refer to the wind-turbine manufacturer's specifications for fastener torque requirements. *Note*: Not all battery packs are located in the hub. Some manufacturers may place them in the nacelle for easier access. This variation should be noted in the manufacturer's PM procedure or electrical schematics. PM of the battery packs should include a test to verify that each battery is charged to the appropriate voltage level and that the internal resistance is within the specifications set by the battery manufacturer. Typical batteries for this application are 12-volt sealed lead storage batteries similar to those used in an uninterruptible power supply (USP) or emergency backup lighting system. **SAFETY: Open circuit breakers or remove any fuses between the charging circuit and the battery packs. Shorting these wires could damage the batteries or the charging circuit. Consult the site safety policy for appropriate PPE requirements when working with battery packs. PPE may include rubber electrical gloves, goggles, face shield, and clothing rated for the associated hazard risk category.** Opening the circuit protection devices will also ensure that the voltage reading is from the battery pack and not the charging system. Each battery may be verified by disconnecting the wire jumpers between the batteries and attaching the instrument test leads to the battery terminals in the correct polarity. Be careful not to short the battery terminals during this step. Shorting a battery can cause an explosion or fire hazard. Follow the battery test instrument manufacturer's guidelines for the instrument setup and interpreting the results.

Figure 14-12 shows an example of a battery inspection and voltage test being performed. Note the individual battery voltages and internal resistance values for each battery

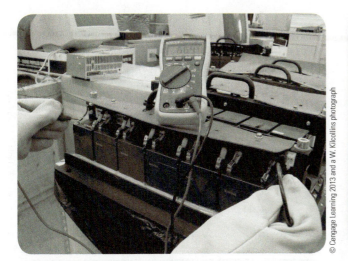

FIGURE 14-12 *Battery inspection and voltage test*

FIGURE 14-13 *Gear-motor assembly used for blade pitch*

in the pack on the PM checklist. If any batteries are out of specification, then notify operations so the pack can be replaced during the maintenance activity or scheduled when a replacement is available. Finish the cabinet maintenance activities by reattaching any wires that may have been disconnected during testing and verify that all components are secure before closing the cabinet covers.

Inspection of Electric Gear Motor Drives

The gear motor assembly will require a visual inspection along with manual activities. Visual inspections include checking the oil level within the gear motor gearboxes; looking for lubricant leaks, chafing and discoloration in wire harnesses, the condition of the gear assembly; and checking for loose or missing mounting hardware. The gearbox assembly or external reservoir will typically have a sight glass or sight tube located on the side of the assembly. To accurately assess the oil level, the gear-drive assembly should be in the upright position. **Figure 14-13** shows examples of gear motor assemblies used for blade pitch. Some drives use a right-angle gearbox while others use an inline assembly. Placing the gear motor assembly in an upright position will require rotating the hub so that each gear motor is located in the bottom position for the inspection. This activity will require multiple trips to the hub for the inspections. Complete this activity in conjunction with other activities that require hub positioning such as torque verification of blade and pitch bearing fasteners.

Planning ahead for activities will save time and money. Add the necessary oil quantity to each gearbox to bring the level back within the specified range. Note the gear-oil quantity added on the PM checklist and note any leaks seen during the inspection. Clean up spilled oil or residue present on the components to prevent damage to paint and cable insulation. Cleaning oil residue will also eliminate dust accumulation and associated problems. **SAFETY: Refer to the gear-oil MSDS for appropriate PPE, handling, and cleanup requirements.**

Maintenance activities for the gear motor assembly include torque inspection of mounting hardware, inspection of the tacho-generator and rotary encoder, along with lubrication of the motor bearings. A **tachometer-generator (tacho-generator)** is a small DC generator used to provide a feedback signal to the motor drive control for rotation speed and direction. The output signal level from the tacho-generator will be proportional to the rotational speed (RPM) of the gear drive. The feedback signal level is used as an analog input by the motor drive control to monitor and adjust motor speed. The polarity of the analog signal is used to determine the motor rotation direction. Rotation of the DC generator in one direction will produce a positive voltage, and rotation in the opposite direction will produce a negative voltage. A **rotary encoder** is an electronic feedback device used by control electronics, such as a PLC, to monitor shaft angular position, distance traveled, and speed. Encoder technology may include contact or noncontact sensors. Encoder contact technology uses brushes and a segmented rotary assembly to produce a coded digital signal that can be interpreted by control electronics for distance traveled, angular position, and speed. Encoder noncontact technology may use a Hall-effect sensor similar to a proximity switch or an optical signal detector assembly. **Table 14-1** shows examples of wind-turbine applications that may use these technologies. An optical encoder may use a metal, glass, or plastic disc etched or machined with lines or slits that are positioned between a light source such as a light-emitting diode and a photodiode receiver. The optical device converts the presence or absence of light as the encoder rotates into a coded digital signal that control electronics can interpret as distance traveled, angular position, and

TABLE 14-1

Wind-Turbine Applications for Noncontact Sensors or Encoders	
Device	**Application**
Proximity switch	Safety interlock detection
	Park brake pad wear
	Park brake status
	Main shaft speed
	Gearbox output shaft speed
	Louver position
	Damper position
	Blade limit positions
	Anemometer
Rotary encoder	Blade-pitch position
	Generator speed
	Anemometer
	Yaw vane

© Cengage Learning 2013

speed. This is the same type of signal output as a contact device, but it is produced with a noncontact technology. **Figure 14-14** shows examples of rotary encoders and a view of some internal components. For more information on motor drive control and feedback devices used on site, refer to the turbine manufacturer's documentation. A torque inspection for mounting bolts that secure the gear motor assembly may be completed with a manual torque wrench or power-bolting system. Consult the manufacturer's maintenance procedure for the recommended torque value or reference a fastener torque chart for the size and material grade used with the application. These bolts are located around the base of the gear motor that secure the assembly to the hub.

Inspection of the tacho-generator or encoder brush assemblies should only be performed by a qualified technician. These brushes are very small in size and may be damaged easily if not handled properly. Consult the manufacturer's procedure for access to the brush assemblies, inspection criteria, and recommended replacement schedule. **CAUTION: Do not attempt this procedure without spare brushes in stock. A damaged brush will prevent returning the wind turbine back into service.** Note inspection findings and components used on the PM checklist for follow-up as necessary. Listing components used for the PM cycle will ensure an accurate adjustment to the site inventory.

Lubrication of the DC motor assembly bearings may be another necessary activity for completing the gear motor maintenance. Refer to the manufacturer's procedure for recommended lubrication grade and the quantity required for the motor bearings. *Note*: Some motor assemblies are designed with sealed bearings that will not require scheduled maintenance. A visual inspection of the motor bearings will indicate their condition. Inspect the bearings for corrosion, discoloration because of elevated operating temperatures, excessive runout, and accumulated wear particles. Each sign indicates that the bearing may be damaged and needs

FIGURE 14-14 *Examples of rotary encoders and a view of some internal components*

replacement. Report damaged bearings to the farm operations team for assessment and replacement if necessary. Make note of findings on the PM checklist for follow-up as necessary.

Inspection of Pitch Gears

The gear motor drive assembly has a spur gear mounted to the end of the gearbox shaft that engages the large ring gear mounted to the root of the blade. Rotation of the gear motor spur gear causes the ring gear to rotate on the pitch bearing moving the blade assembly. **Figure 14-15** shows an example of the gear motor spur and ring gear. Visual inspection of gears should include verification that no mounting fasteners are missing or loose and that the gear teeth are in acceptable condition. Replace any missing fasteners used to mount the ring gear and tighten loose fasteners to the recommended torque level listed in the manufacturer's procedure or site-approved torque chart. Inspection of the gears should include verification of no damage or foreign material wedged between teeth. **SAFETY: Inspection of the gears should be completed using a site-approved LOTO procedure. Accidental running of the pitch assembly such as during an emergency stop would create a pinch or amputation hazard if a technician had a body part jammed between the gears.** Wipe excess grease from the gear teeth and check for cracks, missing teeth, and excess wear. Foreign materials such as loose nuts, washers, and blade resin pieces that are stuck to the gear lubricant can damage the gears if they become wedged between the spur gear and ring gear during operation. Apply a fresh coating of approved grease to the gear teeth after the visual inspection with a brush. **SAFETY: Refer to the lubricant MSDS for appropriate PPE, exposure limits, and cleanup requirements. The site's hazardous-waste policy should provide guidance for disposal of waste grease and soiled rags.** Consult the manufacturer's procedure for recommended type and grade of lubricant.

FIGURE 14-15 *Gear motor spur and ring gear without the blade attached*

Inspection of Hydraulic Pitch Components

Inspections of the hydraulic pitch components are similar to those used for the hydraulic park brake discussed in Chapter 12 on drive-train maintenance. Typical components for this system may include hydraulic cylinders, hoses, fittings, a power unit, valves, accumulators, and a reservoir. Inspection of components should include a visual check for leaks, reservoir oil level, bearing condition, wire connections, and loose or missing mounting hardware. Inspection of hoses and lines for the pitch system should include a check for leaks at fittings and crimp connections, abrasion of the lines, and corrosion. Oil dripping from fittings and crimp connections should be cleaned with an approved cleaner, and the soiled rags should be disposed of according to the site hazardous-waste policy. Check these areas to determine if the fittings require tightening or if there is damage. Tighten the fitting as required to eliminate the leak. If the fitting or hose crimp is damaged, make note of the issue on the PM checklist so it may be replaced on the next schedule service call.

Inspect hoses or lines for abrasions or cuts. Any abrasion of the hose outer jacket that exposes the reinforcing layer can cause a hose failure. Replace the hose assembly if necessary. If no replacement hose assembly is available, then secure the assembly to prevent further damage and advise the farm operations team of the damage for follow-up. Inspect fittings and lines for excess corrosion that can cause the components to fail. A small amount of surface corrosion typically will not cause an issue but should be noted for follow-up. Check for excessive corrosion that has created a significant loss of metal. This level of corrosion may have deteriorated the fitting or line to the extent that it could fail during system operation. Replace the component if a replacement is available or notify the farm operations team to follow-up and make corrections as necessary.

Inspect the actuator mounting assembly for loose or missing hardware. Replace any missing hardware and tighten any components or fasteners that have become loose. Inspect the actuator mounting assembly at both the cylinder cap and rod end connections. **Figure 14-16** shows

FIGURE 14-16 *Hydraulic pitch actuator*

an example of a hydraulic pitch actuator. Wipe off any excess lubricant and inspect the bearing assembly. Check for excess motion, wear, and corrosion. Lubricate the bushing assemblies as recommended by the manufacturer's procedure. Actuate each pitch control and observe that the actuators, linkages, and bearing assemblies move without binding. **SAFETY: If LOTO was performed during a previous step, then this procedure will require removal of the LOTO and reenergizing the system. Notify each technician working around the rotor assembly of the pitch activity and follow the site's safety requirements for the procedure.** This procedure should be one of the last activities before exiting the hub. This inspection may also be accomplished while the pitch bearings are being lubricated. Lubrication of the pitch bearings will be discussed in a later section. Note findings on the PM checklist for follow-up as necessary.

Inspection of Hydraulic Power Unit

Inspection of the hydraulic power unit should include a check for wire connections on valves and sensors along with a check for leaks between valves and the manifold block. Ensure that any wire connections and harnesses on the hydraulic power unit are secure to prevent intermittent signal issues during operation. **SAFETY: Remove power from the electrical circuits before performing this inspection to prevent a possible short circuit or electrical shock hazard. The operating voltage for the valve coils may be from 24 V_{DC} up to 120 V_{AC}.** Refer to the manufacturer's wire schematic or service manual for details on the power-unit operation. Follow the site's safety procedure for required training, PPE, and tools. Inspect the power-unit reservoir for proper fluid level using the sight glass or sight tube. Replenish the reservoir oil level as necessary with approved oil. Refer to the manufacturer's service specifications for the correct type and grade. Note finding on the PM checklist and include the oil quantity added to ensure an accurate inventory adjustment.

HUB PLC, COMMUNICATION, AND CONTROL COMPONENTS

The cabinet enclosing the PLC unit, input and output modules, communication module, control devices, and terminal connections is typically located on the hub bulkhead toward the nacelle. The components in this cabinet act as the coordination center between the wind-turbine control system located in the down tower assembly or nacelle top box and the blade-pitch control system. The wind-turbine control system sends signals through a communication link to the hub PLC communication module to adjust the blades to a specified value. **Figure 14-17** shows an example of a slip ring

FIGURE 14-17 *Slip ring used to connect nacelle PLC to hub PLC*

that would be used to connect the communication link between the hub and nacelle PLC.

A **communication link** consists of small copper conductors or optical fibers that transfer analog or digital signals between control components. Communication link between the nacelle and hub is typically constructed of small copper conductors. The hub PLC in turn sends the required control signals to the pitch controller. The pitch controller converts this command into a signal the individual motor drive units can use to position the blades as required. The motor drive unit commands set gear motor speed, direction of rotation, and final angular position. Feedback to the drive unit from the gear motor assembly ensures that the action is completed within specified parameters such as acceleration, deceleration, speed, and final angular position. After the pitch change is successfully completed, the hub PLC communicates with the nacelle PLC indicating the operation is complete. This feedback process enables the wind-turbine PLC to continue operation. If something in this process is performed outside the specified parameters of blade position, travel time, or activation of any limit switches, the hub PLC will communicate an error code to the wind-turbine PLC and wait for further commands. If the system can diagnose the problem and reset the issue, it will return to service. If it cannot correct the issue, it will remain out of service and send an error code to SCADA for follow-up by a service technician. For all of this to happen correctly, each component has to function properly and electrical connections have to be secure. Any loss of power or communication signal will halt the process. Inspection of all electrical connections to PLC modules, contactors, terminal strips, chargers, overvoltage modules, relays, circuit breakers, transformers, and so on is very important.

SAFETY: Follow the site safety procedure for LOTO activities and PPE required before opening any control cabinet. Follow the maintenance procedure for requirements on torque specifications for fasteners used to mount components. Inspect wire and cable trays for loose or missing covers. Any loose items in this cabinet can cause damage during operation just as the axis cabinets discussed previously. Note findings on the PM checklist for follow-up as necessary.

INSPECTION OF POWER CABLE, GROUNDING, AND BONDING CONNECTIONS

The umbilical assembly from the nacelle to the hub cabinet also consists of larger conductors to supply power and grounding circuits for hub systems. These conductors are terminated to terminal strips in the hub main cabinet for ease of system assembly. Terminal connections are also a great asset when connections need to be changed or components replaced during service activities. There is a downside to these terminal strips: wire connections can loosen from the terminal strips and cause intermittent power loss.

One of the most challenging field service activities is troubleshooting an intermittent problem. Eventually the problem may be solved by checking and tightening each wire connection in the circuit. This does not always work because there may be loose wire connections or contacts within a sealed component. Ensuring that all terminal connections are secure is an important step in the troubleshooting process. One preventative measure to eliminate intermittent problems caused by loose wire connections is to inspect and tighten wire connections during each PM cycle. **SAFETY: Deenergize and LOTO power circuits before inspecting terminal connections in the slip ring and hub cabinets.**

Inspect grounding and bonding connections in hub cabinets and on components to ensure each is secure. A lightning strike to a blade is not a good time to find out whether ground circuits are secure. Passing current from a blade lightning receptor to the tower foundation grounding assembly requires a series of good ground cable connections. Note inspection findings on the PM checklist for follow-up as necessary.

INSPECTION OF SPINNER, DECKING, AND CABINET MOUNTING HARDWARE

Inspection of structural components used to support rotor or hub assemblies should be completed during each PM cycle. Check for loose or missing hardware that secures the spinner, access rails, access covers, interior decking, and cabinet mounting components. Replace any missing fasteners with the same size, grade, and finish. Using lower grade fasteners may break under operating loads and cause structural failure. Replacing galvanized or stainless-steel fasteners with unfinished steel fasteners can create a corrosion issue. Rust formation on fasteners will cause them to decrease in strength, make them difficult to remove, and create rust stains. Rust stains on exterior surfaces of a wind turbine will not impress the customer or neighbors. Tighten any loose fasteners as recommended by the manufacturer's specifica-

tion. Note inspection findings and fasteners used on the PM checklist for follow-up.

INSPECTION AND MAINTENANCE OF BLADE-PITCH BEARINGS

Inspection and maintenance of blade-pitch bearings includes an inspection of surfaces exposed on the interior of the hub along with areas on the exterior. A visual inspection should verify that required seals and covers are present and secure. Missing seals and covers allow moisture to enter the bearing, causing lubricant contamination and corrosion. Moisture contamination of the lubricant will degrade the lubrication properties and cause damage to the moving elements and the race assembly. Wipe off accumulated dust and excess lubricant from exposed bearing surfaces to allow inspection for damage, metal particles, and corrosion. Note inspection findings on the PM checklist and alert the farm operations team of issues that need immediate attention. Issues that require immediate attention include significant rust accumulation, metal or rust particle buildup on the bearing or moving elements, breakdown of the lubricant because of water contamination, and structural damage. Maintenance of the blade-pitch bearings includes addition of lubrication and inspection of mounting fasteners.

Lubricating Blade-Pitch Bearings

The blade-pitch bearings are major components of the rotor assembly because they support the weight of the blades and allow them to be pitched to improve power output during operation. Blade-pitch bearings may have a series of grease fittings around the assembly or a single manifold that feeds multiple locations around the bearing. Lubricating each pitch bearing should include an initial visual inspection of grease fittings, manifold assemblies, and lines connecting the manifold to the bearing. Missing fittings or damaged lines should be repaired or replaced before lubricant is applied. Consult the manufacturer's procedure or specifications to ensure the proper lubricant type, grade, and the quantity to be used.

A recommended practice for lubricating pitch bearings is to slowly rotate each blade while pumping the lubricant into the fittings or manifold assembly. This practice will ensure a uniform distribution of the lubricant around the bearing. Rotating each blade during the lubrication process can be accomplished by using the manual override commands of the blade-pitch system. **SAFETY: The pitch bearing lubrication procedure requires a minimum of two technicians to complete—one to operate the pitch controls and another to**

operate the grease gun. Ensure that clear communication is maintained between technicians and that others in the nacelle and hub are aware of the activity. Blade-pitch–bearing lubrication manifolds and fittings may be located on the exterior or interior of the hub, depending on the wind-turbine model. Applying lubricant to fittings located on the exterior of the hub requires precautions associated with working at heights. Follow the site's safety policy for the use of a full-body harness, a twin-leg lanyard, gloves, safety glasses, and other requirements during lubrication activities on the nacelle and hub. Consult the lubricant MSDS for appropriate PPE, first-aid practices, cleanup, and disposal of waste products. Note inspection findings, lubricant quantities consumed, and components replaced on the PM checklist for further follow-up and accurate adjustments to the site inventory.

Torque Inspection of Blade Bearing Fasteners

Inspection of the blade bearing assembly may require work on the outside of the hub. Inspection and torque verification of the fasteners on each blade bearing will require rotating the hub to gain access to each side of the blades. **Figure 14-18** shows a rotor assembly in a rabbit-eared position. The figure shows that half of the blade bearing fasteners for two blades can be accessed in this position. To access the blade fasteners on the opposite side, the rotor will have to be rotated 120° to the next position. This process would have to be followed again for the last set of fasteners. **SAFETY: Rotor lock LOTO should be performed each time a technician climbs out on the hub. Rotating the hub for each of these inspection and torque activities will require that the technician return to the nacelle, remove the rotor lock LOTO, manually release the park brake, set the park brake after the rotor assembly moves to the next position, and apply the rotor lock LOTO before returning to the hub.**

The torque inspection process for blade bearing fasteners will require power-bolting tools. Consult the wind-turbine manufacturer's specifications for fastener torque value,

FIGURE 14-19 *Torque-tool verification test stand*

number of fasteners to inspect (10% or 100%), and acceptance criteria. Many turbine manufacturers require that power-bolting tools be verified on a test stand at each setup or with a change in tooling. **Figure 14-19** shows an example of torque-tool verification test stand. **SAFETY: Power-bolting tools should be secured to a technician's body harness using a lanyard and carabineers. This precaution will ensure that the tools will not travel any farther than the lanyard if they are accidentally dropped. Dropping an eight-pound tool several feet against a blade or nacelle may cause damage.** A lanyard restraining the tool will minimize impact damage and help with tool handling. Complete the torque inspection for the specified number of fasteners and document the results on the PM checklist. If the test sample fails to meet the manufacture's acceptance criteria, then complete a 100% inspection of all blade bearing fasteners. This step completes the inspection for the outside portion of the blade bearing for the rotor assembly. The last blade bearing assembly PM activity is an inspection and torque verification of the fasteners holding the blade into the pitch bearing.

Inspection and Maintenance of Blade Bolts

Inspection and torque verification of the fasteners securing the blade to the pitch bearing are located within the hub. Access to these fasteners depends on the layout of the pitch-control cabinets. The end of the blade is easily accessed with some hub designs but is limited in others. **Figure 14-20** shows a hub design with easy access to the blade end, and **Figure 14-21** shows a hub with limited access to the blade ends. Inspection of the blade mounting fasteners should include a visual for missing hardware or damaged components. Torque verifications of the blade fasteners should follow the manufacturer's guidelines. These fasteners are locked into the blade assembly, which may include polyester resin, reinforcing fibers, and laminated wood. These materials are

FIGURE 14-18 *Close-up of rotor assembly in rabbit-eared position showing fasteners mounting the blades*

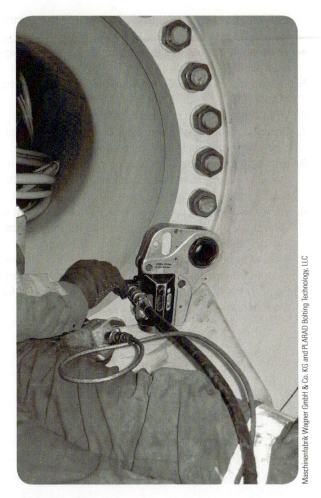

FIGURE 14-20 *Hub design with easy access to blade end*

SAFETY: Access to the end of the blade may require that a technician position a portion of his or her body below the pitch-drive spur and ring-gear assembly. Follow the site-approved LOTO procedure to ensure that any pinch and amputation hazard is eliminated. This activity may require multiple trips to the hub for completion, so follow site safety procedures to ensure everyone's safety. The author suggests a technician don a pair of coveralls or a disposable painter suit to prevent contaminating work clothes with grease from the ring gear if necessary. When the activity is completed, removing the coveralls will prevent contamination of cold-weather clothing and safety gear that will be donned at the end of the day. Follow the manufacturer's specification for number of fasteners to be inspected, torque value, and acceptance criteria. Document all inspection findings on the PM checklist for follow-up later as necessary.

RECOMMENDED CLEANING ACTIVITIES

Recommended cleaning activities for the hub should include wiping grease splatter from interior surfaces and cleaning any lubricants that may have been spilled. The hub has limited ventilation so take care to use only a minimal amount of approved cleaner. **SAFETY: Refer to the cleaner MSDS for appropriate PPE, first-aid requirements, exposure limits, and disposal of soiled rags. Cleaning is typically the last activity before closing up the hub and returning to the nacelle. Take this time to verify that cabinet covers and any safety straps are secure. Ensure that tools, spare hardware, and waste items are packaged up and removed to the nacelle.** The last items on the hub-entry checklist should include a check for loose items along with verification of the SCADA system for errors associated with hub systems.

Make sure all hub systems reset and are ready after closing all cabinets and removing LOTO from power circuits that were applied previously. An extra few minutes now may save hours of work later if something is left behind to tumble within the hub after the wind turbine is brought online. Close up the hub access hatch and return to the nacelle. Complete any open items on the hub-entry checklist and remove the rotor lock LOTO devices.

SUMMARY

The rotor assembly has many PM activities necessary to ensure the hub subsystems operate properly. These activities will vary, depending on the pitch-drive design used for the specific wind-turbine model. It is important that the manufacturer's specifications and maintenance procedures are followed to ensure the systems function as designed during operation. Following the site safety policy will ensure that PM activities will be completed with limited or no

not forgiving when it comes to exceeding the required torque value. A recommendation for this PM activity is to perform the torque inspection on a blade when it is orientated in the hub bottom position. This is another activity that may be coordinated with the inspection of the gear motor oil-level check.

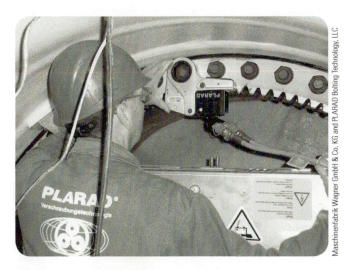

FIGURE 14-21 *Hub with limited access to blade ends*

Maschinenfabrik Wagner GmbH & Co. KG and PLARAD Bolting Technology, LLC

exposure of employees to hazards. Integrating PM activities and safety considerations onto a comprehensive checklist will ensure that no activities are missed and safety risks are managed properly. The morning tailgate meeting is a great time to make sure all of these things come together. Review checklists, MSDS, tool kits, safety devices, safety supplies, procedures, and consumables before leaving the service center to ensure that everything is available for the day. Using this time to coordinate activities and personnel resources will also reduce the overall length of the activity.

REVIEW QUESTIONS AND EXERCISES

1. List several maintenance activities associated with the rotor assembly.
2. List hazards for which you should be prepared to ensure a safe hub entry.
3. List two battery parameters checked for an electric pitch-drive system.
4. What is the leading cause of intermittent electrical signal issues?
5. List attributes that OSHA uses to define a *confined space*.
6. What is the function of a rotary encoder?
7. Describe issues associated with operating a wind turbine with improperly secured components in the hub.
8. List safety considerations for torque inspection of blade bearing fasteners located on the exterior of the hub.
9. What maintenance procedure ensures that lubricant is spread uniformly around a bearing during the application process?
10. Review the hub-entry job-safety analysis and list preventative measures necessary for an internal hub entry. Refer to the blank JSA form listed with Appendix J for more details.

15

External Surfaces

KEY TERMS

aerodynamic force (F_A)
aerodynamic pressure (P_A)
angle of attack
blade root
delamination
drag
Gelcoat
high-pressure region
lift
lightning receptor
low-pressure region
machine head

meteorological (met) mast
nacelle
notice to airman (NOTAM)
parasitic drag
projected area
rime ice
rotor assembly
structural drag
thrust
torque
tower

OBJECTIVES

After reading this chapter and completing the review questions, you should be able to:

- Describe different exterior inspections associated with a wind turbine.
- List safety issues associated with activities on the nacelle assembly.
- Describe the difference between lift and drag.
- Describe issues associated with blade-surface contamination.

- Describe blade defects that may be observed during a visual inspection.
- Describe service and maintenance activities that may require a man basket or suspended personnel platform.

INTRODUCTION

Previous preventative-maintenance (PM) chapters covered maintenance recommendations for systems located within the wind turbine. Wind-turbine designs typically use components to reduce or eliminate direct exposure of control and generation equipment to the environment. The focus of this chapter will be activities used to assess and maintain tower, spinner, nacelle, and blade exterior surfaces that are used to protect wind-turbine systems. These components provide structural support, protection from the environment, improved aerodynamics and aesthetics, along with the ability for the turbine to capture power from the wind. One of the largest wind-turbine structures is the tower assembly.

Tower

The **tower** is designed to position the rotor for optimum power capture height and provide structural support for the machine head. The **machine head** consists of the nacelle, rotor, and interior systems used for control and power generation. Tower designs have included self-supporting lattice structures and monopole (tube) assemblies, depending on the wind-turbine size or manufacturer's requirements. **Figure 15-1** shows examples of a couple self-supporting lattice-tower designs.

Over the past couple of decades, wind-turbine tower designs have shifted from self-supporting lattice structures to a tube style. Technicians may have an opportunity to work on lattice towers if they work at a farm where they are installed. Components for lattice towers were readily

FIGURE 15-2 *Tube tower under construction*

FIGURE 15-1 *Self-supporting lattice-tower designs*

available, easy to ship, and could be assembled on site at remote locations. Improvements in material handling and fabrication of large components have led to the replacement of lattice towers by large tube towers with recent community and utility-scale wind-turbine designs. Tube-tower sections may be formed, machined, assembled, and finished

in a controlled production environment to improve quality and reduce overall tower cost. Construction of a tube tower can be completed within a day with a couple of large cranes and the cooperation of acceptable weather conditions. Tube towers may be constructed of multiple tube assemblies or of interlocking shell sections that make up several cylindrical sections. The shell sections may be fabricated from steel or precast reinforced concrete. These designs reduce the logistical issues associated with transporting long tube sections over roadways to remote sites. **Figure 15-2** shows an example of a tube tower under construction. Other benefits of a tube-tower design include protection of control systems located at the base of the tower and protected access for service technicians from the environment during access to the nacelle. The nacelle, spinner, and tube tower also improve aerodynamic flow around the wind turbine.

Nacelle

The **nacelle** is the housing used to enclose the drive train, generator, and other up-tower control systems. Without these smooth transition shapes, components such as the drive train, generator, and control cabinets would expose multiple surfaces to the oncoming wind that would increase the aerodynamic drag force on the wind turbine compared to a streamlined nacelle assembly. The nacelle assembly also protects the wind-turbine components from exposure to rain, dust, temperature extremes, and other environmental hazards. Enclosing the wind-turbine generator within a nacelle reduces sound emissions from the drive train and generator to the surrounding area and creates an aesthetically appealing finished product. The typical paint scheme for wind turbines is a flat white color that enables them to blend with clouds and reduce the color contrast with the sky. Some manufacturers have gone further and painted lower tower sections with stripes of different shades of green to enable the towers to blend with surrounding grass, shrubs, trees, and other landscape features. Protection of the components, improved aerodynamics, and aesthetics are important for wind turbines, so maintaining these surfaces from the deterioration of environmental exposure becomes an important step in the overall PM process.

Capturing the Wind

The wind-turbine **rotor assembly** is designed to convert mechanical power from the wind into an electrical output. The rotor assembly includes the hub, spinner, blades, and all of the enclosed systems used to control blade pitch. The electrical output process will remain efficient only through diligence, proactive service, and a maintenance plan. A good rotor-assembly maintenance plan should include inspections to ensure blade structural integrity along with activities to maintain clean, smooth exterior surfaces and the aerodynamic profile. Wind-turbine blades are designed to enable the airflow around them to create reaction forces perpendicular and parallel to the oncoming wind. The reaction force perpendicular to the oncoming wind is considered **thrust**, and the force parallel and away from the wind is considered **drag**. The amount of thrust and drag created by an object are a function of shape, surface area, and wind velocity. A thin, flat panel positioned or pitched perpendicular to the wind will create drag without any thrust force. The same flat panel pitched parallel to the wind will create a small amount of drag but no thrust. To create both thrust and drag, the panel must be pitched at an angle to the oncoming wind. **Figure 15-3** shows examples of these three pitch scenarios.

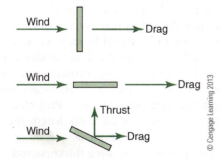

FIGURE 15-3 *Three pitch scenarios*

Because these forces result from the interaction of an object within a moving fluid (wind), they are considered aerodynamic forces.

Aerodynamic Forces

Previously in the book, we described the relationship between wind-turbine instantaneous power captured, swept area, and wind velocity as

$$P_I = \frac{1}{2}\rho A v^3 \eta$$

where ρ (Greek letter rho) represents the density of air (kg/m³, lbs/ft³), A represents the swept area of the rotor assembly (m², ft²), v represents the velocity of the oncoming airstream (m/s, ft/s), and η represents the overall efficiency of the wind turbine. A variation of this formula may be used to determine the aerodynamic force acting on an object positioned in an airstream as

$$F_A = P_A(A)$$

where P_A represents the aerodynamic pressure (psi, pascal, etc.) and A represents the surface area exposed to the airstream (ft², m²). A word of caution: the variables for instantaneous power (P_I) and pressure (P_A) are both represented by the letter P, so it is important to understand the proper use of these equations.

Aerodynamic force (F_A) is a function of aerodynamic pressure created by wind impacting the surface of the object. This force may be calculated using the aerodynamic pressure caused by wind impacting the blade and the blade's exposed surface area. **Aerodynamic pressure** (P_A) is a function of the adjusted air density and the wind velocity squared ($P_A = \frac{1}{2}\rho v^2$). *Adjusted air density* refers to the change in density because of elevation and temperature. Increasing elevation from sea level and an ambient temperature above or below the standard temperature of 15 °C (55 °F) will change the air-density value. An increase in elevation from sea level will decrease air density and so will an increase in air temperature above 15 °C. Decreasing the air temperature below 15 °C will increase the air-density value. Appendix K shows the value of air density with respect to variations in temperature and elevation. This is why cold air blowing in the winter will have more available power than the same velocity wind on a hot summer day. When calculating aerodynamic pressure, ensure that the units of air density are consistent with the units of velocity. For example, use air density in kg/m³ when using wind velocity in m/s.

Thrust, Lift, and Drag

The thrust portion of the aerodynamic force creates a torsion force around the axis of the rotor shaft, which causes rotation of the rotor assembly. The drag portion of aerodynamic force causes the blades to bend away from the oncoming wind. Drag caused by the rotor assembly, nacelle, and tower also causes the tower to bend away from the oncoming wind and sway during changes in the wind. If you have ever spent time working in a tower, you have probably experienced this motion.

© Cengage Learning 2013

SAMPLE PROBLEM

Determine the aerodynamic force created by wind striking a flat panel with area 25m² positioned perpendicular to the oncoming wind. Wind velocity is 12 m/s; use standard atmospheric conditions of sea-level elevation and air temperature of 15 °C. Air density (ρ) = 1.225 kg/m³.

SOLUTION

Aerodynamic pressure (P_A) = ½ρv^2. Units of air density (ρ) are in mass/volume, and the final answer needs to be given in pressure, which would be force/area. Relevant pressure units in the SI system are given in pascals (Pa) or kilopascals (kPa).

$$1 \text{ Pa} = 1 \text{ newton/m}^2$$

$$P_A = \frac{1}{2}(1.225 \text{ kg/m}^3)(12 \text{ m/s})^2$$

$$= \frac{1}{2}(1.225 \text{ kg/m}^3)(144 \text{ m}^2/\text{s}^2)$$

This looks like alphabet soup, but we can simplify the equation by substituting the units: kg(m/s²) = newton (N), and m/m³ = 1/m². Final units would then be N/m² or Pa.

$$P_A = \left[\frac{1}{2}(1.225)(144)\right] \text{N/m}^2$$

$$= 88.2 \text{ N/m}^2$$

$$= 88.2 \text{ Pa}$$

$$= 0.0882 \text{ kPa}$$

Aerodynamic force (F_A) = $P_A \times A$,   A = 25 m²

Units of force are in newtons, so continue to use N/m² to have the equation work:

$$F_A = 88.2 \text{ N/m}^2 (25 \text{ m}^2)$$

$$= \textbf{2,205 N} \text{ or 496 pounds of force}$$

This is why a tornado or hurricane wind can remove relatively heavy material such as wood or sheet metal from a structure with ease and toss it hundreds of feet into the air.

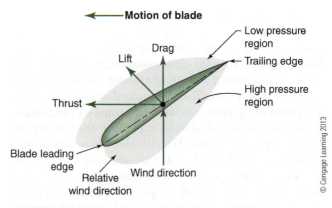

FIGURE 15-4 *Blade showing reaction forces, variations in air pressure, wind direction, and other forces*

rotational speed (RPM), and the numeric constant (9549) were used in previous chapters to relate power (kW) for the rotor-assembly input and generator power output. Varying the angle between the blade center line and the oncoming wind will change the values of thrust, lift, and drag. The angle between the blade center line and oncoming wind is referred to as **angle of attack**. Using blade-pitch adjustments, the wind-turbine control system can maintain consistent generator power output during variations in wind speeds. **Figure 15-4** is a diagram of a blade showing reaction forces, variations in air pressure, and wind direction, along with other useful terms. Rotation of the rotor assembly creates forward motion of the blades through the air. Airflow around moving blades creates a **low-pressure region** on the downwind side or side away from the oncoming wind. The region around the upwind surface exposed to the oncoming wind is referred to as the **high-pressure region**. The resultant **lift** force created by the high- and low-pressure regions around the blade will supplement the thrust generated by the initial wind force that started the process. This phenomenon is the reason why an airplane wing creates lift as it travels through air and why blades are often referred to as *wings*. The combination of these two forces improves the efficiency of a lift-style blade compared to the flat-panel, drag-style blade of the 1930s vintage farm windmills.

Air flowing around a blade or wing not only creates lift but also creates drag. In general, drag is the resistance of a fluid (liquid or gas) to flow around an object. Drag may be structural or parasitic in nature. **Structural drag** (or *profile drag*) is created by air flowing around objects such as the nacelle, meteorological instruments, tower, or blades. The amount of profile drag created by an object is a function of its shape and the square of the fluid velocity. For example, a flat object held parallel to an oncoming wind stream will create more profile drag than a symmetrical object such as a wing with the same projected area. **Projected area** is the cross-sectional product of an area's length multiplied by height exposed to the wind. For example, these objects may have an area created by the same thickness and length, but the smooth transition of the symmetrical object will create

This swaying motion produces the same sensation as being in a boat or ship on open water. **SAFETY: Wind-turbine towers may sway a meter or more with moderate winds, so take precautions if you are prone to motion sickness.**

This swaying action is compounded if the wind turbine is attached to a floating platform located on a large body of water such as the Great Lakes or located several kilometers offshore. On climbing down from a wind-turbine tower after several hours of work, you will notice your balance is off. Use caution until your inner ear adjusts and your balance improves. Torsion force or **torque** is the product of force applied at a distance (i.e., 100 N-m). Typical units of torque are given in newton-meters or foot-pounds. Remember: torque,

less drag. Like friction between two components in contact, drag can be used to benefit or create inefficiencies in a system. Wind-turbine rotor-braking systems such as tip brakes, pitchable tips, and full blade-pitching mechanisms rely on profile drag to stop or slow down.

Parasitic drag (or *skin friction drag*) is resistance to flow created by the amount of contact area exposed to the moving air. This contact area may be referred to as *wetted surface* in some fluid dynamics texts or mechanical engineering reference manuals. Use the mental image of an object being immersed in moving water. The water "wets" all surfaces exposed in the same manner as the air in contact with the object. Decreasing the amount of contact area such as decreasing the width and maintaining the length of the blade or wing will decrease the amount of parasitic drag. Parasitic drag will also increase because of blade-surface imperfections such as Gelcoat erosion, bumps, blisters, surface delamination, and contaminant buildup. Each imperfection will disrupt the smooth flow of air over the blade surface.

A dramatic example of parasitic drag is created by the accumulation of a rime ice layer on the surface of blades. **Rime ice** is formed when supercooled drops of water freeze on contact with a cold blade moving through a cloud or mist. Rime ice has the appearance of small pellets stuck together that create a rough, milky white color. Watching data derived from a supervisory control and data acquisition (SCADA) system during the formation of rime ice will show a dramatic decrease in wind-turbine output even though wind speed does not change. Accumulation of contaminants such as ice also adds weight to the blade that can cause structural damage if the wind turbine is not removed from operation until the ice melts and drops. Allowing ice to build up over the entire length of a blade leading edge may add 1,000 pounds or more to its weight. **SAFETY: Never approach a wind turbine while it is shedding ice. Falling ice can cause serious injury or death. Many pieces may be in excess of 100 pounds.** **Figure 15-5** shows an example of rime ice built up on the

meteorological (met) mast of a wind turbine after a freezing rain shower. This same buildup will form on the external surfaces of the tower, nacelle, and blades. Blade modifications with items such as specialized coatings or electric heater films have been used recently to eliminate ice buildup in an effort to reduce the associated drag and possible structural damage.

Blade-surface defects and buildups of airborne contaminants can take longer to occur, but they can bring on similar reductions in wind-turbine power output. Airborne contaminants include bug splatter, dripping lubricants, dust, salt spray, and other materials. Technicians need to be aware of these issues so they can be corrected before they become structural or power-output issues. **Gelcoat** is the thin finish layer of polyester or epoxy resin that is bonded to the blade, spinner, or nacelle during the molding process. **Delamination** is the separation of bonded layers such as the outer finish coat to the polyester or glass fiber composite layers. This thin layer is used to give the exterior surfaces a smooth, protective finish against moisture, ultraviolet light, and other environmental issues. The outer layer may also include colored pigments to visually enhance the finished product. Gelcoat technology has been used successfully for many years with marine and aircraft applications, so PM and repair techniques for these applications can also be used on wind turbines to detect, prevent, or correct problems. Difference between wind-turbine PM activities and these applications is the elevation above terrain where the activity may be preformed. Quite often it is too expensive to bring wind-turbine components to the ground for cleaning or repair, so they are worked on in place.

Figure 15-6 shows an example of a blade inspection and repair being performed by certified technicians. It takes specialized training and a true desire to hang in a suspended basket or on the end of a rope to perform this type of work. If you are interested in rock climbing, repelling, and high-adrenaline activities and have a great work ethic, this type of career may be for you. Other methods to perform external repair and maintenance activities include a suspended

FIGURE 15-5 *Rime ice built up on met mast of wind turbine after a freezing rain shower*

FIGURE 15-6 *Blade inspection and repair*

First Wind Energy, LLC

Spider

FIGURE 15-7 *Suspended personnel platforms attached to crane load line, boom, and cables attached to rotor or nacelle*

personnel platform attached to a crane load line, a boom, or a cable attached to the rotor or nacelle. **Figure 15-7** shows examples of these methods. Use of a personnel platform is covered under the OSHA construction subsection for cranes, derricks, and construction on lifting personnel (29 CFR1926.1431). This subsection covers regulations on certification, construction, inspection, weather conditions; load-testing criteria; and prelift meeting requirements. Refer to OSHA regulations (http://osha.gov/index.html) and the site's safety plan for approved use and safe operating guidelines.

The following sections will present various inspection and PM activities recommended for external surfaces of the wind-turbine assembly. Wind-turbine manufacturers typically list required exterior inspection activities and schedules in their maintenance manuals, so follow their recommendations and use this text as a guide. Our first discussion will focus on the maintenance activities recommended for the blade exterior surfaces. Access to these surfaces is limited for typical PM cycles, so it is important to use appropriate safety precautions and equipment to observe and document possible issues.

INSPECTION OF BLADE SURFACES

Inspecting blade surfaces is not necessarily an easy task for each PM cycle. The hub is located 80 to 100 meters (260 to 325 feet) above the tower foundation. Even with each blade rotated to its lowest position for inspection, the tip will still be located 40 meters (130 feet) or more in the air. Typical blade inspection may be accomplished in a two-step process. First, the outer blade section can be inspected from the ground using a pair of binoculars or a telescope with a camera. Second, the inner blade section, including the blade root attached to the hub, can be inspected by a technician positioned on the top of the nacelle. The **blade root** is the structural element used to bolt the blade to the pitch bearing located on the hub. The blade root is manufactured of a steel, flanged thimble that is attached to the blade components during the final bonding process. Inspecting exterior surfaces of the blades near the hub can be accomplished while the rotor is stopped in the rabbit-eared position. Manually pitching each of the upper blades will allow for a 360° view around the blade exterior surface. A good time to complete this portion of the blade inspection is when the pitch bearings are being lubricated. The lubrication process requires manually pitching the blades, so why not save time and do the inspection at the same time. **SAFETY: Working from the top of the nacelle requires following the site safety plan for working at heights. This should include appropriate fall-arrest gear, observing maximum wind limits, tracking weather hazards, and using the proper number**

of technicians for the activity. All technicians located in or on the nacelle or hub should be notified of the blade-pitching procedure before initiation. This process will also require manual override of the pitch controls by one technician while other technicians are performing the lubrication and inspection activities. Clear communication between technicians is very important to ensure that no one is placed at risk of a pinching hazard. Surface defects of the blade assembly include cracking of the Gelcoat, delamination of the surface layer, lightning damage, damage because of impacts with ice, and buildup of contaminants. A brief summary of blade construction will be presented before discussing blade defects further. This may improve your understanding of how some defects may cause further damage if not corrected.

Summary of Blade Construction

There are several steps in the construction process of blades. These steps and the construction techniques may vary between manufacturers, but the processes yield a similar assembly. The major sections of the blade include a main spar, leading and trailing edge structures, and outer panels. **Figure 15-8** shows an example of the two outer panels positioned in mold halves. The spar and leading and trailing edge structures are constructed of laminated balsa wood, polyester resin, and reinforcing polyester fiber cloth to achieve strong yet light components. The outer panels of the blade are constructed using a molding process similar to that used with the boat-building and aircraft industries. The inner surfaces of the mold are coated with a releasing agent to enable easy removal of the assembly after the injection process. The inner surfaces of the mold are then sprayed with a thin coating (~0.125") of polyester formulation such as Gelcoat to provide a smooth,

FIGURE 15-8 *Two outer blade panels positioned in mold halves*

FIGURE 15-9 *Blade mold sections being closed for final assembly processing*

protective outer layer for the finished blade. Next, multiple layers of polyester or fiberglass cloth or both are laid up in a controlled orientation to ensure optimal strength of the finished panels. The last steps include closing the mold sections, pulling a controlled vacuum, and then injecting polyester resin to impregnate and bond the layers. **Figure 15-9** shows an example of blade mold sections being closed for final processing of the assembly. Blade assembly is performed by positioning the blade components into specialized fixtures, applying adhesives, and clamping until the adhesives are cured. Final steps in the blade-manufacturing process are to inspect the assembly for defects, make repairs as necessary, and add colored coatings or graphics as required by the customer.

Possible Defects

Defects encountered during the inspection process may include cracks in the outer coating, lifted or missing coating sections, scrapes, impact damage, environmental damage, or contaminant buildup.

Cracks in the coating may be produced by cyclical flexing of the blades during operation, oxidation of the outer coating from prolonged exposure to ultraviolet light, and temperature cycling. Minor cracking of the outer coating typically does not cause immediate issues, but it should be followed during subsequent PM inspections to ensure that surface defects do not become major issues. Minor cracks in the outer coating allow moisture to wick into the coating and possibly between the coating and composite layers. Moisture may cause swelling of wood or fabric layers and a subsequent separation. Moisture present in cracks and between layers will also be aggravated when it freezes during periods of cold

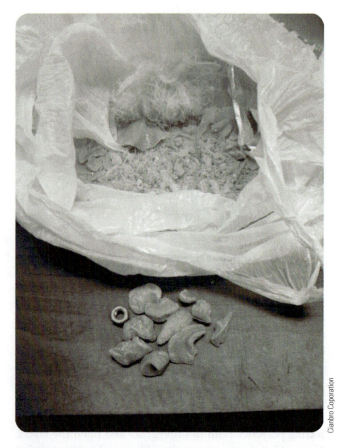

FIGURE 15-10 *Debris left in blade after assembly process*

FIGURE 15-11 *Lightning damage to blade tip*

weather. Expansion of the moisture during the freezing process can cause the layers to further separate.

Observation of areas that exhibit delamination should be brought to the attention of the farm operations group. Continued delamination of the layers may become a structural issue if not repaired by qualified technicians. Some blade designs include a drain hole in the tip of the blade to enable rainwater to drain. This drain may become plugged because of loose polyester resin particles tumbling in the blade or because of an ambitious insect that has constructed a nest. **Figure 15-10** shows an example of debris that may be left in the blade after assembly. If this hole is suspected of being plugged, then it should be cleared to prevent water accumulation that may swell wood structures within the blade or freeze and expand causing damage during cold weather.

Damage to the outer coating from impacts with falling tools or service cranes should be brought to the attention of the farm operations group immediately for review and repair if necessary. Impact damage from ice shedding should be noted and assessed for corrective action as soon as the risk of falling ice has passed. **Figure 15-11** shows an example of lightning damage to the tip of a blade.

Environmental damage to blades that may be observed include lightning strike, pitting of the leading edge, and contaminant buildup. Lightning strikes to a blade typically have minimal damage because of the placement of a grounding circuit within the blade. This circuit is similar to a lightning rod observed on large buildings and other human-made structures. The grounding circuit includes a metal disc known as a **lightning receptor** on the outer surface of the tip with a grounding strap connecting the disc to the hub assembly. This circuit enables the lightning strike to dissipate through the wind-turbine structure to the ground. Depending on the intensity of the lightning strike, the damage may range from discoloration or burning around the lightning receptor to a breakdown of the entire blade tip structure. **Figures 15-12** shows a variety of blade damage examples created by lightning strikes.

Pitting of the blade leading edge is caused by the blade hitting airborne contaminants such as sand, dust, or hail. Long-term exposure to these conditions will deteriorate the leading edge and increase drag. In extreme cases, damage to the leading edge may create structural issues. Contamination buildup on the leading edge is another cause for increasing aerodynamic drag. Some studies suggest that power-output reduction because of extreme contamination buildup may be as high as 50%. Buildup may include dust, bugs, lubricants, and other airborne contaminants. Regions with very little rain fall during the year will be prone to the buildup of contamination on the blades. Rain hitting the blades tends to clean the blades and reduce contaminants. PM for arid regions may include cleaning the blades with a power-wash system to maintain acceptable power-output levels. **Figure 15-13** shows an example of a boom lift setup for inspecting and cleaning blade surfaces. Document inspection results on the PM checklist and include photos of any suspect areas that will require follow-up at a later service call. This will allow the farm operations team to schedule further inspection and cleaning with contractors that have the required equipment and trained technicians to safely perform the activities.

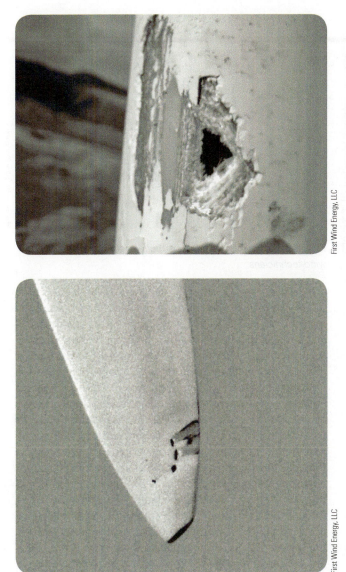

First Wind Energy, LLC

First Wind Energy, LLC

FIGURE 15-12 *Blade damage created by lightning strikes*

© Jim West/The Image Works

FIGURE 15-13 *Boom lift setup for inspecting and cleaning blade surfaces*

INSPECTION OF NACELLE HARDWARE AND SURFACES

Inspections of the nacelle assembly include a visual for damage to the outer surfaces and missing or loose hardware used to mount components to the nacelle. Hardware mounted to the nacelle may include fall-arrest safety anchors, meteorological assemblies, and any structures used for egress or ventilation equipment enclosures. These structures are exposed to harsh environmental conditions throughout the service life of the wind turbine, so they need to be inspected, repaired, and replaced as necessary. Safety equipment mounted on the nacelle assembly used for fall-arrest anchors should be inspected for integrity during each PM activity—preferably before they are used.

Safety Hardware

Fall-arrest anchors include rails, rings, and clevis assembly approved by the wind-turbine manufacturer or site safety plan. Approved tie-off hardware typically is painted or labeled to distinguish it from other hardware that may be mounted to the nacelle, hub, or spinner for service-equipment connections. **Figure 15-14** shows an example of a fall-arrest anchor mounted to the top of a nacelle. Remember, fall-arrest anchors are required to withstand a minimum impact force of 5,000 pounds (22.2 kN), so damaged, corroded, or loose hardware will need to be repaired or replaced. If it still looks like it can support a pickup truck, it probably is acceptable. Consult the

First Wind Energy, LLC

FIGURE 15-14 *Fall-arrest anchor mounted to top of nacelle*

REpower—Jan Oelker Photograph

FIGURE 15-15 *Met mast and obstruction light being serviced by wind technicians*

site safety plan or manufacturer's requirements for approved repair techniques and replacement components. Document findings on the PM checklist and notify the farm operations team of unacceptable safety equipment for immediate repair or replacement. Never expose yourself or other technicians to undue risk by using damaged safety equipment.

Inspection of Meteorological Components

Inspection of the met mast should include a check for missing or damaged hardware or fasteners. The meteorological mast is typically constructed of galvanized steel tubing that is welded or bolted together to support an anemometer, a wind-direction instrument, and obstruction lighting. **Figure 15-15** shows examples of a met mast and obstruction light being serviced by wind technicians. Inspection of the mast should also include verifying that the galvanized fasteners that mount the lights to the nacelle are secure. Any missing or badly corroded steel fasteners should be replaced with comparable fasteners to ensure structural integrity and minimize rust staining to the top and side of the nacelle. Rust particles from corroded hardware may be carried down the side of the nacelle by rain, leaving a visible rust stain after the water dries. This stain can become noticeable from long distance if the source of the rust is not eliminated. It saves time and money to replace rusted fasteners and hardware when they are discovered instead of waiting for a noticeable stain down the side of the nacelle. A stain or contamination on the side of the nacelle means that certified technicians will have to work off the side with specialized equipment for the cleaning process.

Inspection of wind-speed and direction instruments should include verification that mounting hardware is secure and functioning properly. Verification of ultrasonic instrument function can be accomplished by checking recent

SCADA records to determine if the data match other met instruments in the vicinity. Verification of cup anemometers and vane-style instruments may be accomplished by carefully moving the instruments and having a second technician monitor the input signal to the PLC module or observing data shown on the SCADA system screen. These instruments typically have heaters built into the assembly to prevent ice from building up during cold weather. Carefully place your hand on the instruments or use a temperature probe to verify function of the heaters. Inspect obstruction lights to ensure they are functioning properly. Refer to the manufacturer's operation information for details on manual override and inspection criteria. Obstruction lights within the jurisdiction of the Federal Aviation Administration (FAA) in the United States are required to be operational at all times as listed in the approval process documentation for a wind turbine. If an obstruction light will be inoperative for an extended period of time, the nearest FAA office must be notified so it may issue a **notice to airman (NOTAM)**. A NOTAM is the means by which the FAA publishes information on changes to air-space conditions and hazards. Replace inoperative obstruction lights if required by the PM plan or notify the farm operations group for immediate corrective action.

Another inspection requirement for met instruments is to ensure that wire connections and cables are secure to the instruments and met mast. Loose connections or cables may create intermittent signals during operation that can shut down the wind turbine. Follow the cables down the mast to the nacelle and inspect the cable seal assembly located near the base of the mast. The seal is used not only to prevent abrasion of the cables but also to prevent water from entering the nacelle. If a seal is damaged or there are signs that water may have been leaking into the nacelle, apply an approved sealant around the cables to stop the leak. Water leaking in or on the met cables may collect in an electronics junction box and create an electrical short or cause the formation of corrosion on electrical connections. **SAFETY: Refer to the sealant MSDS for appropriate PPE, first aid, and cleanup** practices recommended by the material manufacturer. Make note of each inspection finding along with hardware and materials used on the PM checklist for later follow-up and adjustments to the site inventory.

Ventilation and Egress Assemblies

Access hatches, covers, and ventilation assemblies should be inspected for missing and loose hardware. These components may be held in place with plastic keepers, rubber latches, or galvanized steel fasteners. Carefully check plastic and rubber components for cracking and tears created by exposure to ultraviolet light and mechanical stress of the application. Replace components if they are found damaged to prevent failure and loss of the hatch or cover during operation. If this is not part of the site's PM activity requirements, then ensure the defects are noted on the PM checklist for replacement on the next service call to the tower. Inspect exterior grating, ventilation panels, and composite covers for damaged, loose, or missing hardware. Ensure that loose hardware is secured and replace missing items. Note any components that are damaged or corroded on the PM checklist. Include photos of the damage or corrosion so that components can be assessed later. Photos may help determine whether components can be replaced on the next service call to the turbine.

Exterior access to the hub through the spinner may be covered as part of this inspection activity or during the inspection of the blade root area. Galvanized or stainless-steel fasteners, hinges, and handles should be inspected to ensure they are secure and functioning properly. Loss of mounting hardware for hub access hatches can damage the spinner assembly and create an objectionable noise for neighbors living close to the wind turbine. Having a hatch slam each time the hub rotates can cause considerable stress on mounting hardware and the composite material around the opening. If not corrected, the hatch may break off, creating increased aerodynamic drag and allow for precipitation to enter the spinner or hub interior. Note inspection findings on the PM checklist and include any hardware that may have been used for repairs.

INSPECTION OF TOWER SURFACES

Inspection of the tower's exterior surfaces may be completed from the ground using binoculars or a telescope with a camera. Check for damage to paint and any rust stains that may be forming on the tube sections or at the joints. Note any chips and scratches. Some of these defects may have occurred during the tower-assembly process and some may have occurred after the wind turbine was placed in service. For some reason, teenagers like throwing rocks at wind-turbine tube

FIGURE 15-16 *Personnel basket being used for tower inspection and repair*

towers, they are probably amused by the bell sound that resonates within the tower when rocks strike. It is too bad this amusement comes at a price for the site operator. Each chip or gouge caused by a rock striking the tower is a location for corrosion to form. Coat these surface defects with primer and paint if the PM activity recommends repairs. Some surface defects may require a man basket attached to a boom truck or service crane on the nacelle to position technicians to complete the activity. **Figure 15-16** shows examples of technicians making inspections and repairs to tube towers. If inspecting a lattice tower, this activity would be completed during the torque inspection process of fasteners. Replace any badly corroded fasteners and components as recommended by the turbine manufacturer or site operator. Inspection criteria and corrective action plans should be documented before the PM activity begins. A documented plan will save considerable time and money if maintenance technicians are not required to make judgment calls on what is considered "good" and "bad." Include images of acceptable and unacceptable defects to eliminate field judgment calls. These images will not cover every situation, but they should provide guidance when a judgment call is necessary. Document all inspection findings on the PM checklist for follow-up as needed by the farm operations team.

RECOMMENDED CLEANING ACTIVITIES

Exterior-surface cleaning activities for the tower and blades have been mentioned throughout the chapter. Many of the blade and tower areas cannot be accessed except by certified technicians using specialty equipment from a crane, boom truck, or repelling equipment. These activities should be contracted out to service providers such as Rope Partner, Blade Works, and others that provide certified technicians and specialty equipment to complete these activities safely.

SUMMARY

A variety of PM activities are necessary to maintain the exterior surfaces of a wind turbine. These activities included visual inspections, cleaning, replacement of corroded components, and repair of inoperative equipment. In each case, safety considerations ensure that hazards are understood and that plans are made to eliminate the risks. Some cleaning and repair activities may require working with a man basket or repelling from the nacelle to access the work area. Never attempt these activities without proper training and the required specialty equipment. An attitude that "It will only take a minute" can end tragically. The goal of every service and maintenance activity should be to complete tasks safely and correctly. Never cut corners. The final activity for each PM cycle is to ensure that checklists are completed with appropriate dates, technician names, tower information, part numbers, and quantities of consumed materials.

REVIEW QUESTIONS AND EXERCISES

1. List several maintenance activities associated with the exterior surfaces of a wind turbine.
2. Calculate the aerodynamic force applied to a technician standing on a nacelle ($A = 0.9$ m²) in an oncoming 15 m/s wind velocity. Use atmospheric conditions at sea level and an air temperature of 5 °C. Would the technician experience difficulty in standing or walking because of the wind force?
3. Calculate the aerodynamic force applied to the technician if his or her posture was crouched low to the nacelle. Assume the exposed wind area in this posture is 0.33 m². Would this posture reduce the force on the technician's body and make it easier to move in the wind?
4. Define structural drag and parasitic drag.
5. Give an example where structural drag may benefit a wind turbine.
6. What issues are associated with accumulation of ice on wind-turbine structures?
7. Describe a way to prevent the buildup of rust staining on a nacelle or tower.
8. List three methods in which a personnel platform can be positioned for elevated work activities with a wind turbine.
9. What U.S. government agency sets regulations for the operation of obstruction lighting?
10. What OSHA regulation governs the use of elevated platforms and man baskets?

16

Developing a Preventative-Maintenance Program

KEY TERMS

bill of material (BOM)
commissioning
International Organization for Standardization (ISO)
operations and maintenance (O&M)
power purchase agreement (PPA)

predictive maintenance
preventative maintenance
return on investment (ROI)
service bulletins

OBJECTIVES

After reading this chapter and completing the review questions, you should be able to:

- Describe the need for a comprehensive maintenance program.
- List steps that may be used to develop a preventative-maintenance (PM) program.
- Describe the difference between preventative maintenance and predictive maintenance.
- List information resources that may be used in developing a PM program.
- Describe some of the items included in a maintenance procedure.

- Describe the need for checklists with PM activities.
- Describe the need to document resources necessary to complete a PM cycle.
- Describe the need for accurate documentation of PM activities.
- Describe the importance of tracking a maintenance program's effectiveness.

INTRODUCTION

Purchase and installation of a community or utility-sized wind turbine is a major capital investment. Projects involving wind turbines of this size are typically undertaken by several individuals, a business, or corporate partners with a reasonable expectation of a **return on investment (ROI)**, or the expectation of getting the invested principal back plus a reasonable interest rate for the use of the money. There is not much incentive for an investor to take a loss on an investment or only get the principal back. Capital investment for one turbine may be more than several million dollars, depending on the size. Investment for a wind project may include feasibility studies, permitting, landowner agreements, a power purchase agreement (PPA), marketing, site preparation, equipment purchase, and installation. Site preparation may include construction of access roads, transmission, collector station, and requirements placed on the developer to ensure environmental compliance.

A **power purchase agreement (PPA)** is the purchase contract between the power producer and the consuming organization. The PPA establishes the duration of the contract along

with the fixed price for energy delivered typically in kilowatt-hours(kWh). This effort may be spread over one turbine or several hundred turbines, depending on the wind-farm size. The primary factors for this phase of the project are placing the turbine where it can generate the most energy and completing the installation within the established budget. Once the turbines are installed and commissioned, the goal of the operational phase is to ensure that each turbine is ready whenever the wind blows. If the wind is blowing and the wind turbines are not generating power, the project operator has lost revenue. Revenue obtained from the sale of delivered energy is the primary ROI for the project. Other credit sources for renewable-energy projects include production tax credits and carbon credits. These credits can be used by a renewable-energy project to offset or reduce the expense of doing business.

The typical life expectancy of community and utility wind turbines is around 20 years. To ensure that they produce energy over the expected life, wind turbines require scheduled maintenance and service. Recognizing the need for a comprehensive maintenance program is the first step in ensuring the highest turbine availability and a continuous revenue stream. Developing and implementing a comprehensive maintenance program should be top priority for a wind-farm operator. Whether a wind-farm operator performs the maintenance tasks or sets up an **operations and maintenance (O&M)** contract with another organization, the program should have similar steps: recognize the need, determine requirements, document the program, set up a schedule, perform the maintenance, and track effectiveness. Tracking program effectiveness is as important as determining maintenance requirements. If the maintenance program is not keeping the equipment running, then it is a waste of time, resources, and money. Tracking program effectiveness and continuous improvement activities will be the topic of Chapter 17.

The next step in establishing a maintenance program is to determine maintenance requirements. To determine requirements, one must have a working knowledge of the equipment to be maintained. This may be through previous work experience or through a training program provided by the wind-turbine manufacturer. The training program may also include many of the documents necessary for the development of a maintenance program, so some of the steps in following sections may be completed for you. If that is the case, then this text can provide guidance on documenting your program and tracking maintenance effectiveness. A word of caution: you would not take a car to a technician who does not have relevant training and certification, so do not attempt wind-turbine maintenance activities without proper training and certification. As a wind-farm manager, you should employ trained and certified technicians for maintenance activities and use these technicians as mentors and trainers for less-experienced technicians. Demand the same requirements from your O&M providers.

For those who are new to the business and without the benefit of a developed maintenance program, you will need to start the process. Remember the process steps include recognizing the need, determining requirements, documenting the program, setting up a schedule, performing the maintenance, and tracking effectiveness. It is easy to recognize the need to develop and implement a maintenance program. That is a matter of dollars and good sense with multimillion dollar equipment left in your care. The question may be, where do you start? A good place to start a maintenance program is by determining maintenance requirements. This step requires an equipment inventory of what needs to be maintained.

GETTING STARTED

Wind-farm operators with new equipment may not have to spend considerable time developing an equipment inventory. The wind-turbine manufacturer may supply a **bill of material (BOM)** along with a set of mechanical and electrical drawings. A BOM is an engineering drawing used to list details on subassemblies and individual parts required for the construction of a product. BOM information typically includes quantity, specification numbers, part numbers, and descriptions that are useful in purchasing identical or equivalent items.

Manufacturer-supplied documents can reduce the time and effort needed to complete the inventory step of a maintenance program. If these documents are not available, then the inventory task can be completed by a thorough inspection of the equipment by an experienced technician. Either method will enable a farm operator to begin the process of developing a maintenance program. Maintenance program activities may be divided into two areas: preventative maintenance and predictive maintenance. **Preventative maintenance** is provided through the inspection, cleaning, lubrication, and adjustment of equipment or assemblies to ensure proper operation. **Predictive maintenance** is the repair or replacement of equipment or components based on analysis of data accumulated during the preventative-maintenance process. Trends in wire-insulation resistance and lubricating-oil analysis results are two good examples of using preventative-maintenance data to monitor equipment condition. Comparing these data to manufacturer guidelines, recommendations by experienced test labs, or historical data gathered from similar equipment will help predict present condition and aid in determining if issues may be on the horizon.

Collecting Information

What information should be gathered? Information should include enough detail to determine what, where, and when for each covered item in the maintenance program. *What* includes a description of the equipment, assembly, or part number. *Where* includes the location or locations within the wind turbine. *When* includes estimates on maintenance intervals. An experienced technician with knowledge of similar equipment would have great insight on maintenance interval estimates. Maintenance intervals and scheduling will be a topic for later discussion. An initial inventory table should include a description, model or part number, location, manufacturer, and comments. Using the same table format for the data-collection process will improve consistency of the information. Otherwise, several technicians may collect different information, depending on what they may interpret as important during the inspection. Save time and money by determining what information is important up front and collect that information. **Table 16-1** shows a sample maintenance-inventory table. Customize table headings and layout to meet the needs of your organization, but ensure data collected will be useful to the development process. Use the chapters presented previously on maintenance activities as a guide to equipment that may be encountered during the inspection process.

Equipment description, number, and manufacturer will be valuable when reviewing documentation and finding a manufacturer's contact information. The comments section may include items such as lubrication, fastener size and grade, suggested activities, along with maintenance intervals. Information in the comments section may be brief or very detailed, depending on the technician's field experience. The more detailed the information, the easier conversations will

TABLE 16-1

Sample Maintenance-Inventory Table

Location	Description	Model/Part Number	Manufacturer	Comments
Down tower assembly Control cabinet	PC PLC UPS		Gateway Bachmann Dell	Microsoft Windows 7 Operating system Controller, rack system, optical fiber communication module 120 V_{AC}, 3,000 watts
Tower: Mid deck	Bolts Lighting		Friedberg-Americas Sylvania	M36, Grade 8.8: galvanized Four foot: 40-W fluorescent bulbs
Tower: Yaw deck	Tower hoist Crew lift		Harrington Avanti	¼-ton capacity, chain hoist with remote control ½-ton capacity
Nacelle	Yaw drive Gearbox Generator		Bonfiglioli Moventas Siemens	Four units, 3 ph: 400 VAC w/brake Gearbox 80:1 gear ratio, 100 gallon synthetic oil 3 ph - 600 volt, 2500 kW capacity, brush type
Rotor	PLC Pitch control Pitch drive		Mitsubishi Moog NGC	Controller, rack system, communication module DC drive system Three units, 140 VDC, tacho-generator, encoder

© Cengage Learning 2013

be later with engineers and service technicians in establishing what is important in the maintenance program. Eventually this information will need to be compiled into procedures, checklists, schedules, tooling, and consumables. Use this research time to determine safety and tooling requirements necessary to complete the maintenance activities. Discussions with material suppliers should include requests for material safety data sheets (MSDS).

This information is by no means a complete listing of equipment that may be found in a wind turbine. This should give the reader an idea of how to start the inventory process. Equipment and component manufacturers will vary, depending on the wind-turbine manufacturer and model. Manufacturers listed in Table 16-1 are for discussion purposes only.

Further development of the inventory process may be necessary if the manufacturer's documentation is not

available. Another version of Table 16-1 may include the maintenance items for each piece of equipment categorized by location within the wind turbine. This is where an experienced technician can improve the data-collection process. Some equipment may have multiple items that may need to be maintained, and a listing of each of these during the inventory process will be useful in determining maintenance activities later. For example, multiple maintenance items for a gearbox may include lubricating oil type and quantity, oil filters, and desiccant breather assembly. Additional columns may be added to the original table or categorize the tables by wind-turbine section to clarify multiple items. **Table 16-2** shows a sample table developed to include multiple maintenance items by equipment. Part numbers will be specific to your system, so this column is intentionally left blank with examples of manufacturers for reference only. Use as much

TABLE 16-2

Sample Inventory Table with Equipment Details Categorized by Location (Nacelle)

Equipment	Description	Part Number	Manufacturer	Comment
Main bearing	Roller bearing pillow-block assembly Fasteners		SKF Friedberg-Americas	600-mm shaft, low speed and temperature; lubricant Threaded studs, M36, class 8.8—galvanized
Gearbox	Oil filter Breather—filter		HYDAC AGM	Combination 5:50 micron Drum-desiccant breather type
Parking brake	Breather—filter Oil filter		Svendborg Svendborg	Pleated air-filter cap assembly Filter element: 5 micron
Generator	Shaft bearings Phase brush Ground brush		FAFNIR Helwig Carbon Helwig Carbon	100-mm shaft, high speed and elevated temperature; lubricant 12: 25 mm × 50 mm × 150 mm long 2: 12.5 mm × 25 mm × 50 mm long

© Cengage Learning 2013

space as necessary for comments and include photo images during this step to save valuable time and a return trip to the tower later for more information.

The information in Table 16-2 may be gathered by inspecting the equipment, manufacturer's documentation provided with the wind turbine, or from documentation supplied by component manufacturers. This is by no means a complete listing of equipment that may be found in the nacelle. Use this as an example to develop a table to assist your organization in the maintenance process. **SAFETY: Always use an approved lockout–tagout (LOTO) procedure before opening access panels or covers on energized or pressurized equipment. Never work on equipment without appropriate qualifications and approval by your organization. Obtain the MSDS before handling any material to ensure use of approved personal protective equipment (PPE), and follow all safety recommendations.**

Once the inventory process is complete, there will be pages of information and tables listing equipment that should be included in the maintenance program. This will include the part numbers, equipment models, manufacturers, and plenty of comments. What is next? Use this information to research further details on maintenance recommendations from manufacturer Internet sites, direct conversations with suppliers or service technicians, and from experienced technicians.

SOURCES OF INFORMATION

Resources for equipment maintenance information may come in several forms: training and documentation provided by a turbine manufacturer, conversations with component suppliers and manufacturers, Internet sites, and experienced technicians.

Turbine Manufacturers

Turbine manufacturers may include major component identification, service information, and a list of contacts with the equipment folder presented to the farm operator after the commissioning process. **Commissioning** is the final inspection and testing phase of a wind turbine before it is placed into service. Experienced technicians from the wind-turbine manufacturer perform a visual inspection of the systems along with a series of operational and safety tests to ensure that the turbine will perform as designed. This information will be useful in determining major component identification numbers, suppliers, and manufacturing dates. Major component identification numbers will be useful in tracking any changes that may occur with the equipment over its service life. Major components may include blades, hub, nacelle, tower sections, and other assemblies. For example, a farm operator may track the service life of a component such as a blade that required repair or replacement to determine how it performed compared to other similar components. If several blades at the farm exhibit similar leading-edge erosion, the issue may be tracked back

to a production issue instead of an environmental issue at the farm. Without identification numbers and manufacturing details, this would be impossible to accomplish.

Maintenance and service training of farm technicians at the wind-turbine manufacture's training center is a valuable resource for understanding operation and maintenance requirements. This training typically includes several weeks of lecture and hands-on practical experience with equipment that is similar to that installed at the farm. Major benefits to this training opportunity are student's time to connect the training materials with hands-on experience, time for instructors to walk technicians through the systems without the added stress of production obligations, and team building. Technicians should use this opportunity to set up networking connections with instructors and other technicians that will be useful in the future. On completion of the training program, technicians have an improved working knowledge of systems they will be servicing or maintaining along with documentation on the system requirements provided by the manufacturer.

Component Manufacturers

Developing a network of contacts with component manufacturers enables access to information on their products and training seminars that they may provide. Information may include recommendations on maintenance practices, inspection criteria, lubrication specifications, adjustments, tool recommendations, and service bulletins. **Service bulletins** may include product recalls, changes to maintenance requirements, product improvements, and any other information that the supplier considers important for the customer. This information may include ready-made maintenance procedures that can be adapted into the farm maintenance program. Component manufacturers not only are in the business of selling parts but also are looking to build long-term relationships with customers of fleet equipment. Many components today can be cross-referenced between suppliers, so the best price may not always be the determining factor for a repeat sale. Customer support and part availability have become important factors. Gaining access to supplier knowledge resources can become an important asset to improved farm operations.

Internet Resources

A manufacturer's Internet site is another source of information for components and service recommendations. Manufacturers may include search features on their sites to assist in finding information on equipment or parts. It is typically easier to do a search on a Web site if you have some information to start the process. Model or part numbers are the usual tags a search engine can use to query a database. This is one reason why these numbers were recommended for addition to the inventory table. Typically, if you cannot

Component or Material	Examples of Manufacturers
Bearings	Fafnir
	SKF
	Timken
Lubricants	SKF
	Shell
	Mobil
Fasteners	August-Friedberg Americas
	Portland Bolt
	Bolt Depot
Threaded rod	Williams Form Engineering
Programmable logic Controllers	Siemens
	Rockwell Automation
	Bachmann Electronic
Tower safety equipment	Safe Works LLC
	Capital Safety
	PSA
	Petzl
Oil filters	Parker
	HYDAC
	FRAM
Brake systems	Svenborg

© Cengage Learning 2013

FIGURE 16-1 *Component manufacturers that may be used to research component or material information*

find the information you are looking for through a database search, then you can use the contact information listed on the Web site. This may be through e-mail, instant messaging, or the personal touch of a phone conversation. **Figure 16-1** shows examples of component manufacturer names that may be used for contact information and technical support.

Experienced Technicians

Another resource that can sometimes be overlooked is the experience of a technician. Wind power may be a relatively new industry in the United States and Canada, but many technicians working on wind farms today have had previous power-production experience or experience in industrial settings. Experience in these settings has exposed them to working with generators, motors, gearboxes, fluid power, electrical control systems, and many of the other items used in wind turbines. Maintenance requirements for these systems are the same as those for wind-turbine systems. The difference in the maintenance activity may be that it is located several hundred feet off the ground. Do not underestimate the experience of your workforce.

USING THE INFORMATION

At this point, there should be an inventory of items to be maintained and supplemental information on activities, tooling, consumables, and safety requirements that have been

compiled from the research process. This information may now be assembled into documents for planning and executing the maintenance program. Documents necessary for planning the maintenance program should include schedules, procedures, and resources. Maintenance schedules for a wind turbine are typically set up as semiannual or annual activities. Depending on the size of the wind farm, maintenance may be occurring every day of the year or during a portion of the year. Wind farms with 100 or fewer wind turbines may want to consider running their maintenance cycles to correspond with lower wind months to improve farm power output. For example, high wind months for a geographic area may be November through March. To capitalize on maximum wind-farm output, one maintenance cycle should finish by the beginning of November, and the next cycle should begin during the month of April. Deciding which activities to complete during each maintenance cycle will be determined by information pulled together from the resources.

Procedures

Procedure development should include steps to successfully complete a maintenance activity. Each activity should be broken into sequential steps with notes on safety, requirements, tools, and consumables needed to complete the activity. Documented procedures are a way to ensure that activities are completed to best working practices. There may be a dozen ways to complete an activity, but not all of them may yield the desired results. Consistency is the hallmark of good management. Documenting procedures and practices ensures consistent quality outcomes to work activities and ensures satisfied customers.

Companies that are interested in becoming certified to an international standard such as the **International Organization for Standardization (ISO)** management standard ISO 9001:2008 will need to document and follow consistent work practices from top management down to the frontline service provider. Certification to a standard shows customers that you are dedicated to doing the job right every time.

The following is an example of a procedure with features that are consistent with best working practices. This procedure includes features such as document number, title, revision, and approval, along with information to complete the activity. The document number, title, and revision are used for document traceability within the maintenance system. Document number can be developed using whatever method is suitable to track within your organization. This particular document—"SOP—Standard Operating Procedure, N—Nacelle, MBL—Main Bearing Lubrication, and 1.6 for the Turbine Model. Revision F"—is the sixth revision of this procedure. The date approved is the date that the designated individual approved the revision for general use. Documents are not to be set in stone. They should be revised whenever a discrepancy is found or whenever the process is improved. Having a revision letter or number aids the user in quickly determining if this is the correct revision. These

EXAMPLE OF A WORK PROCEDURE USING INDUSTRY BEST WORKING PRACTICES

Document Number: SOPNMBL1.6
Title: Model 1.6XX Wind Turbine Main Bearing Lubrication
Revision: F
Date Approved: 12 May 2011
Maintenance Cycle: Semiannual and annual
Notes:

- Technician(s): 2
- Lubricant: SKF—LGWM 2 Grease (3 400-g tubes)
- Tools: Portable battery-powered grease gun (extra battery and charger)
- Consumables: Disposable nitrile gloves, paper towels, or cotton rags, all-purpose cleaner, 13 gallon or larger plastic bags

PROCEDURE

- Greasing the main bearing should be completed while the rotor assembly is pinwheeling slowly (approximately 1 RPM). (**Figure 16-2** shows an example of a main bearing assembly with the grease fitting highlighted.)
- **SAFETY: Activity requires two technicians. Task 1 is manual operation of the park brake and task 2 is application of the grease. To prevent injuries, notify all personnel located in and around the nacelle that the parking brake will be released and the rotor assembly will be moving. Review the MSDS for PPE recommended to properly handle the bearing grease. Appendix D shows an example of an MSDS for grease that may be used in this application.**
- Load a 400-g tube of grease into the grease gun. (**Figure 16-3** shows examples of battery-powered grease guns with different adapters for small and large grease fittings.)
- Release the park brake and ensure that the rotor assembly is rotating.
- Attach the grease gun hose to the bearing zerk fitting and inject grease into the main bearing assembly.
- Repeat the process with three tubes of grease to purge existing grease from the bearing.
- Return the park brake to automatic mode and notify personnel that the activity is complete.
- **SAFETY: Remove contaminated grease from the clean-out bucket and dispose of according to the site HAZWOPER procedure SOPHWLUB11.**
- Wipe excess grease from the bearing grease fitting and clean-out bucket assembly. Discard soiled cleaning materials and contaminated grease per SOPHWLUB11.
- Note any discrepancies on the checklist along with the materials consumed.

Maschinenfabrik Wagner GmbH & Co. KG and PLARAD Bolting Technology, LLC

FIGURE 16-2 *Main bearing assembly with grease fitting highlighted*

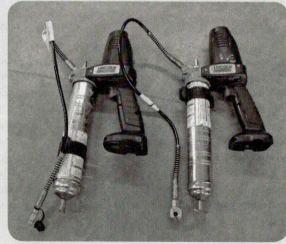

© Cengage Learning 2013 and a W. Kilcollins photograph

FIGURE 16-3 *Battery-powered grease guns with different adapters*

features ensure that the maintenance activity will be completed consistently and produce the desired results.

Procedure information should include information on the activity, requirements, safety considerations, consumables, tools, and steps to clarify the task. Notes serve as a quick reference to important items necessary for the activity. Requirements listed in a procedure should serve to eliminate questions or judgment calls. Listing safety considerations will aid technicians in determining appropriate PPE and tools to reduce the associated risk. Other listed items will help determine what is needed to efficiently complete the activity. Include images of criteria as necessary to further clarify acceptable and unacceptable items such as physical defects. Adding dimensional limits and the required type of measuring device such as a dial caliper can also be effective. Requirements for the sample procedure include type, quantity, and recommended methods to add lubricant to the bearing. The steps of the procedure should serve as a reminder for experienced technicians and may be valuable for training less-experienced technicians. The added information on an acceptable grease gun further clarifies the activity and may help with planning. This procedure may be further enhanced by a supplemental checklist. A maintenance checklist is a great place to write notes, list consumables, and discrepancies and to ensure that each step is completed. The completed checklist will serve as proof that activities were completed and may be used as a reference source for follow-up on the next wind-turbine service call.

Checklists

A checklist may serve as a supplement for a procedure or as a stand-alone document covering several tasks. The purpose of a checklist is to block the steps or tasks into a sequence that can be easily followed and checked off as the steps are completed. Adding notes, images, criteria, and listing materials will add to the utility of the checklist. For example, a checklist including the main bearing lubrication may be presented as the following document.

DOCUMENT NUMBER: CLNDTM1.6

Title: Checklist—Model 1.6XX Wind Turbine Drive Train Maintenance

Revision: C

Date Approved: 22 August 2009

MATERIALS

- Castrol Optigear Synthetic X320 Oil (as needed)
- Mobil SHC 460 WT Grease (3 400-g tubes)
- Mobil Hydraulic 10 W (as needed)
- Klüberplex BEM 41-132 Grease (2 400-g tubes)
- Oil filter XZT18-5/50 m (2)
- HP oil filter 5 m (1)

RECOMMENDED TOOLS:

- Plarad MX-EC/MSX with 60-mm and 55-mm sockets
- Lincoln battery-powered grease gun (2 guns with extra batteries and charge stand)
- Oil canister spanner wrench
- Paint pen: color designated per SOPMCD
- 12-mm ratchet wrench
- 19-mm ratchet wrench
- Feeler gauge, 25 blade, metric
- 5-gallon pail (1)

CONSUMABLES:

- Nitrile gloves (box)
- Paper towels (as needed)
- All-purpose cleaner (as needed)
- 13-gallon plastic bag (as needed)
- Oil-Dri 3RPP8, absorbent mats (as needed)

MAINTENANCE CYCLE:

- A = Annual
- S = Semiannual

Task	Cycle	Description	Initials/Date
1	S/A	Inspect hub adapter: missing, loose, or damaged hardware	
2	S/A	Torque hub adapter fasteners 1,900 ft-lbs (2,570 N-m)	
3	S/A	Inspect main bearing assembly	
4	S/A	Lubricate main bearing assembly (SHC 460 WT)	
5	S/A	Torque main bearing fasteners 1,900 ft-lbs (2,570 N-m)	
6	S/A	Inspect compression adapter: missing, loose, or damaged hardware	
7	S/A	Torque compression adapter fasteners 1,700 ft-lbs (2,300 N-m)	
8	S/A	Inspect gearbox oil level (Synthetic X320 Oil)	
9	S/A	Obtain gearbox oil sample (4-oz. sample size). Indicate: Turbine no., sample date, and technician initials on the sample container	
10	S/A	Replace gearbox oil filters (XZT18-5/50 m)	
11	S/A	Inspect park brake assembly: missing, loose, or damaged hardware	

This represents a partial checklist that may be used for a series of drive-train tasks. Adding further detail to the lines may be useful if the checklist is intended as a standalone document. Adding reference materials or a comments space for notes on the reverse side of the sheet will help keep the information together for future reference.

Checklists are a useful tool to keep track of tasks or steps necessary to complete an activity. The preceding example uses the technician's initials to check off the steps and provide a record of the person who completed the task. It may be disastrous to miss a step and time consuming if repeated. Repeated steps not only may be time consuming but also are costly when extra time, materials, and lost productivity are considered. Resources are another important part of a maintenance program to consider. Understanding resources necessary to complete a maintenance cycle will enable activities to be completed efficiently.

Resources

Maintenance resources come in the form of staffing, materials, equipment, and tools. Staffing includes trained technicians to perform the maintenance and knowledgeable administrative personnel to coordinate teams and finalize documentation for activities. Materials include parts and consumables necessary to change wear items or replenish fluids necessary for continued operation of the equipment. Equipment may include the necessary cranes, trailers, trucks, and safety gear to perform the activities. Tools may include torque wrenches, power-bolting systems, electrical test meters, and other test items. Part of the maintenance development process should be identifying all of the resources necessary to complete the activities. This master resource list should include all items noted on the individual procedures and checklists to ensure that someone planning the activities can determine quickly what is needed. This will ensure that materials and consumables are present in the site inventory or are ordered so that they arrive ahead of the activities.

The master resource list should also include calibration information on tools and test instruments. If a tool or test instrument is out of calibration, it should not be used. These items should be calibrated and ready to use before the maintenance activity begins. If an item will be due for calibration during the maintenance activity, then make sure a suitable replacement is ready for service when needed. As you can see, scheduling becomes an important part of the maintenance program. Having the necessary documents available for review makes the process flow smoothly. Lost wind-turbine production time and teams waiting around for tools and materials become expensive and frustrating, not to mention giving a poor impression to others such as customers or contractors who are on site to supplement O&M staff.

At this point, an equipment inventory, maintenance procedures, checklists, and other documentation to assist in planning activities have been established. The next step in the process is to implement the program so the organization can benefit from reliable and consistent equipment performance.

IMPLEMENTING A MAINTENANCE PROGRAM

Implementing a maintenance program does not mean compiling all of the information into several binders or electronic files and setting them aside to collect dust. When all of the material is prepared and packaged, there is still more work to be done. The next step in the process is to present the program to all employees who will be responsible for using the materials.

Training

Set up a series of training sessions dedicated to discussing the materials and how to use the materials. If you did your homework earlier, then you solicited input from key individuals who will be using the documents. This makes the training process easier because employees—technicians, administration, and managers—all have a stake in a successful program. Stress that the materials are considered the best working practices that the group has already been using. Also stress that the documents are not set in stone. If there are changes in equipment or a better work practice is established, then the documents can be changed. Do not change the process without changing the documents! This leads to confusion and defeats the purpose of a standardized program.

Document Use

Package documents so that they may be easily used in an office or field environment. If your organization uses laptops or other electronic devices to access forms, then develop a format for easy data entry. Advances in electronic devices and networks over the past few years have made access to electronic media seamless between the office and field environments. For this reason, most field-service organizations today use electronic devices for data entry, document retrieval, and electronic-system troubleshooting. If the wind farm is remote or the environment too harsh for electronic devices, then package the documents in binders or folders that are weather resistant. Employees will not use documents if access is limited or unusable when soiled or wet.

Collecting Information

The procedures, checklists, and other forms available for fieldwork should have designated locations to include all information requirements. Stress to field-service technicians that forms should be completed carefully to ensure proper documentation of time, activities, materials, consumables, and comments. Accurately documented checklists will simplify data entry later for inventory, billing, and payroll updates. Thorough descriptions of issues encountered or problems requiring follow-up in the comments sections will make after-action reports easier to complete and direct to appropriate individuals or suppliers.

Tracking Results

Tracking maintenance results and implementing improvements are the final steps in developing a maintenance program. As mentioned previously in the chapter, if the maintenance program is not keeping the equipment running, then it is a waste of time, resources, and money. Use comments from maintenance checklists, supervisory control and data acquisition (SCADA) information, and feedback from service calls to determine if wind-turbine availability issues are the result of personnel, procedures, materials, or environment. Personnel issues may be in the form of attitude, training, or inadequate numbers. Procedure issues may be in the form of improper activities, improper scheduling, or undocumented performance. Material issues may be because of poor workmanship, component design, or formulation. For example, a hydraulic hose breaks during service without previous signs of a problem. This would be a material issue and one not created by ineffective maintenance. If a hose breaks and there were signs of a problem, then this would be a maintenance issue. In this case, a thorough inspection of the hose by an experienced technician may have discovered the problem and comments on the PM checklist would have enabled a scheduled replacement before the failure. Environment issues come in the form of lightning, severe winds, floods, and other natural phenomena. Use this available information on turbine availability to determine the underlying cause of the problem and make adjustments to the maintenance program to prevent a reoccurrence. Tracking program effectiveness and continuous improvement will be discussed further in the next chapter.

SUMMARY

Developing a maintenance program requires several steps. Recognizing that a comprehensive maintenance program will improve wind-turbine output and increase revenue through availability is a first step. Other steps include inventorying equipment, determining requirements, documenting the program, setting up a schedule, performing the maintenance, and tracking program effectiveness. Recognizing that each step in the process is equally important will prevent the development of an ineffective program. Developing accurate procedures and forms that may be easily accessed and understood by personnel will ensure they are used to produce consistent results.

Do not stop the process after the program is implemented. There is always room for improvement! Use comments from maintenance checklists, SCADA information, and feedback from service calls to analyze problems and prevent reoccurrence. Use the program to optimize equipment for reliability and maximum power output. An effective program will improve the organization's ROI and ensure customer satisfaction.

REVIEW QUESTIONS AND EXERCISES

1. List some of the steps in developing a maintenance program.
2. What is the difference between preventative maintenance and predictive maintenance?
3. List information resources that may be used to establish maintenance requirements.
4. Should safety requirements be considered during the development process of a maintenance program? Why?
5. Name an international standards organization and explain why being certified to a standard may be important for you and your customers.
6. Name items that should be included in a typical procedure and explain their importance.
7. What is the purpose of a checklist for a maintenance activity?
8. Name some resources necessary to complete a maintenance activity.
9. What are the last steps in developing a maintenance program?
10. Why are the last steps in the maintenance program as important as determining requirements?

17

Wind-Farm Management Tools

KEY TERMS

attribute data
availability
check sheets
histogram
kilowatt-hours (kWh)
operations and maintenance (O&M)
Pareto chart
plan, do, check, and act (PDCA)

power purchase agreement (PPA)
root cause
run chart
scatter diagram
stakeholders
supervisory control and data acquisition (SCADA)
variable data

OBJECTIVES

After reading this chapter and completing the review questions, you should be able to:

- Describe the importance of continuous improvement.
- Describe the importance of customer satisfaction.
- Describe information sources used to gauge organizational performance.
- Describe the role of safety in organizational performance.
- Describe the use of SCADA and other wind-farm information for trend analysis.
- List the five analysis tools for continuous improvement.

- Describe the use of continuous-improvement tools.
- Describe the four basic steps to continuous improvement.
- Describe the importance of a review process and following up on process changes.

INTRODUCTION

The goal of an organization is to be profitable and to ensure customer satisfaction. Profitability can be gauged by comparing operating expenses to income generated by services or products. If you are a wind-farm owner, then your product is energy in **kilowatt-hours (kWh)** delivered to a customer. This may be through a **power purchase agreement (PPA)** with individual customers or a local electrical utility. A PPA is an agreement between an energy supplier—wind farm, biomass generator, nuclear, or coal power plant—and a customer that may include energy delivered, price, and contract duration. Customer satisfaction would then be determined by the amount of energy delivered to the customer compared to the requirements of the agreement. Not being able to supply a customer's energy needs may require the customer to purchase energy on the open market at a higher price than the PPA. This would not be a way to maintain customer satisfaction.

Products or services from an **operations and maintenance (O&M)** organization would be wind-farm percent availability value or completed service calls requested by the owner. **Availability** is the percent of time the wind turbine is listed as operating or in *ready status*, according to the **supervisory control and data acquisition (SCADA)** system, compared to time available. For example, SCADA may list a wind turbine as operating or in ready status for 725 hours out of a possible 744 hours in a month for 97.4% availability. SCADA is the operating system used to control wind-turbine functions and collect operation data. A wind turbine in ready status means it is functioning without any faults that would prevent normal operation when there is sufficient wind. It may not be producing power because of low wind speed, but it is able to operate when the wind speed increases. If the contract with a wind-farm operator is 98% availability and availability is running at 94%, then the owner would be dissatisfied. After all, the owner has agreements with customers and a return on investment to consider.

Quality service and maintenance practices may be another area that will ensure customer satisfaction. Proper scheduling of resources can improve activity flow and reduce the time equipment is not operational. Productive use of time during a maintenance activity can also reduce the time a wind turbine is not available for production. These are some ideas that may be considered for organizational improvement. No organization is perfect, so striving to improve customer satisfaction should be a priority. If you are thinking this type of philosophy is not necessary for wind-farm operations or services, guess again. Customers have choices. PPA customers can change energy suppliers when a contract term expires. Dissatisfied wind-farm owners can certainly choose another O&M organization or demand a reduction in fees when a contract is due for renegotiation. Improving customer satisfaction is very important for an organization to compete in the global economy. If you cannot do it better for less, then a competitor will certainly try. If you have been in business for awhile, you know it is easier to keep a customer than to win the customer back. The key to improving customer satisfaction and organizational performance is through understanding information available from your processes. Process information can show where performance is going well and where it may need improvements.

WHAT INFORMATION IS IMPORTANT?

Any process information that is useful in determining organizational performance is important. Performance information can be obtained from areas such as equipment operation, service response time, service and maintenance quality, cost control, and safety records. Equipment operation such as production output, availability, and fault data may be obtained from SCADA information, service logs, and production reports.

Service Response Time

Service response time is another variable that may be monitored through SCADA information. In this case, response time would be the time interval from a wind-turbine system fault until the system is returned to operational ready status. Other methods through which service response time can be tracked include:

- production reports received on time,
- projects completed on time,
- time required for technicians to prepare and deploy for assignments,
- time interval from part or material request until received in inventory,
- time to complete a scheduled service activity,
- time to complete a maintenance activity, and
- resource scheduling.

Service and Maintenance Quality

Service and maintenance quality may be gauged from resources such as SCADA information, supervisor follow-up, and customer inspections. SCADA information may show the number of times a wind turbine faulted because of the same problem before the root cause was corrected. This information may also be used to calculate the number of hours a wind turbine was down for the same fault. This accumulated time reduces wind-turbine availability. Supervisor follow-up may be a good practice to determine work quality—workmanship—by both a team and an individual. Workmanship may be gauged as quality and appearance of the repair, condition of the work area, and thoroughness of documentation. Constructive feedback from follow-up should be shared with technicians to help them understand expectations. Share the information after the inspection. Do not wait several months for their performance review because by then no one remembers the issue, and it has become a moot point. Constructive customer feedback on service and maintenance expectations is also a good way to gauge effectiveness. Positive feedback is great, but receiving images of poor quality work and trash left in the wind turbine does not make anyone's day. It creates an uncomfortable experience for the operations supervisor and reduces customer confidence in other work performed by the team.

Cost Control

Effective cost control or profitability is a factor that may be used to gauge an organization's performance. Cost information is tracked from the site supervisor up to the corporation's top executive. Information gathered on operating cost is a means for mangers and stakeholders to determine performance. **Stakeholders** are groups or individuals who have an investment in the operation. Comparison of operating cost to income revenue is a basic measure of profitability. If an organization is not profitable, then it is not using resources effectively and typically cannot stay in business long. Cost may be associated with hours to complete a service call or a maintenance activity, cost of components replaced during a service call, excess inventory levels, hourly wages, overtime, salaries, operation of service trucks, replacement of damaged tools, and shipping—and the list goes on.

Safety Records

The safety record of an organization is another measure of performance. An organization with an effective safety program performs well through reduced costs and the maintenance of good employee morale. A poor safety record increases the costs associated with injuries, reduces the effectiveness of employees through lost time, and affects employee morale.

Safety data may be tracked by several different methods such as days without injuries, lost days for injuries, insurance premium expense, and types of injuries. OSHA 29 CFR 1904.7 regulations define injury tracking and reporting as mandatory functions for many organizations. OSHA regulations for collecting injury data are twofold: enable the employer to analyze and improve its workplace-safety record and allow OSHA to track compliance for the Department of Labor. Considerable information can be captured and analyzed to determine the performance of an organization. The goal of capturing information is not to accumulate and store data but to review the data to gain an understanding of processes. Understanding processes will enable adjustments to improve performance.

ANALYSIS OF AVAILABLE INFORMATION

Organizations collect data continuously on service and production processes. Wind farms are no different than other organizations. The goal in analyzing data is to gain useful insight into how the operation is performing. Organizations typically understand the end results of their service and production activities, but many do not consider the internal steps for these activities. For example, a maintenance activity has an end result of a turbine ready for operation. What are some of the steps in completing an annual maintenance cycle? Think about this for a minute.

Maintenance Information

You may think this is an easy question. Your answer may be: Just send a team to the tower with tools and supplies for a day. Certainly that is part of the maintenance activity, but did you consider these items? Verifying inventory levels, ordering materials, scheduling maintenance teams, verifying equipment and tool requirements, verifying safety requirements, scheduling wind-turbine downtime with the customer, or review of the wind-turbine service log. These are some of the many activities before the team ever gets to the tower. If you thought of these—great. You probably have had previous management experience at a wind farm or other industrial setting. Most technicians only consider their part of the activity.

This is a good point to review examples for these preliminary steps.

- *Verify inventory levels.* What do you need, and what do you have? **Figure 17-1** shows an example of an inventory analysis to determine materials for a maintenance activity.
- *Ordering materials.* What do you need, what is the minimum quantity (case, pail, drum, etc.), and when is the delivery required if the maintenance cycle starts on Monday, May 7? Ordering all of the components for delivery before starting the maintenance cycle is a possibility, but what if there is not enough inventory space or the supplier does not have inventory to complete the order? In either case, possible delivery dates may be determined using the start date and quantity consumed per day. **Figure 17-2** shows an example using inventory levels and maintenance consumption rates to determine delivery dates.

If you are wondering, why inventory levels are drastically reduced for some items at the end of the maintenance cycle, consider the previous statement on excess inventory. If money is tied up in inventory, it cannot be used for other activities. Another consideration: what is the true cost of inventory? Most businesses pay for inventory on credit with suppliers or a financial partner such as a bank or with other lending institutions, so there is an added interest fee for the inventory. That $1,000 of inventory might actually cost $1,250 if it has been on the shelf for an extended period of time. Occasionally, inventory items become obsolete before they are used, so they must be scrapped. That may mean $1,250 in inventory would be worth scrap value to the organization after paying the interest expense for an extended period of time.

- *Scheduling maintenance teams.* The requirement is to have five maintenance teams of four technicians, each available for May 7. Eight technicians may be available on the wind farm, so the other 12 will have to be scheduled from qualified contractors. This would not be a

Description	Current Inventory Levels	Annual Maintenance Cycle Requirement for 75 Wind Turbines	Material Order to Complete Requirements
Gearbox oil filter	68	150	82
HP hydraulic filter	16	75	59
General-purpose grease	52 × 400-g tubes	525 × 400-g tubes	473 × 400-g tubes
High-temperature grease	32 × 400-g tubes	150 × 400-g tubes	118 × 400-g tubes
Gearbox oil	30 gal	55 gal (suggested)	25 gal
Brake fluid	3 gal	5 gal (suggested)	2 gal
Phase brushes	4	900	896
Ground brushes	2	300	298

FIGURE 17-1 *An inventory analysis to determine materials for maintenance activity*

Description	Materials to Complete Requirements	Minimum Order Quantity	Order Quantity	Quantity Consumed per Day	Possible Delivery Schedule for Five Maintenances per day and 25 per Week
Gearbox oil filter	82	Each	82	10	50: May 11 32: May 18
HP hydraulic filter	59	Each	59	5	10: May 4 25: May 11 24: May 18
General-purpose grease	473 × 400-g tubes	10 × 400-g tubes/cs*	48 cs	35	13 cs: May 4 18 cs: May 11 17 cs: May 18
High-temperature grease	118 × 400-g tubes	10 × 400-g tubes/cs	12 cs	10	2cs: May 4 5cs: May 11 5cs: May 18
Gearbox oil	25 gal	5-gal pail or 55-gal drum	5 pails	As required	May 4
Brake fluid	2 gal	5-gal pail or 55-gal drum	1 pail	As required	May 4
Phase brushes	896	Each	896	60	300: May 4 300: May 11 300: May 18
Ground brushes	298	Each	298	20	100: May 4 100: May 11 100: May 18

*Container size

FIGURE 17-2 *Material ordering analysis example*

scheduling task left for May 1. These contractors may be scheduling resources over a six- or 12-month period, so the request for supplemental technicians should be submitted as soon as a maintenance start date is established. This will ensure that technicians will be available as they are needed.

- *Verifying equipment and tool requirements.* For our example, the process would determine the equipment and tools necessary to support five teams. A review of maintenance documentation would establish tools necessary for each team. If equipment is stocked at the wind farm, then the requirement of having it on site is complete. The second portion of the task is to determine the condition of the equipment and tools. Do they need to be serviced or sent out for calibration? Torque wrenches, test instruments, and other tools may require calibration on an annual basis. If the calibration due date is a week into the maintenance cycle, the tool or test equipment will have to be removed from service during the maintenance activity. This is not an acceptable practice if there are no other tools to replace them. Ensure that equipment and tools are ready before they are needed. Expensive maintenance equipment and tool inventories are typically warehoused and maintained at a service center operated by the O&M organization. Scheduling

maintenance equipment and tools should be done several months in advance of the maintenance cycle. This will ensure the items are delivered to the wind farm at least a week before maintenance starts. Using a service center for tools can reduce the cost of tools per maintenance cycle. Having a tool or equipment setup that costs $30,000 (€20,880) sit idle for six months out of the year is not a wise investment. Sharing this expense over several wind farms is a better use of capital. This logic also applies to a centralized component and material inventory location.

- *Verifying safety requirements.* Safety-equipment requirements constitute another topic for review before starting a maintenance cycle. Many safety items have a designated service life, criteria for inspection acceptance, or requirements for functional testing after a period of service. Items such as rubber gloves for electrical activities require dielectric strength verification six months after being placed into service. Even new rubber gloves require dielectric strength testing after sitting unused in inventory for 12 months. Purchasing excessive inventory of safety items with a shelf life such as rubber electrical gloves should be avoided. Ensure that safety gear is available and approved for service so that maintenance is not held up because of safety resources.

These are examples of internal steps that are necessary to ensure that a maintenance activity will run smoothly. Understanding there are several important steps in completing any service or product is beneficial to an organization during the improvement process. The end result may be the same, but there may be ways to reduce cost, time, and materials along the way.

Understanding the connection between inventory levels and farm profitability is a first step toward implementing an improvement. The use of a planned activity such as maintenance to reduce excess inventory levels will show up as an immediate reduction in operating cost. Always review proposed adjustments with the management team before implementing changes to ensure everyone involved understands the goal. Setting minimum and maximum inventory levels for critical items in inventory is another exercise that would be beneficial in cost reduction. Setting minimum levels for critical parts with long lead times will ensure that equipment will not be down waiting for parts to arrive.

Wind-Turbine Availability Information

Now let's analyze information to determine how well an operations team is performing by gauging wind-turbine availability. Maximum wind-turbine availability shows that the operations team is effectively servicing the systems so that they are consistently available to operate. SCADA information and service reports may be used to determine issues that lead to reduced turbine and farm availability. For example, if a wind turbine changes from online to shutdown mode, the SCADA status log will list date, time, and new status. If status change is the result of a fault occurring, then the log also lists fault information. **Figure 17-3** shows an example of a SCADA status log that might be viewed from a wind-farm system terminal, and **Figure 17-4** shows compiled comments from technician service logs. The SCADA status log format may vary between wind-turbine model

Date	Time	Turbine	Status	Condition
11 Nov 2010	0001:23	006	Online	O113
11 Nov 2010	0133:01	073	Fault	E277
11 Nov 2010	0134:01	073	Shutdown	O114
11 Nov 2010	0134:36	073	Ready	O111
11 Nov 2010	0135:22	005	Fault	E337
11 Nov 2010	0135:37	005	Shutdown	O114
11 Nov 2010	0211:16	056	Fault	E277
11 Nov 2010	0212:16	056	Shutdown	O114
11 Nov 2010	0213:01	056	Ready	O111
11 Nov 2010	0245:24	073	Run-up	O112
11 Nov 2010	0245:26	010	Run-up	O112
11 Nov 2010	0250:32	073	Online	O113
11 Nov 2010	0250:44	010	Online	O113
11 Nov 2010	0355:12	056	Run-up	O112
11 Nov 2010	0359:23	033	Run-up	O112
11 Nov 2010	0400:15	056	Online	O113
11 Nov 2010	0404:26	033	Online	O113
11 Nov 2010	0534:22	012	Fault	E080
11 Nov 2010	0534:37	012	Shutdown	O114
11 Nov 2010	0730:57	015	Stop/Reset	O110
11 Nov 2010	0733:34	015	Ready	O111
11 Nov 2010	0735:55	015	Maintenance	M111
11 Nov 2010	0756:01	012	Service	S111
11 Nov 2010	0810:20	005	Service	S111
11 Nov 2010	1210:23	005	Stop/Reset	O110
11 Nov 2010	1213:07	005	Ready	O111
11 Nov 2010	1223:16	005	Run-up	O112
11 Nov 2010	1228:32	005	Online	O113
11 Nov 2010	1411:44	012	Stop/Reset	O110
11 Nov 2010	1418:17	012	Ready	O111

Status information (S), operation (O), and error (E) codes are listed for reference only. Codes and format will vary, depending on the SCADA system provider.
Fault: E277—No wind speed
Fault: E337—Gearbox over temperature
Fault: E080—Generator speed not plausible

FIGURE 17-3 *Example of wind-farm SCADA status log*

© Cengage Learning 2013

Date	Turbine	Fault or Error Code	Time (Hours)	Comments
2 Nov 2011	005	E337	4.25	Gearbox over temperature: Troubleshoot and replace PT100 temperature probe.
4 Nov 2011	016	E127	4.00	Battery voltage out of tolerance low: Test battery pack voltage level (157 V_{DC}) and tighten wire connections to F31 fuse holder.
5 Nov 2011	005	E337	3.50	Gearbox over temperature: Troubleshoot and replaced overvoltage module OVM12.
8 Nov 2011	075	E077	4.25	Generator brush-wear indicator: Inspected brush assemblies and slip ring. Switched status key to maintenance selection. Replaced all phase and ground brush sets. Completed annual generator maintenance requirements. See maintenance checklist for details.
11 Nov 2011	012	E080	3.50	Generator speed not plausible: Inspected connections for generator encoder and high-speed shaft proximity switch. Tightened connections to high-speed shaft proximity switch.
11 Nov 2011	005	E337	3.75	Gearbox over temperature: Troubleshoot entire feedback circuit. Tightened loose connections on PLC input module.
13 Nov 2011	001	E301	2.75	Top box under temperature: Inspect top box heater assembly and found inoperative. Replaced heater assembly.
26 Nov 2011	062	E131	5.50	Blade 0° limit switch not functioning: Inspect limit switches for all 3 axis and found axis 1 switch damaged. Replaced limit switch.
27 Nov 2011	035	E277	2.75	No wind speed: Presence of wind-speed measurement on adjacent wind turbines indicates possible issue with anemometer. Inspected anemometer and found encased in ice. Heater assembly coil found as open circuit. Replaced anemometer assembly.
29 Nov 2011	066	E338	3.75	Main bearing over temperature: Inspected main bearing for indications of lubrication and damage. Bearing appears fine. Troubleshoot temperature feedback circuit and found loose connections on PLC input module.
29 Nov 2011	003	E328	3.25	Yaw time-out: Inspect yaw components and manually operated system. Yaw system functions manually. Signal not present at PLC to indicate motion. Found loose wire connection on terminal strip X10 connection 6U. Tightened connection.

© Cengage Learning 2013

FIGURE 17-4 *Technician service log information*

and manufacturer, depending on the company programming the software code. Information presented in the example is for reference only and does not depict any particular manufacturer's format.

The SCADA status log is a continuous list of operational data that may be reviewed for a specific time or for information over an extended period. Information from this screen along with other system screens may allow initial troubleshooting analysis before a technician ever travels to the wind turbine for a service call. This aspect becomes valuable when deciding what tools, test equipment, and personnel resources may be necessary to complete the call. Using the status log as a historical record in combination with the technician service log records will enable management to determine what was found during the visit to the wind turbine. For example, wind turbine 005 was serviced three times for the same fault over a one-month period. The fault did not return after the third visit, which leads to a conclusion that the **root cause** may have been an intermittent wire connection on the programmable logic controller

(PLC) input module. A root cause is the underlying issue that creates the problem; it is not a symptom of the problem. For example, the symptom for the fault was an intermittent elevated gearbox temperature. The root cause was the loss of electrical continuity through the temperature probe circuit, which showed up as an infinite resistance or elevated PLC temperature. If this is the case, then the two components replaced, 7.75 hours of service time for two technicians (15.5 labor-hours), and 11.75 hours of downtime waiting for service may have been avoided. The 7.75 hours is taken from the technician service log, and the combined downtime for the previous service calls is from the SCADA logs for those days.

How did the multiple service calls impact the availability of wind turbine 005 for the month of November?

Hours for the month of November: 24 hours $\times$ 30 days = 720 hours

Availability time: $720 - 11.75 - 19.75 = 688.5$ hours

Availability percentage: $688.5/720 = 0.956$ or 95.6%

How would availability of wind turbine 005 been impacted if the fault root cause was determined during the first service call?

Availability time: $720 - 7.75 - 11.75 = 700.5$ hours

Availability percentage: $700.5/720 = 0.973$ or 97.3%

The availability improvement would have been $97.3 - 95.6 = 1.7\%$ if the root cause was determined during the first service call.

Analysis of the multiple service visits for the same fault or similar availability situations should be used as a learning tool for technicians and not just as a time for criticism. Remember—the goal is to learn and improve from experience. Troubleshooting should include a review of all components and wire connections within the faulty circuit. A service-call tailgate meeting should include a review of available system documentation to develop a list of possible causes. The goal is to find and correct the root cause to prevent further equipment downtime.

These are just a couple of examples that show how available wind-farm information may be used to start the improvement process.

IMPROVEMENT PROCESS

Deciding which method of analyzing available wind-farm information may sometimes be a challenge. Information from SCADA may include data such as production output, wind speed, availability, and fault, among many other forms of data that indicate operational conditions. Other resources may include inventories, service, maintenance, and safety records, along with any other historical information maintained on the wind turbines. When choosing an improvement project, do not attempt to tackle every perceived issue at once. Choose small projects such as steps within an activity that the team understands. Tackling a project with several steps and multiple data resources can take a lot of time and create frustration if results are hard to understand. Use smaller projects to gain experience with the improvement process and analysis tools. Useful analysis tools for an improvement project include check sheets, Pareto charts, histograms, scatter diagrams, and run charts. Each of these tools allows data to be reviewed and displayed in a way that helps to explain issues.

Before we go any further, it would be useful to briefly explain data sources. Data comes in two forms: variable and attribute. **Variable data**, as the term implies, vary in magnitude. This format makes it easier to see subtle or dramatic variations over time. The following are examples:

- instantaneous power output of a wind turbine—1,233 kW;
- energy produced during a day—36,210 kWh;
- voltage between phases—375.7 VAC;
- current output per phase—1,500 amps;

- generator rotational speed—1,400 RPM;
- wind speed—12 mph (5.4 m/s);
- temperature—212 °F (100 °C);
- pressure—500 psi (345 kPa);
- hours worked per week;
- number of injuries; and
- hours of lost time.

Attribute data are typically gathered in one of two states. For example,

- on or off,
- functional or broken,
- yes or no,
- present or absent, or
- accepted or rejected.

Analysis Tools

Data format available for the wind-farm information resource will determine the type of analysis tool necessary for the review process. Some analysis tools work well with variable data, whereas others are better with attribute data. Choose the analysis tool that will improve the team's understanding of the process.

Check Sheet
A **check sheet** is a simple way to keep track of the frequency of an activity or event. Tracking how often the over-temperature fault occurred on wind turbine 005 would have been a good use of a check sheet. Other examples include:

- number of faults,
- number of times a component is used from inventory,
- maintenance days lost because of weather,
- number of flat tires,
- number of injuries requiring first aid, and
- number of sick days.

Figure 17-5 shows a sample check sheet.

Pareto Chart
A **Pareto chart** is a way to display the occurrence of multiple events. **Figure 17-6** shows an example. It is a way to distinguish the important few issues from the many trivial ones. A Pareto chart could compare occurrence frequency of different fault codes over a period of time. The fault code with the highest occurrence would help identify which fault may be reducing availability and thus prioritize it as the one to tackle first to improve availability. Other examples include:

- number of faults by wind turbine,
- number of parts used by a technician,
- number of parts used by part number,
- injuries by type, and
- safety equipment defects by type.

Date _____

Site Name _____

Service Call Time _____

Turbine Number _____

Lead Technician _____

Fault Code _____

Fault Description _____

_____

_____

Corrective Action _____

_____

Parts Required for Service Call _____

_____

_____

Return Visit for Same Fault (Y/N)

If Yes: Date of Last Service Call _____

Fault Code for Last Service Call _____

Corrective Action Last Service Call _____

_____

_____

Other Information _____

_____

© Cengage Learning 2013

FIGURE 17-5 *Sample service-call check sheet*

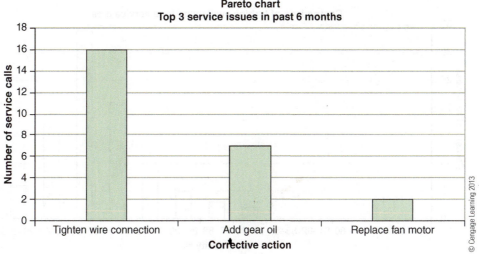

FIGURE 17-6 *Example of pareto chart*

Histogram

A **histogram** can be used to display the frequency of occurrence for a measurement in a process. **Figure 17-7** shows an example of a histogram comparing wind-turbine availability for a service area over a month. This information can help a service manager determine the effort required to improve overall availability. Breaking this down to a wind-farm level would be useful in showing a farm manager the number of wind turbines performing below the contractual agreement. Other examples include:

- average hourly wind speed each day,
- average daily wind speed each month,
- on-time deliveries for shipments,
- service-report errors,
- production-report errors, and
- shipment errors.

Scatter Diagrams

Scatter diagrams are a graphical way to display relationships between two variables, whether positive, negative, or neutral.

Figure 17-8 shows examples of the three scatter-diagram relationships. A positive relationship may be shown by the increase in one variable causing an increase in the second variable. For example, an increase in instantaneous wind-turbine output with respect to an increase in wind speed shows a positive relationship. In another example, increased torque applied to a bolt will increase the amount of clamping force. A negative relationship would be represented by an increase in one variable creating a decrease in the second variable—for example, an increase in temperature will decrease wire-insulation resistance. Another example may be increasing the number of technicians assisting in a maintenance activity will decrease the time to complete the activity. A neutral relationship would indicate that there is no relationship between the

two variables. A neutral relationship may be displayed by the number of wind farms in a state compared to the number of cars parked at a shopping mall. If your team is not sure a relationship exists between two variables, then plot the data on a scatter diagram. The type of relationship or lack of will appears in the distribution of data points. Other examples include:

- wind-turbine output with respect to blade-pitch angle,
- wind-turbine output with respect to yaw angle and the wind direction,
- generator current output with respect to RPM,
- cable temperature with respect to current flow,
- cooling system temperature with respect to fan rotational speed, and
- hours spent exercising each week with respect to muscle strain injuries.

Run Chart

A **run chart** is a means of representing process performance data over a time period. **Figure 17-9** shows an example of a run chart. This analysis tool may be used to monitor wire insulation-resistance variation with service life. Other examples include:

- increase in oil viscosity with service life,
- increase in metal wear elements present in oil samples,
- decrease in troubleshooting time with technician field experience,
- increase in service cost with wind-turbine service life, and
- decrease in maintenance cycle time with technician field experience.

Analysis tools are a great way to track and understand data relationships. Deciding which tool is appropriate for processing data may be challenging at first but will become

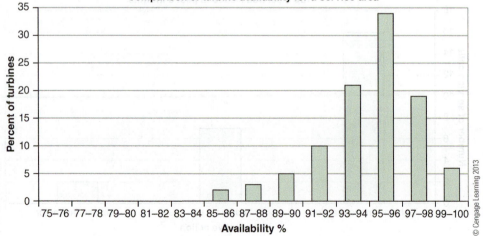

FIGURE 17-7 *Histogram comparing wind-turbine availability for service area over a month*

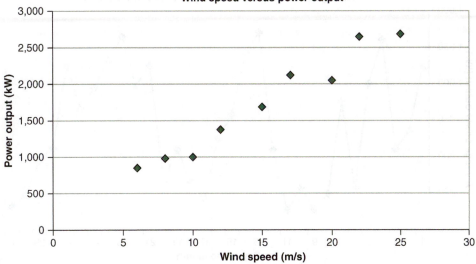

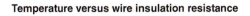

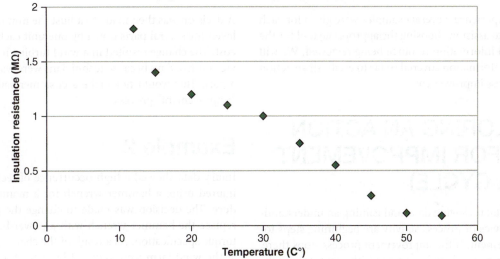

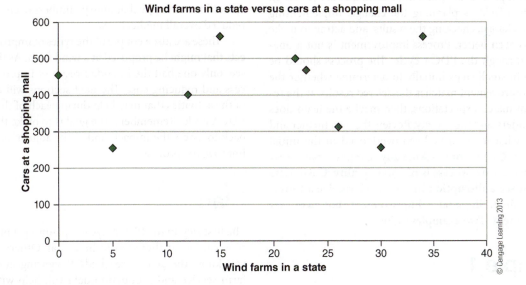

FIGURE 17-8 *Three scatter-diagram relationships*

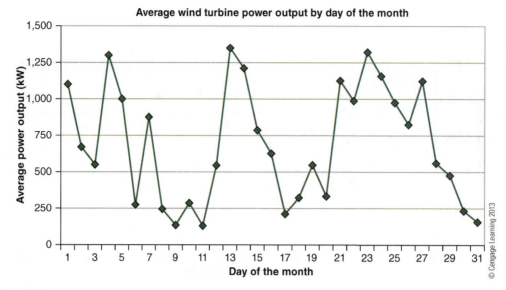

FIGURE 17-9 *A run chart*

easier with experience. Several examples were given for each analysis tool to assist in choosing the appropriate tool for the data type and information resource being reviewed. We will now use the information covered so far to establish an action plan for process improvement.

DEVELOPING AN ACTION PLAN FOR IMPROVEMENT (PDCA CYCLE)

Collecting data, reviewing data, and gaining an understanding of the service or process activity are beginning steps toward improvement. If the improvement process stops there, however, nothing will have been gained. The next steps in the improvement process are **plan, do, check, and act (PDCA)**. Simply put, PDCA is planning the change, implementing (doing) the change, checking the results, and acting to make the process even better. Process improvement is not a one-time pass through the PDCA cycle. The process should be completed in small steps initially to determine whether the intended improvement matches the desired results. If the results do not match expectations, then maybe the team does not fully understand the activity. Review the data further and determine what might have been overlooked in the initial steps. **CAUTION: Before making any changes, ensure that the final service or process is of equal quality. Customers should not see a disruption in service or receive a lower-quality product. The goal is improvement, not customer disappointment. Two examples follow.**

Example 1

Reviewing inventory cost levels showed that savings might be realized by decreasing minimum levels of expensive parts.

A decision was then made to adjust the minimum inventory levels for several parts down by one unit each to reduce site cost. The change resulted in a wind turbine being out of service for five days because technicians were waiting to receive a part. This would not make a customer satisfied with the "improvement" process.

Example 2

Injury data showed a high occurrence of technicians being injured using a hammer wrench for a maintenance procedure. The decision was made to change the procedure and replace the hammer wrench with a power-bolting tool and torque specification. As a result of the change, pinch injuries at the wind farm were reduced by 95%. The customer was satisfied that maintenance time did not change. The corporate office was happy that injury numbers decreased and thus reduced overall medical costs.

These are just a couple of the types of improvement projects that might be proposed at a wind farm. As the reader can see, only one had the intended result of improving the process and reducing cost. The next sections will discuss some of the activities that may help during each of the steps in the PDCA cycle. Remember—the goal is to make the change, go back to assess the impact, and make adjustments to further improve the process.

Plan

The first step in the PDCA cycle is planning, but before planning can occur there has to be a goal. Otherwise, how will you know the goal is reached? Reviewing available wind-farm service and production data will help with determining a goal. Using a Pareto chart for this step may help define the issue. For example, wind-turbine fault issues appear to

PLANNING SUGGESTIONS

Item	Example
Define the objective (be specific)	• Increase turbine availability 2% • Decrease operating cost by 5% • Decrease muscle strain injuries by 25%
Determine actions	• Partner experienced technicians with rookies, update troubleshooting procedures to improve clarity, develop flow charts to aid in troubleshooting • Consult engineering on inventory requirements, work with suppliers to determine component delivery times • Implement stretching exercises to improve flexibility
Assign responsibilities	• Site lead technician to adjust work schedules • Site administrator to coordinate inventory levels and supply procurement schedules • Site safety coordinator to develop stretching exercise requirements
Set a time frame	• Work schedule to be adjusted within a week • Schedule time with engineering within two weeks, set up procurement schedule with supplier within four weeks • safety coordinator to contact local physical therapy (PT) clinic within a week and set up stretching requirements within three weeks
Monitor results	• Use check sheets, histograms, and run sheets to track downtime issues by team • Use inventory level run sheets to track cost and histogram to track on-time deliveries • Use a Pareto chart and run sheets to track injuries
Define the end result	Objective levels to be maintained over a calendar quarter

© Cengage Learning 2013

FIGURE 17-10 *Suggestions to use in the planning process*

be high for several wind turbines. Tackling all faults at once would be overwhelming. Plug the fault occurrence data into a spreadsheet program and determine which fault has the highest occurrence.

Recommendation

A working knowledge of computer software such as spreadsheets, word processing, and Internet browsers is a plus for a wind technician. Use the Pareto chart results to narrow the focus for the improvement process. Does this sound similar to the previously presented issue of an intermittent connection? Remember—choose a product, service, or process that is not meeting expectations. And use all information resources available to narrow the focus to one item that will reduce cost, time, injuries, or that will increase output. Suggestions from customers, managers, and technicians are also good resources for narrowing the focus. Maybe there is one thing that would make someone's job easier or safer.

Now that you have a goal, how do you get there? Planning is the road map to negotiate the turns along the path. **Figure 17-10** shows suggestions to use in the planning process. Activities may vary, depending on the focus of the improvement project.

Do

The next step in the PDCA process is to implement the proposed actions to address the issues. The personnel assigned the tasks will need to develop specific short-term goals that

will enable them to track their progress toward the overall site objective. For example, the site safety coordinator will need to develop a specific stretching regiment that targets the muscles typically injured during work activities. Consulting with physical-therapy specialists will ensure that the exercises improve employees' flexibility and reduce muscle injuries.

Intermediate steps for this goal might include:

• evaluating employees for their initial flexibility,
• determining the initial level for stretching exercises (excessive stretching can also cause injuries),
• determining the types of stretching exercises,
• providing training to employees on benefits and proper exercise methods,
• monitoring employees to ensure they follow the program, and
• continuing to track injuries by type and frequency.

These are some of the intermediate items that would be implemented to ensure that the program objective is attained. If intermediate steps cannot be obtained, then the project will not be completed with the desired results. If you are wondering why flexibility might be an important factor for wind technicians, consider that the work area is around 300 feet (90 meters) straight up a ladder. Wind-turbine manufacturers have provided options such as crew lifts and climb-assist equipment to project developers over the past several years to assist technicians in getting to the nacelle. These options can reduce injuries, but they are not available in every wind turbine. These options also do not exist during construction or in periods when utility power is not available to the wind turbine, so maintaining flexibility should still be considered important.

Implementing intermediate steps for each of the other projects would have different goals but would be approached in the same manner. Establishing and meeting intermediate goals will ensure that the end result is attained for the site project.

Check

Check the results of the project against the objectives set previously in the planning process. Was the project completed within the established time frame? Has the team met the objective? Did the team get close to the objective set previously? If so, what were some of the obstacles encountered along the way? What strategies were used to overcome the obstacles? How effective were the strategies? Were the objectives set to high for the first pass? Share these findings with managers in the organization and determine how they may be addressed to improve the process the next time around. Did the team exceed the objective? If so, great. Congratulate everyone for their effort. The job is not complete. Remember—this is continuous improvement.

Act

Develop a new objective or adjust the original objective with what was learned on the first PDCA cycle. Use lessons learned from obstacles encountered and effective strategies implemented in order to develop a new action plan. Share the new plan with the team and determine what steps need to be adjusted to achieve the objective. Continue the process until the objective is met and can be maintained as a new standard for the operation.

CONTINUOUS IMPROVEMENT

Continuous improvement is not a new philosophy for managers. The continuous-improvement philosophy has been around for many years operating under other names or with different buzz words. Major corporations around the globe use this philosophy to empower their employees to help improve their organizations. Empowering employees to use their skills and knowledge for process improvements has increased organizational efficiency, reduced overall cost, and improved worker satisfaction.

Doing more with less seems to be the standard operating procedure for companies today. Allowing employees to monitor processes and make adjustments to daily activities can free

a manager's time to develop strategic planning. Developing effective plans to navigate the global market will ensure an organization's longevity. It does not matter if the organization is Enercon, Vestas, Iberdrola, or Acciona—strategic planning is a necessity. Managers getting bogged down in daily activities will prevent an organization from moving ahead. The sign of a good manager is to hire the right individuals, give them the right tools and the liberty to do their jobs. Give individuals guidance on the organization's objectives and get out of the way. This philosophy should be the same for a wind farm. Hire the right individual, ensure that he or she has the proper training, tools, and motivation and get out of the way. Providing workers with organizational goals and tools to monitor performance will enable them to make adjustments to achieve those goals. This strategy will enable the organization to improve and compete in today's economy.

SUMMARY

Continuous improvement is a way of life in the global economy. If an organization cannot supply its customer with the services and products they need and want, then the customer will find an organization that can do the job. Using an organized approach such as management tools suggested in this chapter will enable employees and managers to review issues and determine solutions for improving their process. It does not matter if the process is changing gearbox oil or ordering components for the next maintenance cycle—there may be ways to improve the process. Always review proposed process changes before implementation to ensure that the outcome will provide a service or product of equal or better quality. The goal is process improvement, not customer disappointment.

REVIEW QUESTIONS AND EXERCISES

1. List some data sources used to analyze performance.
2. What role does cost play in determining performance?
3. What role does safety play in determining performance?
4. What role does collecting injury data play in an organization?
5. What wind-farm information can help identify availability issues?
6. Explain the difference between variable and attribute data.
7. How is a Pareto chart useful in determining priorities?
8. List some uses for a run chart.
9. List the four steps of the PDCA cycle.
10. What are some benefits of a continuous-improvement process?

Monthly Inspection Form

Safety Item	Identification Number	Manufacturer	Inspection Date	Accept/Reject
Full-body Harness*	PSA111678	PSA	07 July 2011	Accept

*Example of information that may be entered in the columns

Safety Gear Inspection Forms

Full Body Harness

Annual Inspection Checklist

Harness Model/Name:_____

Serial Number:_____ Lot Number:_____

Date of Manufacture:_____ Date of Purchase:_____

Comments:_____

_____

_____

General Factors	Accepted/Rejected	Supportive Details/Comments
1) **Hardware:** includes D-rings, buckles, keepers, and back pads. Inspect for damage, distortion, sharp edges, burrs, cracks, and corrosion.	Accepted Rejected	
2) **Webbing:** Inspect for cuts, burns, tears, abrasions, frays, excessive soiling, and discoloration.	Accepted Rejected	
3) **Stitching:** Inspect for pulled or cut stitches.	Accepted Rejected	
4) **Labels:** Inspect, making certain all labels are securely held in place and are legible.	Accepted Rejected	
5) Other:	Accepted Rejected	
6) Other:	Accepted Rejected	
7) **Overall Disposition:**	Accepted Rejected	**Inspected By:** **Date Inspected:**

Form No: AAE-017   Harness and Lanyard Inspection

Lanyards

Annual Inspection Checklist

Lanyard Model/Name:_____

Serial Number:_____ Lot Number:_____

Date of Manufacture:_____ Date of Purchase:_____

Comments:_____

_____

_____

General Factors	Accepted/Rejected	Supportive Details/Comments
1) **Hardware:** (includes snaphooks, carabiners, adjusters, keepers, thimbles, and D-rings) Inspect for damage, distortion, sharp edges, burrs, cracks, corrosion, and proper operation.	Accepted Rejected	
2) **Webbing:** Inspect for cuts, burns, tears, abrasions, frays, excessive soiling, and discoloration.	Accepted Rejected	
3) **Stitching:** Inspect for pulled or cut stitches.	Accepted Rejected	
4) **Synthetic Rope:** Inspect for pulled or cut yarns, burns, abrasions, knots, excessive soiling, and discoloration.	Accepted Rejected	
5) **Energy Absorbing Component:** Inspect for elongation, tears, and excessive soiling.	Accepted Rejected	
6) **Labels:** Inspect, making certain all labels are securely held in place and are legible.	Accepted Rejected	
Overall Disposition:	Accepted Rejected	**Inspected By:** **Date Inspected:**

Form No: AAE-017   Harness and Lanyard Inspection

Capital Safety

Snaphooks/Carabiners

Annual Inspection Checklist

Hook/Carabiner Model/Name:_____

Serial Number:_____   Lot Number:_____

Date of Manufacture:_____   Date of Purchase:_____

Comments:_____

_____

_____

General Factors	Accepted/Rejected	Supportive Details/Comments
1) **Physical Damage:** Inspect for cracks, sharp edges, burrs, deformities, and locking operations.	Accepted Rejected	
2) **Excessive Corrosion:** Inspect for corrosion, which affects the operation and/or the strength.	Accepted Rejected	
3) **Markings:** Inspect and make certain marking(s) are legible.	Accepted Rejected	
4) Other:	Accepted Rejected	
5) Other:	Accepted Rejected	
6) Other:	Accepted Rejected	
Overall Disposition:	Accepted Rejected	**Inspected By:** **Date Inspected:**

Form No: AAE-017   Harness and Lanyard Inspection

Capital Safety

Example of an Energized Electrical Work Permit

Wind Turbine Site ABC				
Energized Electrical Work Permit				
Date (dd/mm/yyyy):				
Turbine number or facility location				
Circuit description				
Circuit number				
Justification for energized work activity				
Shock hazard analysis results				
Shock protection boundary determination [NFPA Table 130.2(C)]	Nominal Voltage	Limited Approach	Restricted Approach	Prohibited Approach
Arc-flash hazard analysis results				
Arch-flash protection boundary				
HRC requirement and list required PPE [NFPA Table 130.7(C)(9)]	HRC		0, 1, 2, 2,* 3, 4	
Method to prevent unauthorized access				
Tailgate attendance	1.		4.	
	2.		5.	
	3.		6.	
Specific job hazards:				
Approvals:				

Title	Print	Signature
Site Manager		
Lead Technician		
Safety Coordinator		
Team Leader		

Refer to site safety plan for analysis method details and latest edition of National Fire Protection Association 70E Standard as necessary.

D

MSDS for SKF LGWM2 Grease

SAFETY DATA SHEET
MATERIAL SAFETY DATA SHEET

Last changed: 12/11/2008	Internal No:	Replaces date:

LGWM 2

---

1. IDENTIFICATION OF THE SUBSTANCE/PREPARATION AND OF THE COMPANY/UNDERTAKING

TRADE NAME   LGWM 2
TYPE OF USE   SKF extreme pressure wide temperature range grease

National manufacturer/importer

Enterprise	SKF Maintenance Products
Address	Postbus 1008
Postal code	NL-3430 BA Nieuwegein
Country	The Netherlands
Telephone	+31 30 6307200
Fax	+31 30 6307205

Name		**E-mail**	**Tel. (work)**	**Country**
David	Sébastien	sebastien.david@skf.com	+ 31 30 6307200	The Netherlands

Emergency Phone	**Type of assistance**	**Opening Hours**
(+44) 08 45 46 47	NHS Direct	Available 24 hours

---

2. HAZARDS IDENTIFICATION

HEALTH
May be irritating to eyes.
Prolonged or frequently recurrent contact will degrease the skin and may cause skin irritation.

FIRE AND EXPLOSION
Not flammable, but combustible.

ENVIRONMENT
Oil products may cause ground and water pollutants.

---

3. COMPOSITION / INFORMATION ON INGREDIENTS

No	Ingredient name	Reg.No	EC No.	CAS No.	Conc. (wt%)	Classification
1	synthetic hydrocarbon oil		-	-	60 - 100 %	No classification required
2	alkylated diphenylamines		270-128-1	68411-46-1	< 1 %	N,R51/53

Explanation of symbols: T+=highly toxic, T=toxic, C=corrosive, Xn=harmful, Xi=irritant E=Explosive, O=Oxidising, F+=Extremely flammable, F=Highly flammable, N=Dangerous for the environment, Cancer=Carcinogenic, Mut=Mutagenic, Rep=Toxic for reproduction, Conc.=Concentration

INGREDIENT COMMENTS
The complete text of the R-phrases mentioned in section 3 is stated in section 16.

---

4. FIRST AID MEASURES

INHALATION
Provide rest, warmth and fresh air. Seek medical advice if symptoms develop.

SKIN CONTACT
Wash skin with soap and water. Use appropriate hand lotion to prevent degreasing and cracking of skin. Seek medical advice if irritation persists.

EYE CONTACT
Immediately flush eyes with plenty of water while holding eye lids open. Remove any contact lenses and continue rinsing with water for at least 15 minutes (keep the eyelids open). Seek medical advice if irritation persists.

INGESTION
DO NOT INDUCE VOMITING! If vomiting occurs, keep head low so that stomach contents do no enter lungs. Thoroughly rinse the mouth with water. Drink a few glasses of water or milk. Seek medical advice.

5. FIRE-FIGHTING MEASURES

EXTINGUISHING MEDIA
Fire may be extinguished using powder, foam or carbon dioxide (CO2).

SAFETY DATA SHEET
MATERIAL SAFETY DATA SHEET

Last changed: 12/11/2008   Internal No: Replaces date:

LGWM 2

IMPROPER EXTINGUISHING MEDIA
Do not use a direct water stream which may spread the fire.

EXTINGUISHING METHODS
Remove containers from fire area if this can be done without risk. Use water to cool exposed containers. If possible, fight the fire from a protected location.

FIRE AND EXPLOSION HAZARDS
Not flammable, but combustible. Carbon monoxide and hydrocarbons are released in a fire situation.

PROTECTIVE EQUIPMENT FOR FIRE FIGHTERS
General: Evacuate all personnel, use protective equipment for fire fighting. Use a portable breathing apparatus when the product is involved in a fire.

6. ACCIDENTAL RELEASE MEASURES

PERSONAL PRECAUTIONS
Provide good ventilation. Avoid contact with the skin, eyes, and airways. Wear necessary protective equipment. Keep the emission away from sources of ignition.

SAFETY ACTIONS TO PROTECT EXTERNAL ENVIRONMENT
Must not be released into the sewage system, drinking water supply, or ground.

METHODS FOR CLEANING UP
Dry, scrape or absorb the product with inert material and place it in a refuse container. Collect and dispose of materials in closed and labelled containers at an approved disposal or waste collection facility. Clean with soap and water or detergent. Flush clean with plenty of water. Be aware of the risk of slippery surfaces.

7. HANDLING AND STORAGE

HANDLING ADVICE
Provide good ventilation. Avoid contact with the skin and eyes. Wear necessary protective equipment.

STORAGE
Keep in a dry, cool, and well-ventilated area. Keep containers tightly closed. Keep in original container. Keep away from sources of ignition. Store separately from oxidising materials.

8. EXPOSURE CONTROLS/PERSONAL PROTECTION

EXPOSURE CONTROL
Provide good ventilation. Wash the skin at the end of each work shift and before eating, smoking, and using restroom facilities. Eye wash facilities and safety shower must be available when handling this product.

RESPIRATORY PROTECTION
Respiratory protective equipment is usually not necessary. In case of insufficient ventilation, wear respiratory protective equipment with A/P2 filter.

EYE PROTECTION
Wear approved safety goggles if there is a risk of eye splash.

HAND PROTECTION
Protective gloves (chloroprene rubber, nitrile rubber).

PROTECTIVE CLOTHING
Wear suitable protective clothing.

9. PHYSICAL AND CHEMICAL PROPERTIES

Physical State   Paste.
Colour   Brown.
Odour   Characteristic.
Water solubility   Insoluble in water.

Physical and chemical parameters

Parameter	Value/unit	Method/reference	Observation
Melting point	> 290 °C		
Flash point	> 300 °C	ASTM E 659	

SKF Maintenance Products

SAFETY DATA SHEET
MATERIAL SAFETY DATA SHEET

Last changed: 12/11/2008	Internal No:	Replaces date:

LGWM 2

Density	0,9 g/cm³	15 °C	

10. STABILITY AND REACTIVITY

STABILITY
Stable under recommended storage and handling conditions.

CONDITIONS TO AVOID
Avoid heat, sparks, and naked flames.

MATERIALS TO AVOID
Avoid contact with oxidising substances.

11. TOXICOLOGICAL INFORMATION

INHALATION
On warming/heating, the vapours emitted may cause irritation to the respiratory organs.

SKIN CONTACT
Prolonged or frequently recurrent contact will degrease the skin and may cause skin irritation. May cause inflammation of the skin (dermatitis). In case of injection under the skin caused by spraying pistol, wounds must be cleaned thoroughly; if necessary surgical excision.

EYE CONTACT
May cause moderate irritation/smarting.

INGESTION
May cause irritation of the mucous membranes, nausea, and vomiting.

12. ECOLOGICAL INFORMATION

MOBILITY
Insoluble in water. The product floats on water.

DEGRADABILITY
Biodegradable.

ACCUMULATION
Bioaccumulation improbable.

OTHER EFFECTS
Oil products may cause ground and water pollutants.

CONCLUSION
Biodegradable.
Bioaccumulation improbable.

Prevent discharge to the sewage system, waterways, or lakes.

13. DISPOSAL CONSIDERATIONS

GENERAL REGULATIONS
Treat as hazardous waste. Collect and dispose of materials in closed and labelled containers at an approved disposal or waste collection facility. Empty packages must be handled in an environmentally safe manner in accordance with applicable laws and regulations.

CATEGORY OF WASTE
The EWC code is a suggestion only; the end user will select a suitable EWC code. 12 01 12* spent waxes and fats

14. TRANSPORT INFORMATION

Classified as Dangerous Goods: ☐ Yes   ☒ No   ☐ Not assessed

OTHER INFORMATION

SKF Maintenance Products

SAFETY DATA SHEET
MATERIAL SAFETY DATA SHEET

Last changed: 12/11/2008	Internal No:	Replaces date:

LGWM 2

Not classified as dangerous goods.

---

15. REGULATORY INFORMATION

EC-Label   ☒ No   ☐ Yes   ☐ Not assessed

COMPOSITION
  synthetic hydrocarbon oil (60 - 100 %), alkylated diphenylamines (< 1 %)

S-PHRASES
S24/25 Avoid contact with skin and eyes.
S37 Wear suitable gloves.

REFERENCES
Directive 1999/45/EC.
Annex I to Directive 67/548/EEC on Classification and Labelling of Dangerous Substances.
Regulation on Registration, Evaluation, Authorisation and Restriction of Chemicals (REACH) (2008).
European Waste Catalogue (EWC 2002).
Occupational exposure limits (EH40/2005).
Transport of dangerous goods: ADR, RID, IMDG and IATA.

---

16. OTHER INFORMATION

LIST OF RELEVANT R-PHRASES

Nr.	R-Phrase text
R51/53	Toxic to aquatic organisms may cause long-term adverse effects in the aquatic environment.

ISSUED: 12/11/2008

APPENDIX

E

Example of an Oil Analysis Report

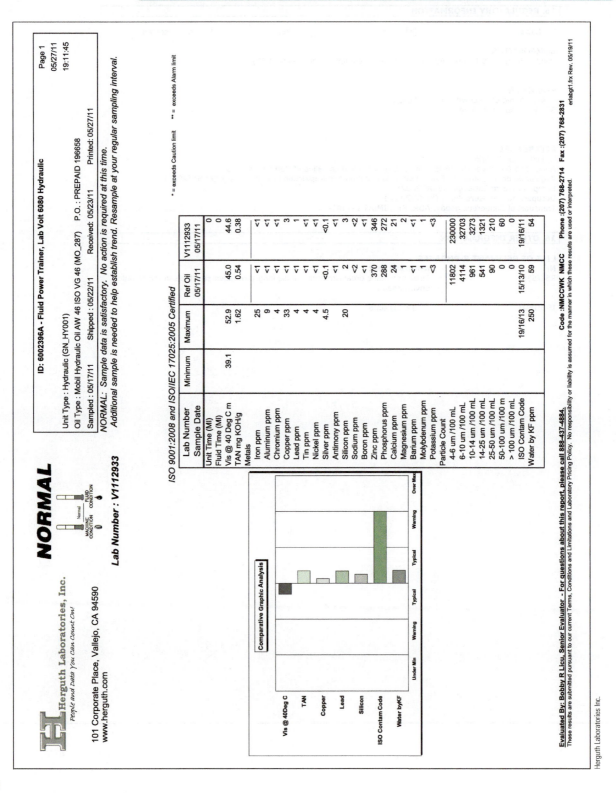

Reference Torque Values for ASTM & SAE Grade Fasteners

Suggested Starting Torque Values

(800) 547-6758 | www.portlandbolt.com

ASTM A307

Bolt Size (in)	TPI	Proof Load (lbs)	Clamp Load (lbs)	Tightening Torque (ft lbs)		
				Waxed	Galv	Plain
1/4	20	1,145	859	2	4	4
5/16	18	1,886	1415	4	9	7
3/8	16	2,790	2,093	7	16	13
7/16	14	3,827	2,870	10	26	21
1/2	13	5,108	3,831	16	40	32
9/16	12	6,552	4,914	23	58	46
5/8	11	8,136	6,102	32	79	64
3/4	10	12,024	9,018	56	141	113
7/8	9	15,200	11,400	83	208	166
1	8	20,000	15,000	125	313	250
1 1/8	7	25,200	18,900	177	443	354
1 1/4	7	32,000	24,000	250	625	500
1 3/8	6	38,100	28,575	327	819	655
1 1/2	6	46,400	34,800	435	1,088	870
1 3/4	5	68,400	51,300	748	1,870	1,496
2	4 1/2	90,000	67,500	1,125	2,813	2,250
2 1/4	4 1/2	117,000	87,750	1,645	4,113	3,291
2 1/2	4	144,000	108,000	2,250	5,625	4,500
2 3/4	4	177,480	133,110	3,050	7,626	6,101
3	4	214,920	161,190	4,030	10,074	8,060
3 1/4	4	255,600	191,700	5,192	12,980	10,384
3 1/2	4	299,880	224,910	6,560	16,400	13,120
3 3/4	4	347,760	260,820	8,151	20,377	16,301
4	4	398,880	299,160	9,972	24,930	19,944

*This chart of estimated torque calculations is only offered as a guide. **Use of its content by anyone is the sole responsibility of that person and they assume all risk**. Due to many variables that affect the torque–tension relationship like human error, surface texture, and lubrication the only way to determine the correct torque is through experimentation under actual joint and assembly conditions.*

Portland Bolt and Manufacturing

Suggested Starting Torque Values

(800) 547-6758 | www.portlandbolt.com

SAE Grade 2

Bolt Size (in)	TPI	Proof Load (lbs)	Clamp Load (lbs)	Tightening Torque (ft lbs)		
				Waxed	Galv	Plain
1/4	20	1,750	1,313	3	7	5
5/16	18	2,900	2,175	6	14	11
3/8	16	4,250	3,188	10	25	20
7/16	14	5,850	4,388	16	40	32
1/2	13	7,800	5,850	24	61	49
9/16	12	10,000	7,500	35	88	70
5/8	11	12,400	9,300	48	121	97
3/4	10	18,400	13,800	86	216	173
7/8	9	15,200	11,400	83	208	166
1	8	20,000	15,000	125	313	250
1 1/8	7	25,200	18,900	177	443	354
1 1/4	7	32,000	24,000	250	625	500
1 3/8	6	38,100	28,575	327	819	655
1 1/2	6	46,400	34,800	435	1,088	870

ASTM A325

Bolt Size (in)	TPI	Tension		Tightening Torque Range (ft lbs) (Min - Max)		
		Min	Max	Plain	Galv	Waxed
1/2	13	12,000	14,000	100 - 117	125 - 146	50 - 58
5/8	11	19,000	23,000	198 - 240	247 - 299	99 - 120
3/4	10	28,000	34,000	350 - 425	438 - 531	175 - 213
7/8	9	39,000	47,000	569 - 685	711 - 857	284 - 343
1	8	51,000	61,000	850 – 1,017	1,063 – 1,271	425 - 508
1-1/8	7	56,000	67,000	1,050 – 1,256	1,313 – 1,570	525 - 625
1-1/4	7	71,000	85,000	1,479 – 1,771	1,849 – 2,214	740 - 885
1-3/8	6	85,000	102,000	1,948 – 2,338	2,435 – 2,922	974 - 1,169
1-1/2	6	103,000	124,000	2,575 – 3,100	3,219 – 3,875	1,288 – 1,550

*This chart of estimated torque calculations is only offered as a guide. **Use of its content by anyone is the sole responsibility of that person and they assume all risk**. Due to many variables that affect the torque–tension relationship like human error, surface texture, and lubrication the only way to determine the correct torque is through experimentation under actual joint and assembly conditions.*

Portland Bolt and Manufacturing

Suggested Starting Torque Values

(800) 547-6758 | www.portlandbolt.com

ASTM A449 / SAE Grade 5

Bolt Size (in)	TPI	Proof Load (lbs)	Clamp Load (lbs)	Tightening Torque (ft lbs)		
				Waxed	Galv	Plain
1/4	20	2,700	2,025	4	11	8
5/16	18	4,450	3,338	9	22	17
3/8	16	6,600	4,950	15	39	31
7/16	14	9,050	6,788	25	62	49
1/2	13	12,050	9,038	38	94	75
9/16	12	15,450	11,588	54	136	109
5/8	11	19,200	14,400	75	188	150
3/4	10	28,400	21,300	133	333	266
7/8	9	39,250	29,438	215	537	429
1	8	51,500	38,625	322	805	644
1 1/8	7	56,450	42,338	397	992	794
1 1/4	7	71,700	53,775	560	1,400	1,120
1 3/8	6	85,450	64,088	734	1,836	1,469
1 1/2	6	104,000	78,000	975	2,438	1,950
1 3/4	5	104,500	78,375	1,143	2,857	2,286
2	4 1/2	137,500	103,125	1,719	4,297	3,438
2 1/4	4 1/2	178,750	134,063	2,514	6,284	5,027
2 1/2	4	220,000	165,000	3,438	8,594	6,875
2 3/4	4	271,150	203,363	4,660	11,651	9,321
3	4	328,350	246,263	6,157	15,391	12,313

ASTM A490

Bolt Size (in)	TPI	Tension (lbs)		Tightening Torque Range (ft lbs) (Min - Max)	
		Min	Max	Plain	Lubricated
1/2	13	15,000	18,000	125 - 150	63 - 75
5/8	11	24,000	29,000	250 - 302	125 - 151
3/4	10	35,000	42,000	438 - 525	219 - 263
7/8	9	49,000	59,000	715 - 860	357 - 430
1	8	64,000	77,000	1,067 – 1,283	533 - 642
1-1/8	7	80,000	96,000	1,500 – 1,800	750 - 900
1-1/4	7	102,000	122,000	2,125 – 2,542	1,063 – 1,271
1-3/8	6	121,000	145,000	2,773 – 3,323	1,386 – 1,661
1-1/2	6	148,000	178,000	3,700 – 4,450	1,850 – 2,225

*This chart of estimated torque calculations is only offered as a guide. **Use of its content by anyone is the sole responsibility of that person and they assume all risk**. Due to many variables that affect the torque–tension relationship like human error, surface texture, and lubrication the only way to determine the correct torque is through experimentation under actual joint and assembly conditions.*

Portland Bolt and Manufacturing

Suggested Starting Torque Values

(800) 547-6758 | www.portlandbolt.com

ASTM A193 Grade B7

Bolt Size (in)	TPI	Proof Load (lbs)	Clamp Load (lbs)	Tightening Torque (ft lbs)		
				Waxed	Galv	Plain
1/4	20	3,350	2,513	5	13	10
5/16	18	5,500	4,125	11	27	21
3/8	16	8,150	6,113	19	48	38
7/16	14	11,150	8,363	30	76	61
1/2	13	14,900	11,175	47	116	93
9/16	12	19,100	14,325	67	168	134
5/8	11	23,750	17,813	93	232	186
3/4	10	35,050	26,288	164	411	329
7/8	9	48,500	36,375	265	663	530
1	8	63,650	47,738	398	995	796
1 1/8	7	80,100	60,075	563	1,408	1,126
1 1/4	7	101,750	76,313	795	1,987	1,590
1 3/8	6	121,300	90,975	1,042	2,606	2,085
1 1/2	6	147,550	110,663	1,383	3,458	2,767
1 3/4	5	199,500	149,625	2,182	5,455	4,364
2	4 1/2	262,500	196,875	3,281	8,203	6,563
2 1/4	4 1/2	341,250	255,938	4,799	11,997	9,598
2 1/2	4	420,000	315,000	6,563	16,406	13,125
2 3/4	4	468,500	351,263	8,050	20,124	16,100
3	4	567,150	425,363	10,634	26,585	21,268
3 1/4	4	674,500	505,875	13,701	34,252	27,402
3 1/2	4	791,350	593,513	17,311	43,277	34,622
3 3/4	4	917,700	688,275	21,509	53,771	43,017
4	4	1,052,600	789,450	26,315	65,788	52,630

*This chart of estimated torque calculations is only offered as a guide. **Use of its content by anyone is the sole responsibility of that person and they assume all risk**. Due to many variables that affect the torque–tension relationship like human error, surface texture, and lubrication the only way to determine the correct torque is through experimentation under actual joint and assembly conditions.*

Portland Bolt and Manufacturing

Suggested Starting Torque Values

(800) 547-6758 | www.portlandbolt.com

ASTM A354 Grade BD / SAE Grade 8

Bolt Size (in)	TPI	Proof Load (lbs)	Clamp Load (lbs)	Tightening Torque (ft lbs)	
				Lubricated	Plain
1/4	20	3,800	2,850	6	12
5/16	18	6,300	4,725	12	25
3/8	16	9,300	6,975	22	44
7/16	14	12,750	9,563	35	70
1/2	13	17,050	12,788	53	107
9/16	12	21,850	16,388	77	154
5/8	11	27,100	20,325	106	212
3/4	10	40,100	30,075	188	376
7/8	9	55,450	41,588	303	606
1	8	72,700	54,525	454	909
1 1/8	7	91,550	68,663	644	1,287
1 1/4	7	120,000	90,000	938	1,875
1 3/8	6	138,600	103,950	1,191	2,382
1 1/2	6	168,600	126,450	1,581	3,161
1 3/4	5	228,000	171,000	2,494	4,988
2	4 1/2	300,000	225,000	3,750	7,500
2 1/4	4 1/2	390,000	292,500	5,484	10,969
2 1/2	4	480,000	360,000	7,500	15,000
2 3/4	4	517,650	388,238	8,897	17,794
3	4	626,850	470,138	11,753	23,507
3 1/4	4	745,500	559,125	15,143	30,286
3 1/2	4	874,650	655,988	19,133	38,266
3 3/4	4	1,014,300	760,725	23,773	47,545
4	4	1,163,400	872,550	29,085	58,100

Notes:

1. Values calculated using industry accepted formula T = KDP where T = Torque, K = torque coefficient (dimensionless), D = nominal diameter (inches), P = bolt clamp load, lb.
2. K values: waxed (e.g. pressure wax as supplied on high strength nuts) = .10, hot dip galvanized = .25, and plain non-plated bolts (as received) = .20.
3. Torque has been converted into ft/lbs by dividing the result of the formula by 12.
4. All calculations are for Coarse Thread Series (UNC).
5. Grade 2 calculations only cover fasteners 1/4"-3/4" in diameter up to 6" long; for longer fasteners the torque is reduced significantly.
6. Clamp loads are based on 75% of the minimum proof loads for each grade and size.
7. Proof load, stress area, yield strength, and other data is based on IFI 7th Edition (2003) Technical Data N-68, SAE J429, ASTM A307, A325, A354, A449, and A490.

*This chart of estimated torque calculations is only offered as a guide. **Use of its content by anyone is the sole responsibility of that person and they assume all risk**. Due to many variables that affect the torque–tension relationship like human error, surface texture, and lubrication the only way to determine the correct torque is through experimentation under actual joint and assembly conditions.*

Portland Bolt and Manufacturing

Power Tool Pressure versus Torque Conversion Chart

PRESSURE / TORQUE CONVERSION CHART

MXT SERIES

Pressure in PSI	P7MXT	1MXT	3MXT	5MXT	10MXT	20MXT	25MXT	35MXT
				Torque in Ft. Lbs				
1500	118	200	480	835	1755	2960	3960	5400
1600	126	214	512	890	1864	3158	4216	5772
1800	143	242	576	1000	2082	3555	4728	6516
2000	160	270	640	1110	2300	3950	5240	7260
2200	176	297	704	1222	2526	4345	5752	8021
2400	193	324	768	1334	2752	4740	6264	8782
2600	209	351	832	1446	2978	5135	6776	9543
2800	226	378	896	1558	3204	5530	7288	10304
3000	242	405	960	1670	3430	5930	7800	11065
3200	259	431	1024	1782	3656	6325	8318	11817
3400	275	457	1088	1894	3882	6720	8836	12569
3600	292	483	1152	2006	4108	7115	9354	13321
3800	308	509	1216	2118	4334	7510	9872	14073
4000	325	535	1280	2230	4560	7905	10390	14825
4200	342	562	1346	2342	4792	8300	10898	15572
4400	359	589	1412	2454	5024	8695	11406	16319
4600	375	616	1478	2566	5256	9090	11914	17066
4800	392	643	1544	2678	5488	9485	12422	17813
5000	409	670	1610	2790	5720	9880	12930	18560
5200	425	697	1674	2902	5948	10275	13450	19306
5400	442	724	1738	3014	6176	10670	13970	20052
5600	458	751	1802	3126	6404	11065	14490	20798
5800	475	778	1866	3238	6632	11460	15010	21544
6000	491	805	1930	3350	6860	11860	15530	22290
6200	508	832	1994	3462	7094	12250	16040	23027
6400	524	859	2058	3574	7328	12645	16550	23764
6600	541	886	2122	3686	7562	13040	17060	24501
6800	557	913	2186	3798	7796	13435	17570	25238
7000	574	940	2250	3910	8030	13830	18080	25975
7200	591	967	2316	4022	8264	14225	18602	26716
7400	608	994	2382	4134	8498	14620	19124	27457
7600	625	1021	2448	4246	8732	15020	19646	28198
7800	642	1048	2514	4358	8966	15415	20168	28939
8000	659	1075	2580	4470	9200	15810	20690	29680
8200	675	1101	2646	4582	9432	16200	21214	30422
8400	691	1127	2712	4694	9664	16600	21738	31164
8600	706	1153	2778	4806	9896	16995	22262	31906
8800	722	1179	2844	4918	10128	17390	22786	32648
9000	738	1205	2910	5030	10360	17785	23310	33390
9200	755	1232	2974	5142	10592	18180	23826	34132
9400	772	1259	3038	5254	10824	18575	24342	34874
9600	788	1286	3102	5366	11056	18970	24858	35616
9800	805	1313	3166	5478	11288	19365	25374	36358
10000	822	1340	3230	5590	11520	19760	25890	37100

Hytorc-Wind, LLC

PRESSURE / TORQUE CONVERSION CHART

MXT SERIES

Pressure in PSI	1MXT-SA	3MXT-SA	5MXT-SA	10MXT-SA
	Torque in Ft. Lbs			
1500	200	480	835	1755
1600	214	512	890	1864
1800	242	576	1000	2082
2000	270	640	1110	2300
2200	297	704	1222	2526
2400	324	768	1334	2752
2600	351	832	1446	2978
2800	378	896	1558	3204
3000	405	960	1670	3430
3200	431	1024	1782	3656
3400	457	1088	1894	3882
3600	483	1152	2006	4108
3800	509	1216	2118	4334
4000	535	1280	2230	4560
4200	562	1346	2342	4792
4400	589	1412	2454	5024
4600	616	1478	2566	5256
4800	643	1544	2678	5488
5000	670	1610	2790	5720
5200	697	1674	2902	5948
5400	724	1738	3014	6176
5600	751	1802	3126	6404
5800	778	1866	3238	6632
6000	805	1930	3350	6860
6200	832	1994	3462	7094
6400	859	2058	3574	7328
6600	886	2122	3686	7562
6800	913	2186	3798	7796
7000	940	2250	3910	8030
7200	967	2316	4022	8264
7400	994	2382	4134	8498
7600	1021	2448	4246	8732
7800	1048	2514	4358	8966
8000	1075	2580	4470	9200
8200	1101	2646	4582	9432
8400	1127	2712	4694	9664
8600	1153	2778	4806	9896
8800	1179	2844	4918	10128
9000	1205	2910	5030	10360
9200	1232	2974	5142	10592
9400	1259	3038	5254	10824
9600	1286	3102	5366	11056
9800	1313	3166	5478	11288
10000	1340	3230	5590	11520

Hytorc-Wind, LLC

PRESSURE / TORQUE CONVERSION CHART

MXT SERIES

Pressure in 100 x Kpa	P7MXT	1MXT	3MXT	5MXT	10MXT	20MXT	25MXT	35MXT
				Torque in NM				
100	154	261	626	1088	2288	3858	5162	7039
120	188	317	756	1312	2731	4664	6203	8548
140	220	371	880	1527	3163	5432	7206	9984
160	254	426	1010	1754	3618	6231	8234	11544
180	285	478	1134	1971	4060	7000	9237	13009
200	317	531	1259	2189	4501	7768	10238	14475
220	351	584	1388	2416	4956	8574	11276	16019
240	382	635	1513	2633	5397	9343	12285	17475
260	414	685	1636	2849	5830	10103	13280	18932
280	447	736	1760	3067	6271	10871	14289	20388
300	482	791	1895	3294	6743	11670	15309	21903
320	513	843	2023	3511	7192	12439	16303	23354
340	546	895	2151	3727	7642	13199	17274	24795
360	579	950	2282	3956	8108	14007	18335	26317
380	611	1002	2405	4172	8546	14767	19337	27756
400	644	1055	2530	4389	8990	15535	20348	29205
420	675	1107	2654	4607	9434	16310	21358	30654
440	709	1162	2783	4834	9911	17103	22384	32141
460	741	1214	2908	5052	10364	17871	23381	33579
480	775	1269	3037	5278	10840	18670	24408	35066
500	808	1321	3165	5496	11293	19439	25420	36508
520	841	1374	3293	5712	11747	20206	26429	37933
540	874	1426	3421	5930	12199	20974	27441	39376
560	907	1479	3555	6156	12673	21766	28503	40875
580	938	1530	3683	6374	13123	22542	29519	42319
600	967	1580	3811	6590	13571	23302	30532	43747
620	1000	1633	3945	6819	14044	24109	31599	45263
640	1033	1686	4070	7036	14494	24878	32604	46707
660	1065	1738	4192	7252	14942	25638	33596	48135
680	1098	1790	4317	7470	15393	26407	34601	49579
690	1114	1817	4379	7578	15617	26787	35096	50293

Hytorc-Wind, LLC

PRESSURE / TORQUE CONVERSION CHART

MXT SERIES

Pressure in 100 x Kpa	1MXT-SA	3MXT-SA	5MXT-SA	10MXT-SA
		Torque in NM		
100	261	626	1088	2288
120	317	756	1312	2731
140	371	880	1527	3163
160	426	1010	1754	3618
180	478	1134	1971	4060
200	531	1259	2189	4501
220	584	1388	2416	4956
240	635	1513	2633	5397
260	685	1636	2849	5830
280	736	1760	3067	6271
300	791	1895	3294	6743
320	843	2023	3511	7192
340	895	2151	3727	7642
360	950	2282	3956	8108
380	1002	2405	4172	8546
400	1055	2530	4389	8990
420	1107	2654	4607	9434
440	1162	2783	4834	9911
460	1214	2908	5052	10364
480	1269	3037	5278	10840
500	1321	3165	5496	11293
520	1374	3293	5712	11747
540	1426	3421	5930	12199
560	1479	3555	6156	12673
580	1530	3683	6374	13123
600	1580	3811	6590	13571
620	1633	3945	6819	14044
640	1686	4070	7036	14494
660	1738	4192	7252	14942
680	1790	4317	7470	15393
690	1817	4379	7578	15617

Hytorc-Wind, LLC

SKF Laser Alignment Report

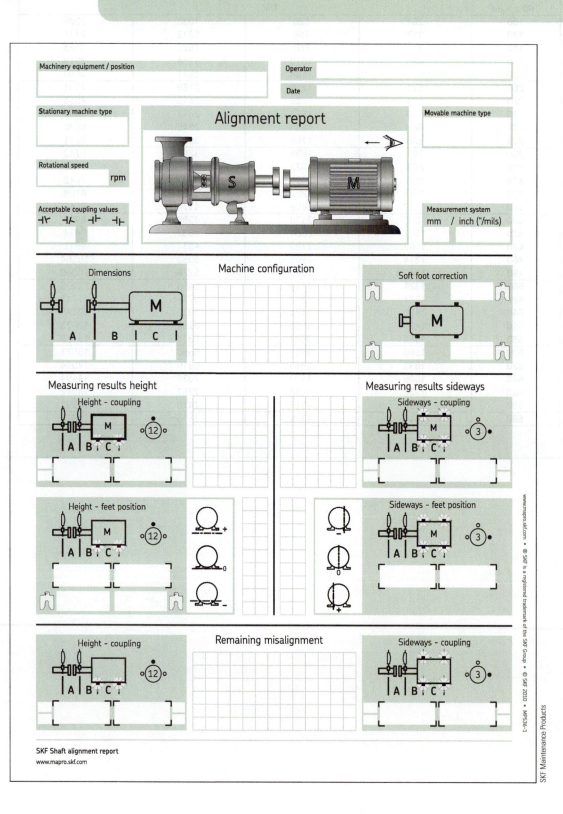

MSDS for SKF LGWM1 Grease

SAFETY DATA SHEET

LGWM 1

Last changed: 06/10/2010   Replaces date: 01/11/2007

1. IDENTIFICATION OF THE SUBSTANCE/PREPARATION AND OF THE COMPANY/UNDERTAKING

TRADE NAME   LGWM 1
APPLICATION AREA   Fat.

NATIONAL MANUFACTURER/IMPORTER

Enterprise	SKF Maintenance Products
Address	Postbus 1008
Postal code	NL-3430 BA Nieuwegein
Country	The Netherlands
Telephone	+31 30 6307200
Fax	+31 30 6307205

CONTACT PERSONS

Name	E-mail	Telephone	Country
David Sébastien	sebastien.david@skf.com	+ 31 30 6307200	The Netherlands

2. HAZARDS IDENTIFICATION

HEALTH
May be irritating to eyes.
Prolonged or frequently recurrent contact will degrease the skin and may cause skin irritation.

FIRE AND EXPLOSION
Not flammable, but combustible.

ENVIRONMENT
Oil products may cause ground and water pollutants.

3. COMPOSITION / INFORMATION ON INGREDIENTS

Ingredient name	Reg.No	EC No.	CAS No.	Conc. (wt%)	Classification
baseoil-unspecified	-	-	-	60 - 100 %	No classification required

Explanation of symbols: T+=highly toxic, T=toxic, C=corrosive, Xn=harmful, Xi=irritant E=Explosive, O=Oxidising, F+=Extremely flammable, F=Highly flammable, N=Dangerous for the environment, Cancer=Carcinogenic, Mut=Mutagenic, Rep=Toxic for reproduction, Conc.=Concentration

4. FIRST AID MEASURES

INHALATION
Remove victim immediately from source of exposure.Provide rest, warmth, and fresh air.Seek medical advice if symptoms develop.

SKIN CONTACT
Immediately flush contaminated skin with soap or mild detergent and water. Immediately remove soaked clothing and flush the skin with water.Use appropriate hand lotion to prevent degreasing and cracking of skin.Seek medical advice if irritation persists.

EYE CONTACT
Immediately flush eyes with plenty of water while holding eye lids open.Remove any contact lenses and continue rinsing with water for at least 15 minutes (keep the eyelids open).Seek medical advice if irritation persists.

INGESTION
DO NOT INDUCE VOMITING!If vomiting occurs, keep head low so that stomach contents do no enter lungs.Thoroughly rinse the mouth with water.Drink a few glasses of water or milk.Seek medical advice.

1 / 4

5. FIRE-FIGHTING MEASURES

SAFETY DATA SHEET

EXTINGUISHING MEDIA

Extinguishing agent: carbon dioxide, powder, foam, water spray, or water mist.

IMPROPER EXTINGUISHING MEDIA

Do not use a direct water stream which may spread the fire.

EXTINGUISHING METHODS

Remove containers from fire area if this can be done without risk.Use water to cool exposed containers.If possible, fight the fire from a protected location.

FIRE AND EXPLOSION HAZARDS

Not flammable, but combustible.Carbon monoxide and hydrocarbons are released in a fire situation.

PROTECTIVE EQUIPMENT FOR FIRE FIGHTERS

General: Evacuate all personnel, use protective equipment for fire fighting. Use a portable breathing apparatus when the product is involved in a fire.

6. ACCIDENTAL RELEASE MEASURES

PERSONAL PRECAUTIONS

Provide good ventilation.Avoid contact with the skin, eyes, and airways.Wear necessary protective equipment.Keep the emission away from sources of ignition.

SAFETY ACTIONS TO PROTECT EXTERNAL ENVIRONMENT

Must not be released into the sewage system, drinking water supply, or ground.

METHODS FOR CLEANING UP

Dry scrape, or absorb the product with inert material and place it in a refuse container.Collect and dispose of materials in closed and labelled containers at an approved disposal or waste collection facility.Clean with soap and water or detergent.Flush clean with plenty of water. Be aware of the risk of slippery surfaces.

7. HANDLING AND STORAGE

HANDLING ADVICE

Provide good ventilation.Avoid contact with the skin and eyes.Wear necessary protective equipment.Keep away from sources of ignition.

STORAGE

Keep in a dry, cool, and well-ventilated area.Keep in original container.Keep containers tightly closed.Store at temperatures below 50°C. Avoid contact with oxidising substances.Keep away from sources of ignition.

8. EXPOSURE CONTROLS/PERSONAL PROTECTION

EXPOSURE CONTROL

Provide good ventilation.Wash the skin at the end of each work shift and before eating, smoking, and using restroom facilities.Eye wash facilities and safety shower must be available when handling this product.

RESPIRATORY PROTECTION

Respiratory protective equipment is usually not necessary.In case of insufficient ventilation, wear respiratory protective equipment with A/P2 filter.

EYE PROTECTION

Wear approved safety goggles if there is a risk of eye splash.

HAND PROTECTION

Protective gloves (nitrile rubber, Viton®).

SAFETY DATA SHEET

LGWM 1

Last changed: 06/10/2010   Replaces date: 01/11/2007

PROTECTIVE CLOTHING
Wear suitable protective clothing.

9. PHYSICAL AND CHEMICAL PROPERTIES

PHYSICAL STATE   Paste.

COLOUR   Orange.

ODOUR   Mineral oil.

WATER SOLUBILITY Insoluble in water.

Physical and chemical parameters

Parameter	Value/unit	Method/reference	Observation
Flash point	> 150.00 °C		
Density	< 1.00 g/cm³	25 °C	

10. STABILITY AND REACTIVITY

STABILITY
Stable under recommended storage and handling conditions.

CONDITIONS TO AVOID
Avoid heat, sparks, and naked flames.

MATERIALS TO AVOID
Avoid contact with oxidising substances.

11. TOXICOLOGICAL INFORMATION

INHALATION
On warming/heating, the vapours emitted may cause irritation to the respiratory organs.May cause coughing and breathing difficulties.

SKIN CONTACT
Prolonged or frequently recurrent contact will degrease the skin and may cause skin irritation.

EYE CONTACT
May cause moderate irritation/smarting.

INGESTION
May cause irritation of the mucous membranes, nausea, and vomiting.

12. ECOLOGICAL INFORMATION

MOBILITY
Insoluble in water.The product floats on water.

DEGRADABILITY
Not readily biodegradable.

OTHER EFFECTS
Oil products may cause ground and water pollutants.

SAFETY DATA SHEET

LGWM 1

Last changed: 06/10/2010   Replaces date: 01/11/2007

CONCLUSION
Not readily biodegradable.
Oil products may cause ground and water pollutants.

Must not be released into the sewage system, drinking water supply, or ground.

OTHER INFORMATION
German water pollution classification (WGK): 1 Water pollutant.

---

13. DISPOSAL CONSIDERATIONS

GENERAL REGULATIONS
Treat as hazardous waste.Collect and dispose of materials in closed and labelled containers at an approved disposal or waste collection facility.Empty packages must be handled in an environmentally safe manner in accordance with applicable laws and regulations.

CATEGORY OF WASTE
The EWC code is a suggestion only; the end user will select a suitable EWC code.12 01 12* spent waxes and fats

Watse product number: 7021 Oily and greasy waste.

---

14. TRANSPORT INFORMATION

Classified as Dangerous Goods: **No**

OTHER INFORMATION
Not classified as dangerous goods.

---

15. REGULATORY INFORMATION

EC-Label: No

COMPOSITION
baseoil-unspecified (60 - 100 %)

S-PHRASES
S24/25 Avoid contact with skin and eyes.
Not translated to English

REFERENCES
Directive 1999/45/EC.
Annex VI to Regulation (EC) No 1272/2008 on Classification, Labelling and Packaging of Dangerous Substances.
Regulation (EC) No 1907/2006 of the European Parliament and of the Council of 18 December 2006 concerning the Registration, Evaluation, Authorisation and Restriction of Chemicals (REACH).
European Waste Catalogue.
Not translated to EnglishTransport of dangerous goods: ADR, RID, IMDG, and IATA.

---

16. OTHER INFORMATION

ISSUED: 22/03/2006

4 / 4

SKF Maintenance Products

Reference Job-Safety Analysis (JSA)

GENERIC WIND-TURBINE SERVICE WITH HUB-ENTRY ACTIVITY

The specific service activity or hub-entry procedure will vary, depending on the wind-turbine manufacturer and model. Always follow the specific procedure developed by the wind-farm safety coordinator where you are working.

Job-Safety Analysis

A job-safety analysis (JSA) is a tool to aid in preparation for a task or work activity. The purpose of a JSA is to highlight hazards associated with an activity and allow team members the opportunity to review relevant safety procedures, select required PPE, technical information, and tools to complete the activity safely. The JSA below is set up in a checklist format to aid in ensuring each activity is reviewed and appropriate action is considered before the activity is started.

Tailgate Meeting Date _____

Wind-Farm Name _____

Step	Description of Activity	Possible Injury	Corrective Action	Y/N
1	Collect tools, equipment, and materials for the scheduled work activity	Cuts, strains, trip hazards, and chemical exposure	Use of gloves	—
			Use proper lifting technique	—
			Good housekeeping techniques	—
			Review appropriate material safety data sheet (MSDS)	—
2	Travel to the turbine site	Bodily injury because of automobile accident, and weather hazards	Vehicle safety inspection	—
			Review of road map to determine best route of travel	—
			Review of traffic advisories for roadway issues	—
			Limit driver distractions during trip	—
			Notify safety coordinator or farm manager of intended route, estimated travel time, and contact information	—
			Obtain weather forecast for work activity duration and include travel time round-trip	—
			Driving technique should be appropriate for weather conditions	—
3	Arrive on site and unload tools, equipment, and materials	Cuts, strains, trip hazards, overhead hazard, chemical exposure, and weather hazards	Use gloves	—
			Use hardhat	—
			Use safety glasses	—
			Use proper lifting technique	—
			Good housekeeping techniques	—
			Review appropriate MSDS	—
			Use appropriate weather gear	—
			Notify safety coordinator or farm manager of arrival at turbine	—
4	Enter wind-turbine tower and remove from service	Shock, electrocution, and arc-flash hazard	Use appropriate flame-resistant (FR) clothing	—
			Do not enter cabinets without appropriate shock and arc-flash analysis and required personal protective equipment (PPE)	—
			Do not perform the activity unless qualified	—

Step	Description of Activity	Possible Injury	Corrective Action	Y/N
5	Disable remote access to the wind turbine	Shock, electrocution, and arc-flash hazard	Use appropriate FR clothing	—
			Do not enter cabinets without appropriate shock and arc-flash analysis and required PPE	
			Do not perform the activity unless qualified	—
			Follow the manufacturer's procedure or approved company procedure	—
6	Climb the tower (technician 1)	Fall hazard, strains, cuts, overhead hazard	Use appropriate fall-prevention equipment	—
			Use appropriate fall-protection equipment	—
			Use hardhat, safety shoes, safety glasses, and climbing gloves	—
			Use appropriate stretching exercises before the climb	—
			Have rescue equipment inspected and ready for use as necessary	—
			Ensure technicians are trained for possible rescue scenarios	—
			Do not perform the activity without a weather brief on possible lightning in the area during the activity	—
			Never climb a tower with the risk of a lightning strike	—
			Never climb the tower with the risk of high winds	—
7	Inspect the hoist assembly, lower the cable or chain, and prepare to bring materials to top of tower	Fall hazard, pinch, cuts, electrocution, overhead hazard	Use appropriate fall-arrest equipment	—
			Use gloves	—
			Use safety glasses	—
			Keep body parts away from moving parts	—
			Ensure safety covers are over moving components	—
			Ensure all electrical safety guards are in place	—
			Use hardhat	
			Follow recommended safety inspection for hoist, cable, chain, and connectors	—
8	Load lift bag and raise load (repeat as necessary)	Cuts, strains, trip hazards, chemical exposure, overhead hazard	Use gloves	—
			Use hardhat	—
			Use safety glasses	—
			Use proper lifting technique	—
			Good housekeeping techniques	—
			Review appropriate MSDS	—
			Do not overload lift bag and hoist assembly	—
			Never stand under load during a lift	—
			Use hand signals or other appropriate communication technique during the lift	—
9	Climb tower (technician 2)	Fall hazard, strains, cuts, overhead hazard	Use appropriate fall-prevention equipment	—
			Use appropriate fall-protection equipment	—
			Use hardhat, safety shoes, safety glasses, and climbing gloves	—
			Use appropriate stretching exercises before climb	—
10	Prepare wind turbine for hub entry	Pinch, crush, shock, electrocution, and arc-flash hazard	Keep body parts away from moving components	—
			Do not perform the activity unless qualified	—
			Follow manufacturer's procedure or approved company procedure	—
			Use hand signals or other appropriate communication technique during manual operation	—

© Cengage Learning 2013

(Continued)

Step	Description of Activity	Possible Injury	Corrective Action	Y/N
11	Disable and secure rotor assembly for entry	Pinch, crush, and entanglement	Keep body parts away from moving components	—
			Do not perform activity unless qualified	—
			Follow manufacturer's procedure or approved company procedure	—
			Perform approved lockout–tagout (LOTO) procedure	—
12	Disable and secure electrical and mechanical systems for work activity	Shock, electrocution, arc-flash hazard, pinch, and entanglement hazard	Keep body parts away from moving components	—
			Do not perform activity unless qualified	—
			Follow manufacturer's procedure or approved company procedure	—
			Perform approved LOTO procedure for required systems	—
13	Enter hub (external) Ensure technician 3 is available and ready for support as necessary	Fall hazard, strains, cuts, trip hazard, and weather hazards	Use appropriate fall-prevention equipment	—
			Use appropriate PPE	—
			Use safety shoes, climbing gloves, safety glasses, and hardhat as directed by company policy	—
			Follow appropriate confined space procedure for hub-entry activity	—
			Use appropriate weather gear	—
			Do not attempt entry unless wind speed is within guidelines established for activity	—
14	Enter the hub (internal)	Strains, cuts, crush, and trip hazard	Use gloves, safety glasses, and hardhat	—
			Follow appropriate confined space procedure for hub-entry activity	—
15	Return from hub (external)	Fall hazard, strains, cuts, trip hazard, and weather hazards	Use appropriate fall-prevention equipment	—
			Use appropriate fall-protection equipment (PPE)	—
			Use safety shoes, climbing gloves, safety glasses, and hardhat as directed by company policy	—
			Use appropriate weather gear	—
			Notify ground support that entry activity is completed	—
16	Return from hub (internal)	Strains, cuts, crush, and trip hazard	Use gloves, safety glasses, and hardhat	—
			Notify ground support that entry activity is completed	—
17	Reverse procedures to return hub and turbine to service	Follow guide for hazards listed above	Use appropriate corrective actions as listed above for each hazard	—
			Notify safety coordinator or farm manager of departure from turbine site and estimate travel time for return trip	—
			Follow up with a weather briefing and updated traffic report	—

Wind turbine manufacturer & model _____

Team leader _____

Attendance record_____

Safety coordinator review _____

This JSA does not cover all scenarios for all wind-turbine configurations. Refer to the organization safety policy and procedures used at your wind farm.

Reference Job-Safety Analysis (JSA)

GENERIC WORK ACTIVITY

The specific work activity procedure will vary, depending on the wind-turbine manufacturer and model. Always follow the specific procedure developed by the wind-farm safety coordinator where you are working.

JOB-SAFETY ANALYSIS

A job-safety analysis (JSA) is a tool to aid in preparation for a task or work activity. The purpose of a JSA is to highlight hazards associated with an activity and allow team members the opportunity to review relevant safety procedures and select required PPE, technical information, and tools to complete the activity safely. The following JSA is set up in a checklist format to help ensure that each activity is reviewed and appropriate action is considered.

Tailgate Meeting Date _____

Wind-Farm Name _____

Step	Description of Activity	Possible Injury	Corrective Action	Y/N
1				
2				
3				
4				
5				
6				
7				
8				
9				
10				
11				
12				
13				
14				
15				
16				
17				

Wind-Turbine Manufacturer & Model _____

Team Leader_____

Attendance Record_____

Safety Coordinator Review_____

K

Air Density (kg/m³) for Various Temperatures and Elevations from Sea Level

Elevation (Meters)	Pressure (Millibars)	Temperature (°C)									
		−40	−30	−20	−10	0	10	15	20	30	40
0	1,013	1.514	1.451	1.394	1.341	1.292	1.247	1.225	1.204	1.164	1.127
500	955	1.427	1.368	1.314	1.264	1.218	1.175	1.154	1.135	1.097	1.062
1,000	899	1.345	1.289	1.238	1.191	1.148	1.107	1.088	1.069	1.034	1.001
1,500	848	1.267	1.215	1.167	1.122	1.082	1.044	1.025	1.008	0.974	0.944
2,000	799	1.194	1.145	1.099	1.058	1.019	0.983	0.966	0.949	0.919	0.889
2,500	753	1.126	1.079	1.037	0.997	0.961	0.927	0.911	0.895	0.866	0.838
3,000	710	1.061	1.017	0.977	0.939	0.905	0.873	0.858	0.844	0.816	0.789
3,500	669	0.999	0.957	0.921	0.886	0.853	0.823	0.809	0.795	0.769	0.744
4,000	631	0.942	0.903	0.868	0.835	0.804	0.776	0.762	0.749	0.725	0.702
4,500	594	0.888	0.851	0.818	0.787	0.758	0.731	0.718	0.706	0.683	0.661
5,000	560	0.837	0.802	0.771	0.741	0.714	0.689	0.677	0.665	0.644	0.623

Glossary

A

AC *See* Alternating current.

Accumulator Fluid-power component used to maintain a fixed volume of fluid at system pressure.

Active braking Wind-turbine yaw-brake system that uses electrical or hydraulic actuators to maintain nacelle position when the yaw drive is not activated.

Actuator Device used to convert electrical, hydraulic, or pneumatic power to useful work.

Aerodynamic force Force on an object created by fluid flow around the object.

Aerodynamic pressure Force applied to the surface area of an object created by its interaction with a fluid. Units include psi and MPa.

AGMA *See* American Gear Manufacturers Association.

Alkali Chemical with a higher pH value such as household bleach (pH = 13) compared to distilled water, which is considered neutral (pH = 7).

Alternating current (AC) Current amplitude that varies with time, typically as a sine wave.

American Gear Manufacturers Association (AGMA) Organization comprised of industry leaders that promotes uniform standards for material properties, equipment, and manufacturing practices.

American National Standards Institute (ANSI) Organization comprised of industry leaders that promotes uniform standards for safety training, equipment requirements, material properties, component identification, and manufacturing practices.

American Petroleum Institute (API) Organization comprised of industry leaders that promotes uniform standards for lubricant properties and manufacturing practices.

American Society for Testing and Materials (ASTM) Organization comprised of industry leaders that promotes uniform standards for testing, material properties, equipment requirements, component identification, and manufacturing practices.

American Society of Mechanical Engineers (ASME) Organization comprised of industry leaders that promotes uniform standards for material properties, testing, and manufacturing practices.

American Wind Energy Association (AWEA) Association of businesses, research organizations, academic organizations, and government agencies located in the United States with a common interest in promoting wind-energy development, implementation, employment, and education resources.

Amplitude Measured value of signal strength at any given time or frequency level.

Angle of attack Angle formed between the oncoming relative wind and centerline across the wing or blade from leading edge to trailing edge.

Angular misalignment Misalignment of two shafts because of an angle between their centerlines.

Anhydrous Lacking or without water.

ANSI *See* American National Standards Institute.

Antiseize lubricant Lubricant with PTFE, graphite, or molybdenum disulfide added to prevent galling or cold welding of mating threads.

API *See* American Petroleum Institute.

Approach boundaries Established industry best practices used to determine safe working distances from energized conductors. Industry-established distances, also known as *boundaries*, are referred to as *limited*, *restricted*, and *prohibited*.

Arc flash Sudden release of energy from an electrical system because of a short circuit or other fault condition.

Aromatic Crude-oil stock that has varying properties, depending on the amount of paraffinic or naphthenic oil in the raw stock.

ASME *See* American Society of Mechanical Engineers.

Asphaltic Thick tarlike deposits referred to as *crude bitumen* that typically must be heated to flow for processing and use.

ASTM *See* American Society for Testing and Materials.

Attendant Trained, skilled, and authorized individual who provides a safety watch for a confined-space entrant.

Attribute data Data that may be subjective in nature to identify an object, such as color, relative size, or functional status—for example, an object is blue, it fits a specific size hole, it operates when energized, or it is go–no go.

Asynchronous generator Electrical device used to produce electrical output using a squirrel cage rotor and a wound stator assembly. The asynchronous generator will operate as a synchronous AC motor without external mechanical input at the rotor shaft, but when a mechanical input (prime mover) drives the rotor shaft speed above the motor's synchronous speed, it will produce an electrical output. Maximum output for an asynchronous generator is a rotor speed a couple of percent higher than the synchronous speed when run as a motor.

Availability Percentage of operational time compared to total time a wind turbine could be producing power.

AWEA *See* American Wind Energy Association.

Axial misalignment Misalignment of two shafts because of incorrect spacing between the ends.

Axial runout Movement or displacement of the shaft within the equipment along its centerline.

Axis cabinet Hub-control cabinet used to protect control components of a blade-pitch drive system.

B

Bedplate Common structural component or assembly used to mount the drive train and other systems to the tower.

Bill of material (BOM) List of components, associated specifications, and quantities that are used to make up an assembly.

Blade Airfoil used to convert motion of oncoming air into rotational movement of the wind-turbine rotor assembly and drive train.

Blade pitch Controlled movement of wind-turbine blades.

Blade root   Large section of the blade used as an attachment structure to the hub.

Body belt   Belt support system used only for work positioning activities after OSHA regulation changes were implemented in January 1998.

Body support system   Full-body harness assembly with associated connector and anchorage components used to arrest a fall.

Bolt   Type of fastener using threads as a means to create and maintain a clamping force between joined components through the use of washers and a nut to complete the clamping assembly.

BOM   *See* Bill of material.

Bonding   Method of electrically connecting components together to ensure they are maintained at the same potential.

Boundary film   Condition similar to mixed film in that it has inadequate lubricant to separate surfaces when components are at rest.

Brake caliper   Clamp assembly used to close brake pads onto a disc to slow or prevent movement of an assembly such as the output shaft of a gearbox.

Brake pad   Sintered metallic or composite material used in a brake system to clamp the movable brake disc.

Bulk modulus   Measure of a fluid's incompressibility (psi or Pa).

C

Cable keepers   Rubber U-shaped devices placed along the access ladder to secure the safety cable (vertical lifeline) during normal operation of the wind turbine.

Cable tray   Tray assembly used to support and protect cables run between equipment.

Canadian Centre for Occupational Health and Safety (CCOHS)   A not-for-profit federal department corporation established in 1978; governed by a tripartite council representing government, employers, and labor to ensure a balanced approach to workplace health and safety issues.

Canadian Standards Association (CSA)   Not-for-profit member-based association to create standards aimed at enhancing trade and promoting safety and well-being.

Carabiner   Metal loop with a spring-loaded or screw-type gate assembly used to secure safety items to the D-rings on a full-body harness.

CCD   *See* Charge-coupled device.

CCOHS   *See* Canadian Centre for Occupational Health and Safety.

Charge-coupled device (CCD)   Electrical device used to detect electromagnetic radiation and convert the energy into a signal that can be displayed as an electronic image or photograph.

Check sheet   Document that may be used to compare order of events, activities, or attributes. An example may be a list of steps in an activity that need to be followed to achieve a desired result.

Commissioning   Process of visual inspection and system testing of a wind turbine to ensure proper operation before presenting it to the customer for production.

Communication link   Metal wires or optical fibers used to transfer signals between control equipment and systems.

Competent inspector   Person trained to pertinent government regulations and industry standards and authorized by his or her employer to inspect and verify status of safety equipment.

Complex grease   Grease with barium and aluminum salts added to soap thickeners to produce desired properties.

Compressible   Fluid property that enables a reduction in volume because of an external force.

Compression   External force that creates a reduction in a part's length along the axis of applied force. Units are given in pounds-force or newtons.

Compression coupling   Collar assembly used to lock the main shaft to the gearbox input shaft.

Conduction   Transfer of thermal energy to cooler objects through physical contact.

Confined space   Space considered not designed for human occupancy, capable of partial or full entry by an individual, and with limited ingress and egress.

Confined-space permit   Safety policy that includes written procedure and permit process whereby entry into confined space is controlled so that an activity may be performed safely.

Constrictor   Snake that traps, crushes, and kills by coiling its body around prey.

Control system   Electrical equipment used to monitor system conditions and control subsystems in accordance with a preprogrammed routine to enable normal operation.

Convection   Transfer of thermal energy to a surrounding fluid of lower temperature.

Cooling damper   Panel that may be positioned to direct or prevent airflow in a ventilation system.

Cooling system   Equipment used to produce a controlled decrease in temperature for a system.

Corrosion   Chemical reaction between metal and atmospheric oxygen that creates an oxide of the metal. One example would be rust (iron oxide).

Crest factor   Measure of sine wave distortion; $CF = V_p/V_{rms}$.

CSA   *See* Canadian Standards Association.

D

DC   *See* Direct current.

Dehydration   Physical condition created by reduced water levels within tissues.

Delaminating   Separation and subsequent lifting of layers of polyester resin or fabric covering the blade or nacelle.

Department of Environmental Protection (DEP)   State agency tasked with protecting human health and the environment.

Desiccant   Anhydrous salt used to absorb moisture from air.

Digital multimeter (DMM)   Digital electrical test meter with test functions that may include resistance, voltage, current, capacitance, inductance, and continuity.

Direct current (DC)   Current amplitude that does not vary with time.

Direct drive   Wind-turbine design with rotor assembly connected directly to the generator.

Directional control valve   Component used to alter the path of a fluid within a system.

Disc defragment   Computer memory maintenance process used to locate file data stored in multiple memory locations during normal operations and store it in consecutive memory strings. This maintenance process can reduce memory retrieval time and improve computer efficiency.

Distortion Variation of a signal compared to its true form. Distortion of an AC signal may be determined by the relationship between peak voltage value and RMS voltage value.

Ditching Emergency act of landing an aircraft in water.

DMM *See* Digital multimeter.

Doubly fed induction generator An electrical device that uses an active wound rotor and wound stator assembly to produce a controlled AC output through the use of a controlled rotor input signal from an external control system. This generator design can produce a controlled AC output signal, such as 60 Hz, at varying rotation speeds from well below to above the normal synchronous speed. Normal synchronous speeds for generators include 3,600 RPM for two magnetic poles, 1,800 RPM for four magnetic poles, and so on.

Down tower assembly (DTA) The structure of a wind turbine that may include control systems, power conversion, electrical interconnections, switching assemblies, and personnel access equipment.

Drag Force on an object that resists its motion.

D-ring Loop-shaped metal component attached to web straps on a body harness. D-rings may be used as attachment points for carabiners, snap hooks, and other required safety devices for a full-body harness assembly.

Drive system Combination of control, feedback, and actuators used to move or position components or assemblies.

Drip loop Loop placed in power and control cables below the yaw deck used to allow slack for motion of the nacelle and to enable water to drip from the cables.

Drive train Components used to connect the rotor assembly to the generator. Components may include main bearing, main shaft, compression coupling, gearbox, and flexible coupling.

Dry lubricants Lubricants such as lead, graphite, polytetrafluoroethylene (PTFE), and molybdenum disulfide.

DTA *See* Down tower assembly.

Dye-penetrant test Nondestructive test using brightly colored or UV luminescent dye to highlight holes or cracks on the surface of metal structural components.

E

Electrically safe work condition All electrical hazards must be determined and eliminated so a person can work safely.

Electrical safety program Written program to determine electrical hazards associated with a workplace, planning to eliminate employee exposure, training on approved work methods, and feedback mechanism to ensure program is up to date and reviewed as required to ensure compliance with national standards and government regulations.

Electrical work permit Means of ensuring that work with electrical hazards is evaluated and performed in an approved manner to reduce or eliminate possibility of injury. Permit process also ensures management involvement so safety issues may be addressed as necessary if issues arise.

Electrolyte Term given to a solution of water and a salt or an acid. Example of an electrolyte would be the solution of water and sulfuric acid within a lead storage battery.

Electromagnet Magnet formed by passing current through a wire wound around a ferrous material. Magnetic pole orientation will be dependent on the direction of current flow within the wire.

Emergency locator transmitter (ELT) Device used to emit a predetermined emergency radio frequency when activated to aid rescue personnel in location of disabled or crashed aircraft or water vessel.

Emergency medical service (EMS) Organization trained to provide aid during a medical emergency.

Emulsion Two liquids or a solid and a liquid that do not mix, enabling one of them to stay suspended in the other.

Entrant Trained, skilled, and authorized individual who may enter a confined space.

Entry supervisor Trained, skilled, and authorized individual who is knowledgeable about hazards associated with an organization's confined-space entry requirements. Individual is responsible for ensuring that entrants and attendants comply with all company and regulatory standards.

Environmental Protection Agency (EPA) U.S. federal agency tasked with regulatory functions to ensure human health and protection of the environment.

Extreme pressure (EP) Lubricant additive that reacts with the metal surfaces to create a tough protective film.

F

FAA *See* Federal Aviation Administration.

Fall protection PPE used to prevent a fall from occurring or to safely arrest a fall if one occurs.

Fastener Devices used to join components. Examples include screws, bolts, pins, and rivets.

Federal Aviation Administration (FAA) Agency of the U.S. Department of Transportation tasked with regulating air safety practices, pilot and equipment certifications, and airspace requirements.

Fiber optics Communication system using a controlled light source and photodetector connected by an optical fiber cable.

Filter element Component used in a hydraulic system circuit to capture and hold contaminants.

Filtration system System used to prevent contamination of lubricant, pumps, manifolds, actuators, and other components of fluid systems.

Fire resistant (FR) Material capable of surviving a standard heat intensity for a specified time period without structural failure.

Flange Circular lip attached to a tube used as a means of attachment to another component.

Flex coupler Flexible coupling assembly used to allow small amounts of misalignment between connected shafts.

Flexible coupling Component used to transfer power between two shafts with a flexible structure to reduce shaft stress created by minor misalignment.

Flow Motion of a fluid within a system. Units of measure include gallons per minute (gpm) or liters per minute (lpm).

Flow control Device used to adjust the velocity of fluid within a fluid system.

Fluid power Pressurized fluid system used to provide controlled power to remote actuators.

Fluid transfer   Low-pressure system used to move fluid for transportation, processing, cooling, or heating.

Foundation   Structure used to support and stabilize a building or other assembly such as a wind-turbine generator. Generators may use a spread footing or rock anchor for on-shore systems and a piling or floating design for off-shore applications.

FR   *See* Fire resistant.

Frequency   Number of wave oscillations within a second. Units of frequency are in hertz (Hz).

Frequency converter   *See* Power converter.

Friction   Force between bodies in contact that resists their motion.

Frost bite   Freezing of the water within living tissue.

Full film   Condition in which there is sufficient lubricant between mating surfaces to completely separate even their high points while at rest and in motion.

G

Galvanizing   Application of thin layer of molten zinc to a carbon steel component to create a corrosion-resistant surface. Galvanized components have a noticeable crystalline appearance to their surface finish created by the zinc solidification process.

Gas   Form of lubricants that includes air, nitrogen, or other inert pressurized gas to prevent surface contact between objects.

Gearbox   Assembly used to convert rotor input from high torque and low speed to a low-torque, high-speed output for conventional high-speed generators.

Gear ratio (GR)   Typical industrial calculation for a gearbox ratio: the input speed divided by the output speed. Methods to determine the ratio of a gear set include a comparison of the meshing gear's set pitch diameters or number of teeth on each gear in the mating set. Example: GR = driven gear teeth number/driver gear teeth number.

Gelcoat   Thin layer of sprayed polyester resin used to form a protective finish layer for wind-turbine blades, aircraft components, and boat hulls.

Generator   Electrical device used to convert a mechanical input to an electrical output. Device uses either a rotating conductor coil and stationary magnetic field or a stationary conductor coil and rotating magnetic field. Generator requires three things to produce an electrical output: conductor coil connected to a load, magnetic field, and relative motion between a conductor coil and magnetic field.

GFCI   *See* Ground fault circuit interrupt.

GR   *See* Gear ratio.

Grade   Properties of a component or material that meet predefined characteristics as established by representatives of a specific industry.

Grease manifold   Common manifold used to distribute grease to multiple locations on a machine or within a bearing assembly.

Grid   Electrical infrastructure consisting of controls, switches, transformers, and transmission lines used to transfer power from a generation source to an end user.

Ground fault circuit interrupt (GFCI)   Electrical safety device used to monitor current flow within a power circuit that will open whenever the return current flow does not match the supply within a predetermined level.

Grounding   Method of electrically connecting components to earth ground.

Guarding   Physical component or system used to prevent accidental contact with a hazard.

H

Harness classification   Classification of a full-body harness assembly according to the recommended work activity. Classifications include A (fall arrest), D (controlled descent), E (limited access), L (ladder climbing), and P (work positioning).

HAWT   *See* Horizontal axis wind turbine.

Hazard communication (HAZCOM)   Required safety program used to educate employees about chemical hazards in the workplace.

Hazardous material (HAZMAT)   Material such as flammable or poisonous gases, liquids, or solids that would be a danger to life or the environment if released without precautions.

Hazardous waste operation (HAZWOPER)   Method of protecting personnel, environment, and the surrounding public through use of regulations, policies, and procedures for activities involved with chemicals and waste by-products.

Hazard risk category (HRC)   Electrical hazard categories as defined by industrial associations, including methods for addressing hazards such as recommended working practices, appropriate PPE, and tool specifications for typical workplace electrical activities.

HAZCOM   *See* Hazard communication.

HAZMAT   *See* Hazardous materials.

HAZMAT team   Emergency team trained, knowledgeable, and skilled in procedures for recognizing, cleaning, and disposing of hazardous materials.

HAZWOPER   *See* Hazardous waste operation.

Head   Part of a fastener used to supply a clamping force to joined components. A threaded fastener head is formed or machined to provide a means for applying torque to the fastener.

Heat exchanger   Device used to extract heat energy produced by system inefficiencies such as electrical resistance or friction in mechanical systems.

Heat exhaustion   An organism's physical condition created by a combination of an elevated core temperature and dehydration.

Heating system   Equipment used to produce a controlled increase in temperature for a system.

Hemolytic   Chemical that breaks down red blood cells and releases hemoglobin into surrounding tissues.

Hemotoxic   Chemical that breaks down body tissues, creates degeneration of organs, and disrupts normal blood clotting.

High-pressure region   Region of a wing or blade with an increase in air pressure because of compression of the oncoming relative wind.

High-speed shaft   Output shaft of a wind-turbine gearbox.

Histogram   Plotting of data values on a two-dimensional graph that may be used to visually show the distribution of data. The x-axis of the graph indicates the item to be tracked, and the y-axis indicates the occurrence of the tracked item.

Hook's law   Formula that describes the behavior of a ductile material loaded below its proportional limit. Also known as *linear elastic-theory*.

Horizontal axis wind turbine (HAWT)   Wind-turbine design constructed with the axis of the rotor horizontal to the horizon. Typical commercial wind turbines are designed with this configuration.

Horizontal lifeline   Support system used to arrest a fall but allow freedom of motion along a horizontal work surface such as a steel construction project.

Horsepower   Measure of a system's potential to perform useful work. Power of a fluid system may be calculated using parameters such as flow and pressure—for example, $HP_H = (pQ)/1,714$, where p is pressure (psi), Q is fluid flow (gpm), and 1,714 is a constant associated with the unit of measure.

Hot–cold–hot test   Method of verifying the condition of an electrical circuit and the test meter used for the activity. Process starts with verification of the test meter on a known energized circuit, proceeds to the circuit being verified for zero-energy state, and ends with a second verification of the test meter on the same known energized circuit.

Hot work   Work activity capable of producing an ignition source that may create a fire or explosion if not properly planned and executed.

HRC   *See* Hazard risk category.

Hub   Attachment structure used to mount blades to the wind-turbine drive train.

Hub adapter   Disc assembly used to mount the rotor assembly to a wind-turbine main shaft, gearbox input shaft, or generator shaft.

Hydraulic   System that uses an incompressible fluid as a power-transfer medium.

Hydraulic actuator   Device used to convert moving, pressurized, fluid into usable work.

Hydraulic cylinder   Device used in a hydraulic circuit to provide useful work in linear motion; also known as a *linear actuator*.

Hydraulic fluid   Virtually incompressible fluid such as water or oil used in hydraulic systems.

Hydraulic head   Pressure created by the weight of a column of fluid supported above a surface. Units include psi and MPa.

Hydraulic motor   Device used to convert hydraulic system power into rotary motion to produce useful work.

Hydrodynamic   Full film condition created when components are moving at rated velocity and lubrication in a system is allowed to circulate between components.

Hydrometer   Tool used to determine the specific gravity of an aqueous solution such as water and glycol or acid.

Hydrostatic   Full film condition created when an external pump and distribution system is used to maintain a pressurized flow of lubricant between mating components.

Hyperthermia   Elevation of body core temperature because of excessive physical exertion in a hot environment.

Hypothermia   Physical condition created by core temperature dropping below levels required for normal body functions.

Hypoxia   Physical condition created by an oxygen deficiency within a bloodstream.

I

IGBT   *See* Insulated gate bipolar transistor.

Impact-grade socket   Machined steel socket that may be used safely with impact or power-bolting tool.

Incident energy   Measure of thermal energy available at a given distance from an electrical arc fault; measured in units of Cal/cm^2.

Incompressible   Fluid property enabling a volume to resist reduction because of an external force.

Infrared   Wavelength of light located below the red portion of the visible electromagnet spectrum.

Insulated gate bipolar transistor (IGBT)   Solid-state electrical device used to switch DC power on and off at a controlled rate to produce a simulated AC signal. Used typically with high-power applications such as with wind-turbine output.

International Organization for Standardization (ISO)   European organization comprised of international industry leaders that promotes uniform standards for material properties, testing, manufacturing practices, and finished component properties.

Inverter   Electronic circuit used to produce a controlled stepped AC output from a DC input source.

Investment casting   Metal casting process that uses a wax model and plaster mold constructed around the model to produce a detailed cavity for the finished casting.

ISO   *See* International Organization for Standardization.

J

Job briefing   Brief prework meeting to ensure that personnel are aware of workplace hazards associated with the performance of an activity. Meeting should include presentation of appropriate materials to allow assessment and selection of safety items to mitigate workplace hazards.

Job-safety analysis (JSA)   Planning tool for evaluating workplace hazards and determining a means to reduce or eliminate the hazard exposure to employees performing the activity.

K

Kilowatt-hours (kWh)   Unit of energy used by electrical utilities to show a customer's consumption.

L

Ladder safety device   Fall-arrest system comprised of a cable and sleeve or rail and a slider that may be used with ladder access equipment.

Land use and regulatory commission (LURC)   State agency tasked with planning, permitting, and protecting land and natural resources in unorganized territories.

LASER   *See* Light amplification by stimulated emission of radiation.

Lattice tower   Open tower design made of three or four major structural legs tied together with horizontal cross members used to support a wind-turbine machine head.

LED   *See* Light-emitting diode.

Lift   Net force on a wing or blade caused by unequal pressure regions created by differences in airflow around its shape.

Lift bag   Bag used to hoist tools and supplies to upper levels of the wind turbine.

Light amplification by stimulated emission of radiation (LASER)   Coherent source of light that may be used to measure precise distances between components by determining the time required for the light to travel, reflect, and return to the transmitting unit.

Light-emitting diode (LED)   Solid-state device used to produce emitted radiation at a discrete wavelength.

Lightning receptor Conductive disc mounted near the end of a wind-turbine blade connected by a braided conductor or large cable to the hub ground circuit to dissipate a lightning strike.

Linear actuator Device used to convert hydraulic system power into linear motion to produce useful work.

Liquid Form of lubricant; includes animal, vegetable, mineral, and petroleum oils.

Lockout–tag out (LOTO) Recommended set of steps used to determine an equipment's energy source(s), remove the source(s), institute associated isolating practices, and other activities used to eliminate or reduce an employee's exposure to work activity hazards. Refer to federal regulations, industry best practices, and your organization's safety policies for activity specific details.

Low-pressure region Region of a wing or blade with a decrease in air pressure because of increased fluid flow velocity over the surface.

Lubricant additives Additives that can be combined with lubricants to enhance their overall performance.

M

Machine head Top level of a HAWT assembly containing power-generation equipment attached to the rotor assembly.

Magnetic particle test Nondestructive test using magnetic flux lines and iron particles to visualize the integrity of a steel or cast-iron structural component.

Main bearing Large bearing assembly mounted to the bedplate behind the rotor adapter used to support the main shaft.

Main shaft Drive-train component used to connect the wind-turbine rotor assembly to a gearbox or generator, depending on the system design.

Material safety data sheet (MSDS) Document providing chemical safety information for workplace activities.

Mega-ohm meter Electrical test instrument used to measure large values of resistance such as with wire-insulation resistance tests.

Metal coating Coating used for corrosion resistance or to improve surface finish appearance.

Metal inert gas welding Welding process using a consumable metal electrode as a filler metal and an inert atmosphere to prevent oxidation of the molten metal during the joining process; also known as *MIG welding*.

Meteorological (met) mast Structure used to mount the anemometer and wind-direction sensor above the wind-turbine nacelle.

Mixed film Condition where lubricant is present between contacting surfaces, but the layer is inadequate to separate the surfaces.

Modules Discrete components of a PLC system that may include communication, processing, input, and output devices.

Modulus of elasticity *See* Young's modulus.

MSDS *See* Material safety data sheet.

N

Nacelle Wind-turbine machine-head component used to protect control, drive, and generation equipment from the environment.

Naphthenic Crude-oil stock that is less thermally stable than paraffinic and has a lower viscosity index.

National Electrical Code (NEC) Industry best practices recommended by the National Fire Protection Association (NFPA) set forth in its Standard Electrical Practices 70.

National Fire Protection Association (NFPA) Organization comprised of industry leaders who promote uniform standards for safety training, equipment requirements, material identification, and manufacturing practices.

NATO *See* North Atlantic Treaty Organization.

NEC *See* National Electrical Code.

Neurotoxin Chemical that interferes with normal functions of the nervous system, creating partial or full paralysis.

NFPA *See* National Fire Protection Association.

Nondestructive testing Test that may be used to determine integrity of a material without damaging the material.

North Atlantic Treaty Organization (NATO) Military alliance made up of North American and European governments to provide a collective defense against military threats.

Notice to airman (NOTAM) Bulletin used to disseminate safety and operational information to pilots about changes to airspace between the normal publication schedule.

Nut Hex or square profile component manufactured with a threaded center through the hole used to mate with a bolt.

O

O&M *See* Operations and maintenance.

Occupational Safety and Health Administration (OSHA) Division of the U.S. Department of Labor tasked with regulating issues that may impact human health and safety in the workplace.

Oil analysis Process of testing oil to determine condition of a motor, gearbox, or hydraulic system.

Oil cooler Heat exchanger used with an oil-cooling system to shed the buildup of thermal energy created by system inefficiencies.

OPEC *See* Organization of Petroleum Exporting Countries.

Operating system Programmed code of instructions used to control data processing and memory storage and retrieval functions, along with coordination of input and output devices for a computer such as a PC.

Operations and maintenance (O&M) Ongoing effort to ensure optimal operation of equipment.

Organization of Petroleum Exporting Countries (OPEC) Organization made up of Middle Eastern, African, and South American countries that export crude-oil resources.

Orifice Opening or hole of a set diameter in a fitting.

OSHA *See* Occupational Safety and Health Administration.

P

Pad-mount transformer Large transformer assembly mounted in a box-shaped enclosure. This transformer configuration may be used to increase the wind-turbine output to match the local transmission infrastructure.

Paraffinic Crude-oil stock that is thermally stable and has a relatively high viscosity index.

Parallel misalignment Misalignment of two shafts because of horizontal or vertical spacing between their centerlines; also known as *radial misalignment*.

Parasitic drag Resistance to fluid flow created by the length of surface exposed or wetted by the fluid.

Pareto chart A graphic representation used to compare data results from the highest occurrence to insignificant values. Typically a Pareto chart is set up as a bar graph showing a top occurrence item and the next closest two or three that need to be addressed to improve a process.

Parking brake Braking assembly used to prevent rotation of the wind-turbine drive assembly.

PASS *See* Pull, aim, squeeze, sweep.

Passive braking Wind-turbine yaw-brake system that uses a spring-loaded assembly to maintain nacelle position when the yaw drive is not activated.

PC *See* Personal computer.

PDCA *See* Plan, do, check, and act cycle.

Peak-to-peak Measure of a sine wave signal from the positive peak to the negative peak value.

Peak value Amplitude of a sine wave signal measured from the 0 voltage level to peak value.

Permanent magnet Ferrous material with magnetic properties because of formation process.

Permanent magnet generator Generator designed with permanent magnets to produce a magnetic field.

Permit-required confined space Confined space that has an added safety hazard that may cause a serious injury or death. Safety hazards include type of atmosphere, engulfing medium, mechanical configuration, and electrical hazard.

Personal computer (PC) Computer used for standalone operations or as part of a local or wide area network. Typical operations include data acquisition, processing, and storage, along with computing applications for word processing, data analysis, and Internet searches.

Personal flotation device (PFD) Device used as a buoyancy aid for personnel in water.

Personal locator beacon (PLB) Smaller version of an ELT used as an aid in locating lost or injured personnel.

Personal protective equipment (PPE) Last line of defense to reduce or eliminate workplace hazards. Examples of PPE may include fall prevention, fall arrest, eye protection, and hearing protection.

PFD *See* Personal flotation device.

pH Measure used in chemistry to determine the extent an aqueous (water) solution is acidic or alkaline (pH scale runs from 0; strong acid to 14; strong alkali).

Pillow block Block assembly used to mount a bearing assembly.

Pinwheeling Free spinning of the rotor assembly with the blades feathered to 90° and the parking brake disengaged.

Pitch bearing Bearing used to provide movement of blades mounted to the hub.

Pitch control System of control, drive, and feedback components used to position wind-turbine blades.

Pitch diameter Diameter that passes through the midpoint of the teeth on a gear.

Pitch system Combination of control, feedback, and actuators used to position wind-turbine blades for normal operation and emergency braking.

Plan, do, check, and act (PDCA) cycle Management tool used to improve a process through incremental steps of making adjustments, measuring the change, and making further adjustments to refine the process to match desired results. This process is typically used several times to improve or refine the end result of a process, so it is often referred to as a *continuous improvement* philosophy.

Plastic fluid Change in lubricant viscosity that occurs when it is mechanically stressed.

PLB *See* Personal locator beacon.

PLC *See* Programmable logic controller.

Pneumatic System that uses compressed gas as a power-transfer medium.

Positive contact Confirmed contact and established communication with a person who will assist with an emergency situation.

Positive displacement Attribute of pump design to move a fixed volume of fluid with each revolution of the input shaft.

Power-bolting tool Wrench that uses hydraulic or electric power to provide output torque for threaded-fastener assembly.

Power converter Electrical system used to condition wind-turbine power output to match the grid voltage and frequency; also known as a *frequency converter*.

Power purchase agreement (PPA) Agreement between power producer and customer on price, contract duration, and other items of interest to both parties.

Power unit Equipment used to provide fluid power to a hydraulic system.

PPA *See* Power purchase agreement.

PPE *See* Personal protective equipment.

Predictive maintenance Monitoring system changes to flag issues with system components so corrections can be made before failure occurs.

Pressure Force created by combined collisions of fluid molecules against a container's surface. Typical units used are pounds per square foot (psi) and megapascals (MPa).

Pressure control Device used to adjust the pressure of a fluid within a fluid system.

Pressurized fluid Hydraulic fluid supplied at pressure to power a bolting tool.

Preventative maintenance Scheduled maintenance activities to prevent equipment failure.

Prime mover Device used to provide input power to a system. Examples include wind-turbine rotor assembly, electric motor, and gasoline engine.

Production tax credit (PTC) U.S. incentive for wind-energy paid per kilowatt-hour generated for the first 10 years of equipment service.

Programmable logic controller (PLC) Microcomputer used to monitor and control industrial systems through the use of a program code, input, and output devices.

Projected area Cross-sectional area of an object exposed to an oncoming fluid that restricts the fluid's movement.

Proportional limit Stress level a material can endure before permanent deformation occurs. Loading and unloading below this level will return the material to its original length.

PTC *See* Production tax credit.

PTFE Polytetrafluoroethylene.

Public utility commission (PUC) State agency tasked with providing utility industry oversight and regulatory functions to provide fair market practices.

Pull, aim, squeeze, sweep (PASS) Acronym used to describe the use of a fire extinguisher: pull the pin, aim toward the fire, squeeze the trigger, and sweep (side to side).

Pump Device used to increase the power of a fluid system.

Pump unit Hydraulic power unit used to supply pressurized fluid to a power-bolting tool.

Q

Qualified person Trained and skilled individual who is authorized by his or her employer to perform an electrical work activity.

R

Rabbit-eared Positioning of a wind-turbine rotor assembly with one of the three blades directly down and the other two positioned up. Visual appearance looks like ears on a rabbit or the antenna that may be used on portable television or radios.

Radial misalignment *See* Parallel misalignment.

Radial runout Displacement of a shaft in a direction perpendicular to the centerline of the equipment.

Radio frequency (RF) Electromagnetic radiation produced as a discrete frequency or range of frequencies and used to transfer information between a transmitter and a receiver electrical device.

Reaction arm Arm attached to a power-bolting tool that is used to stabilize the tool and prevent rotation when applying torque to a fastener during assembly.

Reaction point Location on an adjacent fastener or the equipment structure where the reaction arm contacts during the torque process.

Rectifier Electronic circuit used to convert an AC input signal to a DC output.

Rescue system Equipment used to retrieve or lower an individual if he or she has fallen during a working-at-height activity or during an emergency situation.

Reservoir Device used to maintain a reserve supply of fluid for a hydraulic system.

Resonance Frequency value at which the vibration amplitude is maximum because of the mechanical configuration.

Retina Portion of the eye containing rod- and cone-shaped structures that provide chemical signals to the optic nerve when stimulated by the visible light spectrum.

Return on investment (ROI) Recovery of capital investment from profit realized from an operational project.

RF *See* Radio frequency.

Rime ice Ice buildup created by supercooled water droplets coming into contact with an object that has a temperature below freezing. The ice appears rough and may be milky in color.

RMS *See* Root mean square.

Rock-anchor foundation Foundation that uses treaded rods to secure the assembly to an underlying rock formation for stability.

Rock-anchor rods Threaded rods that are grouted into the underlying rock formation below a foundation to stabilize the assembly.

ROI *See* Return on investment.

Root cause Cause that may be attributed with creating a problem or failure and not a symptom of the problem or failure. A root cause may be attributed to variation in process, material, or personnel or to environmental factors.

Root mean square (RMS) Value that is the statistical magnitude of a varying quantity such as a sine wave; for example, volts RMS would be the signal peak voltage value divided by the square root of 2 ($V_p/\sqrt{2}$).

Rotary encoder Control system feedback component used to provide precise positioning of a rotating system.

Rotor Moving coil or laminated metal assembly of a motor or generator that is mounted to the central shaft.

Rotor adapter Structural component of drive train used for attachment of the rotor assembly to the main shaft or gearbox input shaft.

Rotor assembly Wind-turbine assembly that includes blades and an attachment structure.

Rotor lock Locking mechanism used to hold the wind-turbine drive assembly during a hub-entry activity.

Run chart Graph used to track variation in data measurements with time.

Rupture strength Stress level at which material failure occurs.

S

SAE *See* Society of Automotive Engineers.

Safety interlocks Method of disabling or eliminating a hazard by use of electrical or mechanical components attached to a safety guard.

Sag Voltage level below specifications caused by a sudden increase in load or decrease in supply conditions.

Saponification Chemical reaction between an oil and alkali that forms a metallic soap.

SCADA *See* Supervisory control and data acquisition.

Scatter diagram Graph of data values used to represent the correlation between two contributing factors. Scatterplots are useful to show a positive, negative, or neutral (no) correlation.

SCBA *See* Self-contained breathing apparatus.

SCR *See* Silicon-controlled rectifier.

Self-contained breathing apparatus (SCBA) PPE that includes a mask, pressure regulator, and breathable compressed air supply; used for entry into an atmosphere that may cause serious injury or death if inhaled.

Self-retracting lifeline (SRL) Vertical lifeline fall-arrest safety system using a spring-loaded, self-retracting cable or synthetic web strap as a connector between a full-body harness assembly and an anchorage system.

Semisolid Emulsified systems such as metallic soaps or solid particles in oil.

Sensor Device used to provide system parameter feedback to a control system. Examples include thermocouplers, pressure transducers, and proximity switches.

Service bulletin   Published document used to provide up-to-date information from a manufacturer to customers on recent system improvements or items that may need special attention to prevent future operation issues.

Shank   Part of the threaded fastener without threads.

Shock-absorbing lanyard   Connecting device designed with a deceleration system to reduce impact forces created by a fall. U.S. Department of Labor and OSHA regulations require that approved deceleration systems reduce impact forces to 1,800 pounds or below, whereas industry associations such as ANSI require that deceleration systems reduce impact forces to 900 pounds or below.

Sight glass   Device used as a visual reference to indicate fluid level or condition.

Silicon-controlled rectifier (SCR)   Solid-state electrical device used as a switch in an electrical control circuit.

Slip ring   Electrical device with movable rings and stationary brushes that can be used to transmit power or electrical control signals between the nacelle and hub.

Society of Automotive Engineers (SAE)   Organization comprised of industry leaders that promote uniform standards for material properties, testing requirements, and manufacturing practices.

Soft foot   Condition in which one of the four motor or generator feet is not located in the same plane as the other three; can create instability or damage to the mounting assembly because of high stresses in the soft foot.

Solenoid   Electromechanical device using a controlled magnetic field that is produced around a coil of wire by electrical current flow; used to move a ferrous component producing useful work.

Solid   Form of lubricant; includes graphite, lead, and molybdenum disulfide.

SOP   *See* Standard operating procedure.

Sound insulation   Foam sheets or panels used to absorb sound waves or dampen vibration to reduce sound propagation from a source.

Specific gravity   Ratio of specific weight of a fluid compared to an equal volume of water. Specific gravity of freshwater equals 1.

Specific weight   Weight per unit volume of a material such as a fluid or solid.

Spinner   Shell or cover used to improve aerodynamic flow around the hub and nacelle.

Spring force   Reaction force created by compressing or extending a ductile material within its elastic range.

SRL   *See* Self-retracting lifeline.

Stakeholders   Person or group who has an physical, emotional, or financial investment in a project or organization.

Standard operating procedure (SOP)   Defined method of activities or procedures developed and adopted by an organization for its employees.

Star pattern   Process of tightening fasteners located on a circular assembly to minimize an uneven clamping force between the components.

Stator   Stationary wire coil of a motor or generator used to form magnetic poles.

Strain (ε)   Change in length (δ) of a material divided by its original length (l); $\varepsilon = \delta / l$.

Stress (σ)   Load (F) applied to a material divided by its cross-sectional area (A); $\sigma = F/A$.

Structural drag   Fluid flow resistance created by the shape of an object. A uniform shape such as a sphere or cylinder will have less resistance than a flat, blunt shape.

Stud   Type of threaded fastener manufactured without a head. Typical configurations include continuous thread and threaded regions on each end separated by a shank.

Supervisory control and data acquisition (SCADA)   System of software and hardware used to monitor and control the functions of a system such as a wind-turbine generator.

Swell   Voltage level above specifications because of sudden decrease in load or increase in supply conditions.

Sympathetic vibration   Induced vibration transmitted through a structure to a location away from the source.

Synchronous generator   Electrical device used to produce an AC output, with the frequency directly proportional to the rotational speed of the device.

Synchronous speed   Shaft speed at which a generator can provide the proper output AC voltage frequency. Synchronous speed (RPM) = 120 (frequency)/number of magnetic poles.

Synthetic compounds   Compounds created from multiple chemical reactions of discrete atoms or molecules to form a molecule with desired characteristics.

T

Tachometer-generator   Control-system feedback component used to provide a proportional voltage signal dependent on a drive motor's direction of rotation and shaft speed. Also called a *tacho-generator*.

Tension   External force that creates an elongation or stretching of a part along the axis of applied force. Units are given in pounds-force or newtons.

Tensioning tools   Tool used to apply a tensile load to a fastener during the assembly process.

Terminal strip   Wiring device used to terminate multiple wire pairs.

Thermography   Use of thermal imaging or measurement of emitted radiation to analyze temperature variations or distribution for a component or system.

Thickener   Oil additive to produce grease such as a metallic soap, clay, or graphite.

Thixotropic   Decrease in grease viscosity with time.

Thread   Helically shaped V profile machined or formed on the threaded fastener used to engage with a mating part with the same profile.

Thread gauge   Gauge machined with a standard thread profile so that it can be quickly compared to an unknown thread. A typical thread gauge has an assortment of blades, each with a specific standard thread profile.

Thread pitch   Spacing measured from the crest of one thread to the crest of the adjacent thread.

Threads per inch (TPI)   Standard measure of thread pitch used by American and Canadian fastener manufacturers to describe thread spacing.

Thrust   Unbalanced force applied to an object that creates motion.

TIG welding   *See* Tungsten inert gas welding.

Top box   Control console located in the nacelle of a wind turbine.

Torque   Force applied at a distance from a point. Units include foot-pounds and newton-meters.

Torque multiplier   Tool that uses a gearbox to multiply the mechanical advantage supplied from the input shaft to the output shaft.

Torque wrench   Wrench with a mechanism that indicates torque value applied to the tool bit or socket.

Tower foundation rods   Threaded rod used to clamp the wind-turbine tower assembly to the foundation.

TPI   *See* Threads per inch.

Transformer   Electrical circuit device used to increase or decrease electrical output voltage through an induced voltage from a primary coil to a secondary coil.

Transient   Short-duration spike in voltage created by a lightning strike on the electrical grid or by a switching activity.

Tribology   Study, application, and principles of friction, lubrication, and wear.

Tube tower   Formed steel tube structure providing support for the wind turbine, protection for control systems, and internal access to the nacelle.

Tungsten inert gas welding   Welding process using a Tungsten electrode, filler metal, and an inert atmosphere to prevent oxidation of the molten metal during the joining process. Also known as *TIG welding*.

U

Ultimate strength   Maximum stress a material can withstand. This stress level corresponds to the highest point on a ductile materials stress–strain curve.

Unified National Coarse (UNC)   Standard coarse-thread spacing agreed on by representatives of the United States, Canada, and Great Britain in 1948.

Unified National Fine (UNF)   Standard fine-thread spacing agreed on by representatives of the United States, Canada, and Great Britain in 1948.

Uninterruptible power supply (UPS)   Backup power unit comprised of several batteries and an inverter to supply AC power to electronic systems during a power interruption.

United States Coast Guard (USCG)   One of the five armed forces of the United States and the only military organization within the Department of Homeland Security.

Universal serial bus (USB)   Computer communication system.

UPS   *See* Uninterruptible power supply.

USB   *See* Universal serial bus.

USCG   *See* United States Coast Guard.

V

Variable data   Data that may be used to show a numeric variation in a population. An example of variable data would be a series of dimensional measurements that could be used to determine an average value or the amount of variation by which one group may be different from another.

Vertical axis wind turbine (VAWT)   Wind-turbine design constructed with the axis of the rotor assembly perpendicular to the horizon. Example of this design configuration would be a Darrieus wind turbine.

Vertical clearance   Distance below the work area taking into account length of connecting devices, deployment of shock-absorbing devices, location of anchorage with respect to the dorsal D-ring on the harness, and the ability of the harness to slide up the back during deceleration to prevent impact with an object.

Vertical lifeline   Fall-arrest system used when working on elevated work surfaces such as roofing activities.

Viper   Snake that injects toxins through its fangs during a bite to disable or kill its prey. Examples include rattlesnakes, copperheads, cottonmouths, and coral snakes.

Viscosity   Measure of force to shear molecular attraction within a fluid and allow it to flow.

Viscosity index   Measure of oil stability with changes in temperature.

Visible spectrum   Portion of the electromagnetic spectrum that stimulates the optic nerve and allows our brains to interpret colors and physical shapes in environment around us.

W

Wear elements   Small particles of metal present in lubricating oil because of abrasion between moving components within a mechanical system.

Wear product   Elements that may appear in an oil-analysis report such as iron, chromium, silver, lead, tin, and copper that indicate abrasion or other damage to components.

Washer   Flat disc manufactured with a hole through the center and used with screws and bolts to distribute the clamping force between the head and nut over a larger area.

Wild AC   AC signal frequency other than that used by utilities; typically, 60 Hz in North America and 50 Hz in Europe and other regions.

Wind-resource study   Collection and analysis of wind data used to determine commercial viability of wind resources. Typical organizations that collect and study wind-resource data include independent meteorological contractors, wind-project developers, and government agencies.

Wire tray   Small tray assembly used to support and protect wires run between equipment or within a control console.

Work-positioning lanyard   Connecting device designed to allow hands-free work from an elevated work surface; not designed as fall-arrest systems and so should be used in conjunction with a fall-arrest system to prevent fall injuries.

Y

Yaw   Motion around the center axis of the tower.

Yaw bearing   Bearing assembly used to support the machine head of a wind turbine.

Yaw brake   System used to hold the nacelle assembly in place during normal operation.

Yaw deck   Deck below the yaw bearing and nacelle assembly.

Yaw ring   Ring mounted to the top of a wind-turbine tower; used as the fixed part of an active or passive braking system.

Yaw system   Wind-turbine control and drive system used to position the rotor assembly into the oncoming wind.

Yield strength   Determined by drawing a line parallel to the Young's modulus slope that intersects the stress–strain curve and the 0.002 strain value (0.2%) on the lower strain axis of the graph.

Young's modulus   Straight-line portion of a stress–strain graph produced during tensile testing of a ductile material. Within this region on the graph, material deformation is proportional to the applied load and will return to its original length when the load is removed. The slope ($E = \sigma/\varepsilon$) of this line will vary for each ductile metal. Also known as the *modulus of elasticity*.

Z

Zerk fitting   Quick-connect device used for adding grease to a component.

Zero-energy condition   State or condition of a system with all sources of energy isolated, deenergized, and disabled to prevent reenergizing.

Index